Paul Fumagalli, Joachim Schoenes

Magneto-optics

Also of interest

Magnetic Hybrid-Materials.
Multi-scale Modelling, Synthesis, and Applications
Odenbach, 2021
ISBN 978-3-11-056877-6, e-ISBN 978-3-11-056963-6

Semiconductor Spintronics
Thomas Schäpers, 2021
ISBN 978-3-11-063887-5, e-ISBN 978-3-11-063900-1

Spintronics.
Theory, Modelling, Devices
Blachowicz, Ehrmann, 2019
ISBN 978-3-11-049062-6, e-ISBN 978-3-11-049063-3

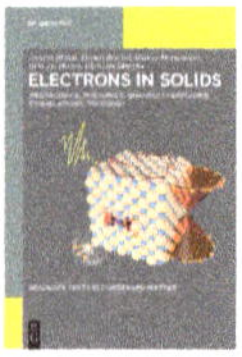

Electrons in Solids.
Mesoscopics, Photonics, Quantum Computing, Correlations, Topology
Hendrik Bluhm, Thomas Brückel, Markus Morgenstern,
Gero von Plessen, Christoph Stampfer, 2019
ISBN 978-3-11-043831-4, e-ISBN 978-3-11-043832-1

Paul Fumagalli, Joachim Schoenes

Magneto-optics

An introduction

DE GRUYTER

Author
Prof. Dr. Paul Fumagalli
Freie Universität Berlin
Institut für Experimentalphysik
Arnimallee 14
14195 Berlin, Germany
Paul.Fumagalli@fu-berlin.de

Prof. em. Dr. Joachim Schoenes
Technische Universität Braunschweig
Institut für Physik der Kondensierten Materie
Mendelssohnstrasse 3
38106 Braunschweig, Germany
Joachim.Schoenes@gmail.com

ISBN 978-3-11-063522-5
e-ISBN (PDF) 978-3-11-063530-0
e-ISBN (EPUB) 978-3-11-063545-4

Library of Congress Control Number: 2021950120

Bibliographic information published by the Deutsche Nationalbibliothek
The Deutsche Nationalbibliothek lists this publication in the Deutsche Nationalbibliografie; detailed bibliographic data are available on the Internet at http://dnb.dnb.de.

Cover image: Paul Fumagalli. The figure shows a ferromagnetic thin film with stripe domains on a nonmagnetic substrate. The orientation of the magnetization is perpendicular to the film surface and points either up (red stripes) or down (blue stripes). The incoming linearly-polarized light becomes elliptically polarized after reflection. The semimajor axis is rotated with respect to the incoming polarization direction and the direction of rotation depends on the orientation of the stripe domains.
Typesetting: Integra Software Services Pvt. Ltd.
Printing and binding: CPI books GmbH, Leck

www.degruyter.com

Preface

This book aims at graduate students. Knowledge of solid-state physics, quantum physics, and magnetism is advantageous but not a precondition. The content of the book is limited to the key aspects of magneto-optics. Therefore, magneto-optics of free carriers in doped semiconductors will only briefly be mentioned, although it has long been a focus area of magneto-optic research.

The foundation of this book is the lectures *Solid-State Optics* and *Magneto-optics*, which were given for many years by one of the authors, Joachim Schoenes, at Swiss Federal Institute of Technology (ETH) in Zurich, and – based thereupon – the lecture *Introduction to Magneto-optics*, which was given by the other author, Paul Fumagalli, at Freie Universität Berlin during many years, as well.

https://doi.org/10.1515/9783110635300-202

Contents

1 Introduction

1.1 Definition of magneto-optics

Magneto-optics (MO) is a wide field in physics and describes, in general, any interaction between electromagnetic radiation and a material which is magnetized. The application of a magnetic field to induce a magnetization is in principle not necessary in case of the existence of spontaneous magnetization. The local magnetic field produced by the spontaneous magnetization will lead to similar MO effects as an external magnetic field. Nevertheless, a magnetic field is usually applied even in the case of spontaneous magnetization to avoid indeterminacies due to the formation of magnetic domains. If the area that is illuminated by the incoming light covers several magnetic domains, the MO effect measured is the average of the signal produced by the individual domains. By applying a magnetic field, the magnetic domains will be aligned. On the other hand, one can take advantage of this behavior when studying MO hysteresis curves.

MO effects are not limited to the energy range of visible light but exist in principle in the whole spectral range of electromagnetic waves: MO phenomena are known in the microwave as well as in the X-ray range and the DC limit relates to the Hall effect. In the microwave range, predominantly free carriers give a contribution. In X-ray range, transitions from inner shells to the conduction band are important. This is known as X-ray magnetic circular dichroism (XMCD) and, for example, has been successfully used for the proof of an induced magnetic moment in Co/Pt multilayers.

1.2 History of magneto-optics

To understand the importance and impact of Faraday's discovery of the first MO effect, one has to know that in the beginning of the nineteenth century the true nature of light was still unknown. Émile Verdet has given an elaborate survey on the history of wave optics in his lectures on optics published in 1869 (1). Light was believed to spread by means of an ether in a very mechanistic way, like sound waves in fluids. But a corpuscular nature was assumed possible, as well. Yet another property of light was not really understood: polarization. This property appears when light is reflected from transparent surfaces at a specific angle (which is known today as Brewster angle), or, when light is transmitted through birefringent materials such as calcareous spar. Here, the light splits into an ordinary and an extraordinary ray, traveling in different directions, which are both linearly polarized with planes of polarization orthogonal to each other. It is the merit of Augustin Fresnel to have contributed in his short life to a very deep understanding of the behavior of linear polarization without the knowledge of the true nature of light. Being an engineer by training, he occupied himself more and

https://doi.org/10.1515/9783110635300-001

more with optics. In 1819, he published with his mentor Arago four rules about how linearly polarized light rays will interfere with each other (2). He was able to calculate for the first time the intensity of reflected light by combining his knowledge about how to add up polarized waves with the Huygens principle (3, 4). In a second note to his note on the calculation of the hues that thin layers of birefringent crystals produce (5), Fresnel developed his well-known equation describing the intensity of a light ray (linearly polarized and perpendicular to the plane of incidence) that is reflected from a transparent material (6): "Je vais faire voir maintenant comment on peut calculer l'intensité de la lumière réfléchie sous une incidence quelconque pour le faisceau polarisé suivant le plan de réflexion." The amazing fact is that Fresnel was using a purely mechanical picture of excited ether molecules in a fluid producing transversal light waves (discounting longitudinal waves completely, which made him feel uncomfortable). Moreover, he did not use boundary conditions at all (6): "Soit *m* la masse d'un élément différentiel du premier milieu, qui, en glissant sur lui-même, met en mouvement l'élément différentiel contigu *m*' du milieu réfléchissant, que je suppose de même élasticité. Dans le premier instant, *m*' était en repos, et *m* avait une vitesse *v*; un instant après, les deux élémens ont la même vitesse, et c'est alors que s'arrête le déplacement du premier par rapport au second; mais, en raison du déplacement effectué, le premier doit recevoir après, en sens contraire, toute la partie de la vitesse initiale qu'il a perdue." In his picture, the ether molecule of the first material accelerates the molecule of the second material at the interface until they reach a common velocity. Then, the second molecule transfers the velocity difference back to the first molecule. The whole process is mathematically equivalent to an elastic collision of two masses. He assumes the intensity of the light to be proportional to the square of the velocity of the ether molecules. Therefore, the intensity ratio between reflected and incoming ray is equal to the ratio of the square of the velocities of the ether molecules after the interaction. To get the intensity of the reflected ray, Fresnel replaced the masses of the ether molecules by a product of volume and density. The volumes he deducted from the width and wavelength of the incoming and the refracted ray and the density he assumed proportional to the inverse square of the propagation velocity of the light (like sound waves in fluids). Putting all this together, he finally derived what is called today the first Fresnel equation. Being an engineer, he carefully tested the equation by comparing with very accurate measurements and found excellent agreement. The same year, Fresnel also published the formula for light polarized in the plane of incidence, this time without presenting a derivation (7). With our knowledge of the correct physics, it seems like a miracle that Fresnel was able to derive the correct formulas with his improper assumptions. Nevertheless, his formulas together with his work on the interference of polarized light prepared the ground for a better understanding of the interaction of polarized light with material and of the role the refractive index plays.

When in 1845 Michael Faraday observed a rotation of the plane of polarization of light passing through lead borosilicate glass, to which he had applied a magnetic field along the light propagation direction (8), his discovery triggered a deeper understanding of the nature of light and its relation to electromagnetism. Faraday notes in his famous paper (8): “I believe that, in the experiments I describe in the paper, light has been magnetically affected, i.e. that that which is magnetic in the forces of matter has been affected, and in turn has affected that which is truly magnetic in the force of light.” Faraday examined a variety of transparent diamagnetic materials, some fluids such as water and alcohol and about 150 different aqueous solutions such as soluble acids, alkalis, and salts. In all of them, he found the new effect. In this way, he could show that a change of state of matter under the influence of magnetic forces is a general property not limited to paramagnetic or magnetically ordering materials. In the same study, Faraday revealed the following qualitative relations for the rotation of the plane of polarization, since then known as Faraday rotation θ_F:

1) $\theta_F \propto l$, where l is the distance of the light traveled in the material;
2) $\theta_F \propto B$, where B is the magnetic field applied to the material;
3) the rotation is strongest when the force lines of the magnetic field are parallel to the light ray, no effect happens when they are perpendicular;
4) in a material (e.g., a sugar solution), showing optical activity, or, as Faraday calls it, a natural rotation, which means the plane of polarization rotates without magnetic field, the Faraday effect just adds to it;
5) a positive rotation $\theta_F > 0$ is defined as a rotation of the plane of polarization in the same direction as the positive current must flow in a coil to produce the magnetic field;
6) an empty coil does not generate a rotation of the plane of polarization.

Faraday thought that all forces have a common origin or are directly related and mutually dependent. With his discovery, he could for the first time interconnect magnetism and light and proof that they are not independent (8): “Thus is established, I think for the first time, a true, direct relation and dependence between light and the magnetic and electric forces; and thus, a great addition made to the facts and considerations which tend to prove that all natural forces are tied together, and have one common origin.” Because Faraday found for all the materials he investigated the same sign of the rotation, he assumed that all materials behave like that. However, as we will see later, there is no direct connection between these quantities. We know today that dia-, para-, and ferromagnetism differ in the temperature dependence and dispersion of the Faraday effect. From Faraday’s last relation it follows that the magnetic force is not acting directly on the light but only through the material. From the third relation follows the possibility of constructing a light trap, the so-called Rayleigh light trap (9), as sketched in Fig. 1.1.

Faraday was very good at finding the fundamental dependencies of the effect in a rather qualitative way. The person who proofed his findings quantitatively was Émile Verdet. He developed an exact measurement of the magnetic-field strength and carefully verified the proportionality between magnetic field and rotation (10). Thereafter, Verdet established the angular dependence between magnetic field

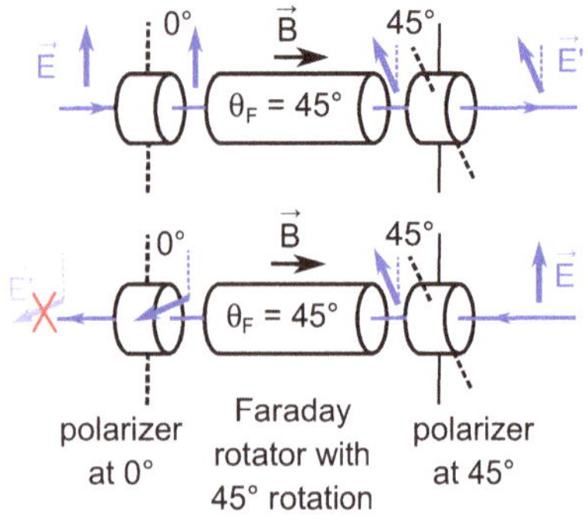

Fig. 1.1: Rayleigh light trap. A piece of glass, with a length such that it rotates the polarization in a magnetic field by 45° (called Faraday rotator), is placed between two polarizers set at 0° and 45°, respectively. Light traveling to the right is transmitted while it is blocked when travelling to the left due to the 45° rotation independent of travelling direction. The system therefore acts as an optical isolator which transmits radiation in one direction only.

and light-propagation direction. In another investigation, he discovered materials with a negative Faraday rotation: iron salts (11). Most importantly, he mentioned for the first time the phenomenological model that the Faraday rotation could originate from the difference of the reciprocal propagation velocities of right and left circularly polarized light (12). This is identical to the difference of the corresponding refractive indices. He wrote that Fresnel had already used a similar argument to explain the rotation of linearly polarized light in optically active materials (13). In a succeeding publication (14), Verdet restates a theory proposed by Airy (15) put forward shortly after the discovery of Faraday. The result is again that the Faraday rotation can be explained by two different velocities for right and left circularly polarized light. However, to achieve the desired result, Airy added to the fundamental wave equation a (non-physical) cross term which mixed the two wave-amplitude components. He, as well, was inspired by the idea of Fresnel. Having established an exact measurement of the magnetic field, Verdet was able to tabulate the relative size of the Faraday rotation for a wealth of liquids setting water equal to 1 (11). The phenomenon of dispersion, that is, that the rotation depends on the wavelength, had been found by Edmond Becquerel immediately after Faraday's discovery (16). He concluded that the rotation varies as the inverse square of the wavelength. Verdet showed through much more accurate measurement that this law is only approximately true (14).

The next big advancement took place in 1865 when James Clerk Maxwell published his dynamical theory of the electromagnetic field where he proposed his famous equations (17). Interestingly, he still assumed a ubiquitous ether as the origin of the propagation of electromagnetic waves (17): "We have therefore some reason to believe, from the phenomena of light and heat, that there is an æthereal medium filling space and permeating bodies, capable of being set in motion and of transmitting that motion from one part to another, and of communicating that motion to gross matter so as to heat it and affect it in various ways." He was able to calculate the velocity of the waves and concluded (17): "This velocity is so nearly that of light, that it seems we have strong reason to conclude that light itself (including radiant heat, and other radiations if any) is an electromagnetic disturbance in the form of waves propagated through the electromagnetic field according to electromagnetic laws." In addition, Maxwell's theory gave an answer to the question why longitudinal waves do not exist.

The behavior of light at interfaces, that is, reflection and refraction, was not discussed by Maxwell at all. This was done twelve years later by Hendrik Antoon Lorentz in his doctoral thesis (18). By refining a derivation published by Helmholtz (19), Lorentz could show that Fresnel's equations follow naturally from Maxwell's equations. He also discussed attenuated total internal reflection, a phenomenon, that already Newton had observed: When light is internally reflected within a prism and a second prim is pressed against the surface, from which the light is reflected, the reflected light is attenuated and a weak ray will be transmitted into the second prism. For easy mathematical treatment, Lorentz introduced a complex representation of the electromagnetic wave, where the real part characterizes the true physical light ray.

The investigation of the Faraday rotation was continued intensively by Henri Becquerel, who is mainly known for the discovery of radioactivity for which he was awarded the Nobel Prize in 1903 together with Marie Curie. Continuing the research his father started (16), Becquerel was interested in establishing a relation between Faraday rotation and refractive index taking also into account the effect of dispersion (20). With this knowledge, Becquerel could estimate the expected size of rotation for materials in the gas condition. So far, a Faraday rotation had not been found in gases. Becquerel succeeded by increasing the sensitivity with a special equipment comprising a 3-m long sealed copper tube where the light passed nine times back and forth yielding a total length of the light path of 27 m (21, 22). August Kundt and Wilhelm Conrad Röntgen had discovered a Faraday rotation in dense carbon disulfide gas under high pressure one year earlier than Becquerel (23–25) but Becquerel was first in measuring Faraday rotation in gases at ambient pressure. In another beautiful experiment, Becquerel was able to prove by direct measurement the model proposed by Verdet that the Faraday rotation originates from a different velocity of right and left circularly polarized light in an MO-active material (26). For his experiment, he split circularly polarized light into two parallel beams and let them travel through identical pieces of heavy flint glass before they were recombined to produce an interference pattern. The trick was that the two flint glasses were positioned in different locations of the horseshoe-shaped electromagnet: one between the two poles and one outside of them along the prolongation of the force lines. That way, the second flint glass experienced an inverted though weaker magnetic field than the first one. The interference pattern then shifted in opposite direction when switching on the magnetic field with opposite polarity. A similar experiment had been made by Augusto Righi already two years before (27). He let a right and a left circularly polarized beam travel in parallel through an MO-active liquid (carbon disulfide and a zinc chloride solution) and observed the superposition of the two beams which resulted in a linearly polarized beam rotating in different directions depending on the sign of the applied magnetic field.

About at the same time, in 1876, a new MO phenomenon was discovered by John Kerr (28). He observed a rotation of the polarization plane of linearly polarized light after reflection from the iron pole of an electromagnet that was polished to become a mirror. This is now called polar Kerr effect. In addition, he noticed that the light was elliptically polarized after reflection. Kerr related the phenomenon to the metallic reflection (29). The best results he got when the angle of incidence was about 75° but he could also show the effect at normal incidence. These findings were confirmed by James Edward Henry Gordon (30). Edwin H. Hall found a similar effect with the same sign in cobalt and nickel (31). However, Kerr did not relate the effect to a difference in the refractive index of circularly polarized light in combination with the Fresnel equations. This was done just a few months after his discovery by George Francis Fitzgerald (32). A year later, Kerr found a rotation of linearly polarized light, with plane of incidence parallel to the magnetic field, when it was reflected from the surface of an iron mirror which was aligned parallel to the magnetic field. This is now referred to as longitudinal Kerr effect (33). In 1884, August Kundt succeeded in preparing iron, cobalt, and nickel films on glass substrates that were thin enough to still be transparent. By weighing, he was able to determine their thickness which was of the order of 50–70 nm. In these films, Kundt found a specific Faraday rotation, which was more than ten thousand times stronger than for the dielectric materials Faraday had investigated (34). In the same investigation, Kundt discovered that in iron the Faraday and the Kerr rotation was larger for red than for blue light. The rotation follows thereby the magnetization and saturates (35).

Of great importance for a deeper understanding of the MO effects was 1896 Zeeman's discovery of the splitting of spectral lines of Na gas in a magnetic field (36). Lorentz had given him the hint that the outmost split lines should be circularly polarized in opposite way which Zeeman could indeed verify. Only two years later, Woldemar Voigt realized the close connection between Faraday and Zeeman effect and set up a phenomenological theory (37). In addition, he published in 1908 the first book on MO (38). At about the same time, Paul Drude formulated his theory of the electrons in metals which led to a deeper understanding of the behavior of metals (39). All theories were based on classical mechanistic models and were made to explain the phenomena seen in experiments. Nevertheless, they are very instructive and will therefore be discussed in this book as well.

From the turn of the century, further progress in understanding the microscopic origin of MO went along with the development of quantum mechanics. Later, the availability of computers led to first band-structure calculations which allowed the computation of MO spectra. A detailed overview of this development can be found in the Habilitation treatise of Peter M. Oppeneer (40) which will be summarized in the following.

Using a Heisenberg model to describe ferromagnetism in a crystal, Otto Halpern and Henry Rainsford Hulme related in 1932 the Faraday rotation to different selection rules for right and left circularly polarized light for transitions between energy levels which are split by spin–orbit interaction. Halpern argued only qualitatively (41), whereas Hulme presented a detailed calculation yielding the correct order of magnitude for the Faraday rotation in iron and cobalt (42). It took another 20 years before Petros N. Argyres was able to develop a more thorough theory of MO treating spin–orbit interaction and spin polarization at the same time (43). In those days, MO was generally understood as an interplay between exchange splitting and spin–orbit coupling.

In experiment, only improvements of the measurement techniques had been made since the turn of the century. No new MO phenomena were discovered except for a MO Kerr effect found by Quirino Majorana in the nonferromagnetic metals silver, gold, platinum, aluminum, and bismuth (44). The rotation was about a thousand times smaller than the rotation which Kerr had found in ferromagnets.

A direct comparison between theoretical and measured MO spectra became possible when Bernard R. Cooper together with Henry Ehrenreich (45) proposed an interband theory of the Kerr effect in nickel (46). The band structure of nickel had just become known at that time. However, they could only estimate the required optical transition-matrix elements. The situation changed when Walter Kohn developed with Pierre Hohenberg (47) and with Lu Jeu Sham (48) density-functional theory and local spin density approximation. With these tools, accurate band structure calculations became feasible. Joseph Callaway together with C. S. Wang (49) and with M. Singh (50) made further progress by employing an expression for the optical conductivity taken from linear-response theory which was developed by Ryogo Kubo (51). This way, they computed the absorptive part of both the diagonal and the off-diagonal optical conductivity of nickel and iron. This can be considered as the first band-theoretical calculation of a MO spectrum because the off-diagonal optical conductivity relates directly to the Faraday and the polar Kerr effect. However, the agreement between theory and experiment was moderate. With the discovery of magneto-absorption in semiconductors in the microwave range by G. Dresselhaus, A. F. Kip, and Charles Kittel (52) and in the infrared by Benjamin Lax together with S. Zwerdling (53), and with L. M. Roth (54), MO spectroscopy developed to a technique for the investigation of the band structure of semiconductors. Moreover, it became customary to relate MO phenomena to the material's band structure.

At the end of the 1980s, several groups started to calculate MO spectra again. Hubert Ebert (55) computed the absorptive part of the diagonal and off-diagonal conductivities of ferromagnets Fe, $FeCo_3$, and $FeCr_3$, and Yu. A. Uspenskii and S. V. Khalilov (56) of iron, nickel, and cobalt with comparable agreement with experiment as the spectra computed by Callaway and coworkers. G. H. O. Daalderop et al. (57) computed the absorptive components of the optical conductivity of UNiSn. Using Kramers–Kronig transformation, they computed the dispersive components as well and, subsequently, the first theoretical MO Kerr spectrum. An explicit calculation of the Kerr spectra of iron and nickel was performed by Peter M. Oppeneer et al. (58), who developed a precise computational technique to solve the difficulties inherent to the evaluation of MO spectra. The MO Kerr effect of iron computed this way provides a precise description of the experimental spectra. The computed Kerr rotation of nickel compares reasonably well to low-temperature data measured by G. Q. Di and S. Uchiyama (59). Since then, ab initio calculations of the Kerr spectra of various compounds have been reported. Altogether, these investigations have supplied a wealth of evidence that precise MO spectra can be accurately calculated based on relativistic band-structure theory. Moreover, it is nowadays feasible to make ab initio predictions of MO spectra.

In experiment, the photon energy range in which MO spectra could be measured was extended to 0.5–5 eV and a wealth of materials were investigated between 1960 and 1990 starting with the ferromagnets iron, cobalt, and nickel by G. S. Krinchik and V. A. Artem'ev (60). Most of the accumulated experimental MO data of this period is collected in the reviews of K. H. J. Buschow (61), Joachim Schoenes (62), and Wolfgang Reim and Joachim Schoenes (63). Two occurrences boosted the interest in MO research again. The first was the discovery that magnetic domains could be observed by means of the MO Kerr effect in polar geometry by H. J. Williams, F. G. Foster, and E. A. Wood (64) and in longitudinal geometry by C. A. Fowler and E. M. Fryer (65). This opened the road to visualizing surface and subsurface magnetic domains. However, the resolution is always limited by the wavelength of the illuminating light.

The second occurrence was the realization that the polar Kerr effect could be employed to read out stored magnetic information by H. J. Williams et al. (66), R. L. Conger, and J. L. Tomlinson (67), and E. J. Supernowicz (68). In the early 1970s, the discovery of amorphous ferromagnetic thin films based on GdCo alloys with square hysteresis loops, perpendicular magnetic anisotropy, and Kerr rotations of approximately 1° by P. Chaudhari, J. Cuomo, and R. J. Gambino at IBM paved the way for MO storage applications (69). Although the patent application existed for these materials as MO storage media as early as 1976, it took another 15 years until the first product appeared on the market. The reasons for this delay are found in the difficulty to sufficiently downsize the elaborate optics. In addition, a major research effort was required to develop a compact, powerful, and durable laser diode. Only the prospect of commercial use in form of an audio CD (compact disk) led to a breakthrough in laser diode development. Another reason might have been that IBM worried about creating with such a new storage device a competition for their own products and risking a loss in the lucrative business with conventional magnetic storage media. Thereupon, traditional storage media, phase-change media (which are inserted into the optical CD-R and CD-RW drives), and, especially, flash memory were improved so much that today it would take a considerable research effort for MO storage media to compete with conventional magnetic and optical storage disks and solid-state memory. The main advantage of MO storage materials over conventional magnetic storage devices – namely the simple replacement of the medium due to the contactless sampling by a laser beam – has become obsolete because of the incredible downsizing of conventional storage media. Even the last remaining advantage over CD-R and CD-RW disks and flash memory, an almost unlimited overwrite capability and much longer lifetime of stored information, has become dispensable due to the tremendous increase in storage capacity in conjunction with an extreme price erosion of flash memory.

A novel technique, extending MO to the X-ray region, evolved during the 1980s. A precondition was the availability of highly polarized X-rays from synchrotron radiation sources. The starting point where theoretical considerations on a possible MO absorption at the $M_{2,3}$ ($3p \rightarrow 3d$) edge of nickel made by J. L. Erskine and E. A. Stern (70). Ten years later, a theoretical paper by B. T. Thole, G. van der Laan, and G. A. Sawatzky predicted strong linear magnetic dichroism in the $M_{4,5}$ X-ray absorption spectra of magnetic rare-earth materials (71). One year later, the predictions could be confirmed in experiment by the same authors and coworkers (72). Yet another year later, Gisela Schütz and coworkers succeeded in measuring the XMCD in iron (73). Since then, especially with the continuous improvement of synchrotron radiation sources, XMCD has developed to an indispensable tool for element-specific investigations of magnetic properties. Because the initial states are core shell levels of the atoms constituting the material, XMCD is a truly element-specific method allowing to measure the properties of any element in a material system. A recent example is depicted in Fig. 1.2, where Faraday rotation and XMCD of a monolayer of graphene (G) are plotted (74). The X-ray energy range has been chosen to be only sensitive to the MO response of the graphene monolayer. The graphene has been grown on a 5.8 nm thick cobalt layer deposited on a tungsten substrate. A very large Faraday rotation and XMCD effect are observed. The measurement clearly indicates that cobalt induces ferromagnetism in graphene.

Yet another new direction was opened through the fast development of ultrashort-pulse lasers. Soon after the demonstration of the first lasers, the precondition for ultrafast optics was found in 1964: mode locking (75), which allowed the simultaneous oscillations of a large number of highly coherent laser modes superposing to very short pulses. In the 1980s, femtosecond-pulsed solid-state laser became commercially available and boosted the field of nonlinear optics (76). Since then, two highly topical MO techniques evolved: time-resolved MO spectroscopy, involving pump-probe techniques, and magnetic second-harmonic generation, involving highly focused laser beams that produce higher harmonics at the surface or at an interface (77).

Of course, the treatment of these novel MO techniques is beyond the scope of an introduction. Nevertheless, this book will still allow the reader to understand the basic principles of the techniques.

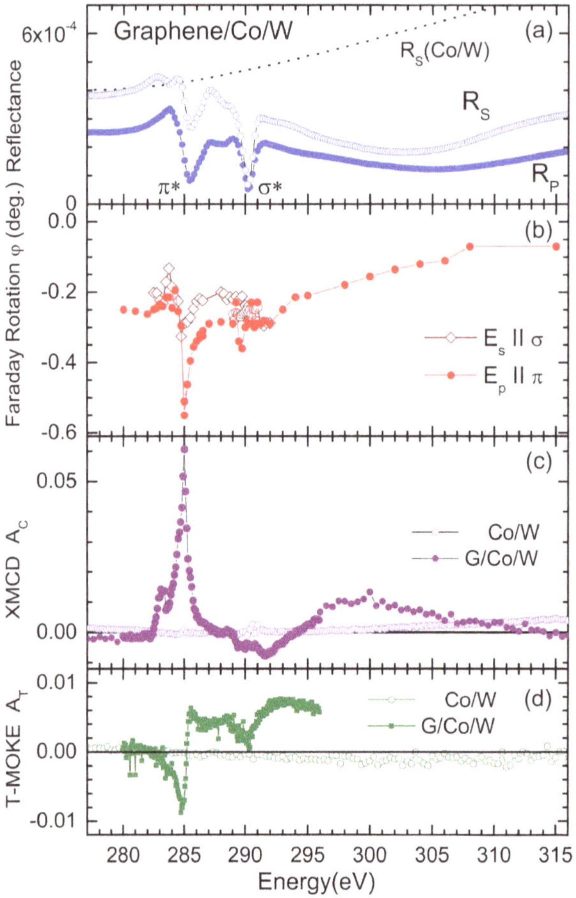

Fig. 1.2: Magneto-optical spectra measured on graphene/Co/W. (a) Absolute reflectance R_p and R_s obtained with orientation of electric field E_p parallel to π* bonds and E_s parallel to σ* bonds, respectively. Calculated R_s for the substrate Co/W (dotted line). (b) X-ray Faraday rotation spectrum measured for p-polarized light (filled symbols) and s-polarized light (open symbols) on in-plane magnetized graphene. (c) XMCD asymmetry obtained with circularly polarized light (filled symbols for G/Co/W). The reference substrate without graphene does not show magnetic signals (open symbols). (d) T-MOKE asymmetry A_T measured with p-polarized light (filled symbols for graphene/Co/W) and open symbols for reference sample Co/W. Figure from Ref. (74).

1.3 Classification of magneto-optic effects

1.3.1 Faraday effect

The oldest and best-known MO effect is the Faraday effect, also called magnetic circular birefringence (see Fig. 1.3). A magnetic field is applied to a sufficiently transparent sample parallel to the propagation direction of linearly polarized light. After having passed through the sample, the plane of polarization of the light is rotated by an angle θ_F, which we call the Faraday rotation. If the sample exhibits absorption the emergent light will generally be elliptically polarized. Then, θ_F designates the rotation of the semimajor axis relative to the initial polarization direction and the ratio of the two semiaxes of the ellipse is defined as the tangent of the Faraday ellipticity, $\tan(\eta_F)$.

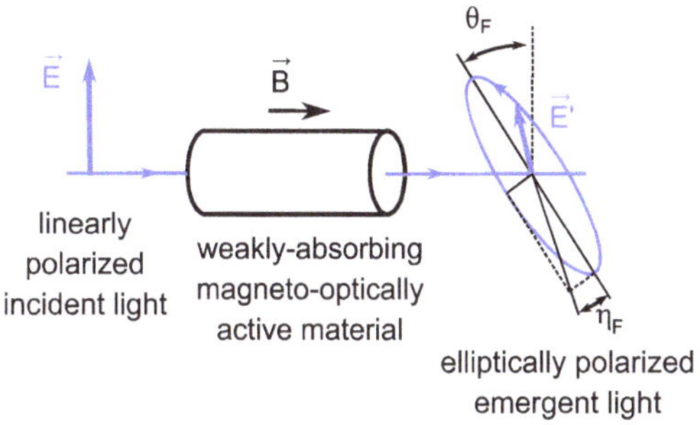

Fig. 1.3: Definition of the Faraday effect. Linearly polarized light is transmitted through a weakly absorbing magneto-optically active material. A magnetic field $\vec{B}$ is applied parallel to the light propagation direction. When leaving the material, the light is elliptically polarized. The direction of the semimajor axis is rotated by an angle θ_F with respect to the polarization direction of the incoming light. The ratio between semiminor and semimajor axis is tan η_F.

As we will learn later, linearly polarized light can be viewed as a combination of a right and a left circularly polarized wave. Keeping that in mind, a rotation θ_F is equivalent to a phase shift between the two circularly polarized partial waves or to a different propagation velocity of the two partial waves. As we shall see in Section 2.4.2, θ_F is proportional to the difference between the refractive index of right and left circularly polarized light. The difference is, in turn, proportional to the applied magnetic field.

In this context, it should be mentioned that there are many molecules that, being in solution, rotate the polarization direction of the light without having a magnetic field applied. This property is called ***optical activity*** and has nothing to do with magnetism. It occurs only in chiral materials or molecules, that is, materials or molecules having no mirror symmetry.

1.3.2 Voigt effect

If the magnetic field is applied perpendicular to the propagation direction of the light, the so-called Voigt effect can be observed, also known as transverse or linear magnetic birefringence (78). It has a sizable magnitude mainly in the immediate vicinity of sharp absorption lines. Through the action of the magnetic field, the refractive index for light polarized parallel to the magnetic field is not altered. For light polarized perpendicular to it, a spectral splitting of the absorption lines is observed. As a result (of a rather lengthy calculation), light which is linearly polarized at an angle of 45° to the magnetic field experiences a phase shift which is proportional to the difference between the refractive index of light polarized parallel and perpendicular to the magnetic field and proportional to the traveled distance. The emerging light is then elliptically polarized with the semimajor axis along the direction of the input polarization. The difference of the refractive index is proportional to the ***square*** of the applied magnetic field. In case of absorption, we find a rotation of the polarization plane, too.

1.3.3 Cotton–Mouton effect

Using the same geometry as in the Voigt effect, Aimé Cotton and Henri Mouton observed 1907 in certain liquids an effect, which is expressed in the same way with respect to change of polarization and field dependence as the Voigt effect (79). The origin of this so-called Cotton–Mouton effect is found in the alignment of magnetically anisotropic molecules within a magnetic field. When the molecules are also optically anisotropic, the alignment leads to a birefringence of the liquid in a homogeneous magnetic field.

The relationship with the ***electro-optical*** Kerr effect is that both have as a condition ***optically*** anisotropic molecules. The difference, however, is that in the first case a ***magnetic*** anisotropy of the molecules is necessary and in the other case an ***electrical***.

1.3.4 Majorana effect

In search of the magnetic analog of the electro-optical Kerr effect, Quirino Majorana discovered 1902 a magnetic birefringence in a transverse magnetic field in colloidal solutions (80). The experimental arrangement is identical to that of Cotton and Mouton. The effect differs from the Cotton–Mouton effect only in the fact that instead of molecules optically anisotropic microscopic particles are aligned in the magnetic field.

1.3.5 Magneto-optic Kerr effect

If linearly polarized light is reflected from the specular surface of magnetic materials, the reflected light will in general be elliptically polarized as shown in Fig. 1.4. This effect is named Kerr effect after its discoverer. Here, we define the deviation of the semimajor axis of the polarization ellipse from the direction of the incident polarization as the Kerr rotation θ_K and the ratio of the two semiaxes as the tangent of the Kerr ellipticity η_K.

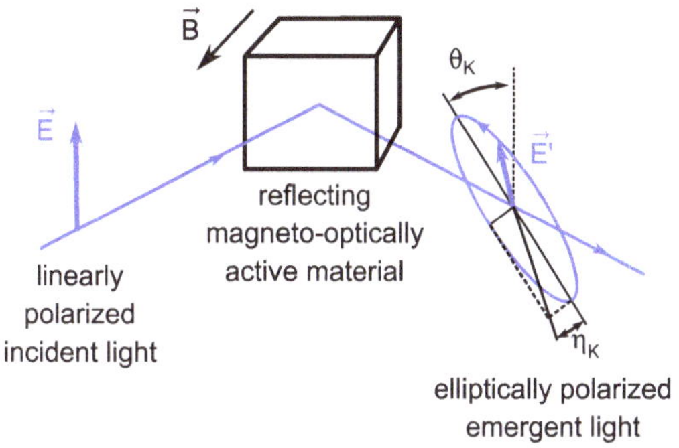

Fig. 1.4: Definition of the polar Kerr effect. Linearly polarized light is reflected from a polished magneto-optically active material. A magnetic field $\vec{B}$ is applied perpendicular to the reflecting surface of the material. After reflection, the light is elliptically polarized. The direction of the semimajor axis is rotated by an angle θ_K with respect to the polarization direction of the incoming light. The ratio between semiminor and semimajor axis is tan η_K.

Depending on the geometry, three different Kerr effects can be distinguished as plotted in Fig. 1.5. For the ***polar*** Kerr effect, the magnetic-field direction or the magnetization is ***perpendicular*** to the sample surface and ***within*** the plane of incidence of light. For the ***longitudinal*** Kerr effect, the magnetic field direction is also ***within*** the plane of incidence of light but ***parallel*** to the surface of the sample. For the ***transversal*** (also called transverse or equatorial) Kerr effect finally, the magnetic field direction is ***perpendicular*** to the plane of incidence of light and ***parallel*** to the sample surface.

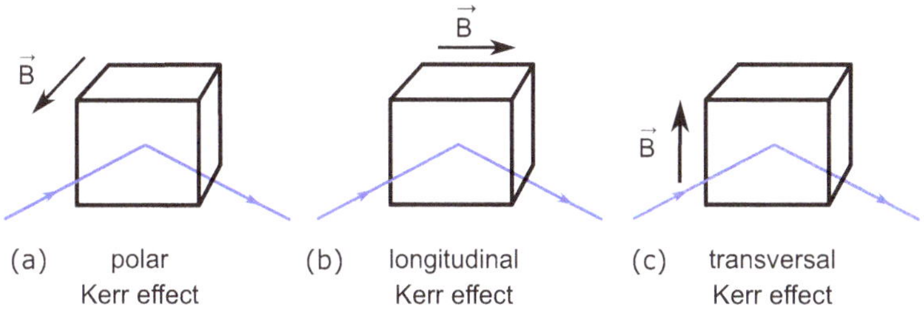

Fig. 1.5: The three types of Kerr effect: (a) polar Kerr effect, (b) longitudinal Kerr effect, and (c) transversal Kerr effect.

The polar geometry is most widely used since it establishes a direct connection with the Faraday effect and because it is the largest effect of the three. If no need for wavelength-dependent measurements exists, for example, taking MO hysteresis curves at a fixed photon energy of the light, then the other geometries are often advantageous since they are much easier accomplished experimentally. The longitudinal Kerr effect is, despite being a small effect, very useful for studying hysteresis curves in thin films because the easy axis of magnetization is usually in plane. A polar Kerr hysteresis curve would then not show a hysteresis because it is only sensitive to the hard axis component of magnetization.

For normal incidence of the light, the longitudinal Kerr effect merges toward the transversal. Then it occurs that light polarized parallel to the magnetic field is reflected without change. At grazing incidence, the longitudinal Kerr effect resembles the polar one at perpendicular incidence because the light direction becomes parallel to the magnetic field direction. In the problem in Section 2.6.1, we will use a simple argument to show these similarities between polar and longitudinal Kerr effect. Of course, the argument is very heuristic because of the different reflectivity at grazing incidence as opposed to normal incidence. We will discuss reflectivity in more detail in Sections 2.2 and 2.3. Furthermore, as we shall see in Section 3.2, the magnetic field introduces a true anisotropy in the material in a plane perpendicular to the magnetic field direction. Therefore, the anisotropy of the surface in the polar Kerr geometry differs from the anisotropy in the longitudinal and transversal Kerr geometry. This requires a different treatment for each Kerr effect. In this book, we limit ourselves to the discussion of the Faraday effect and the polar Kerr effect at normal incidence.

2 Basics of solid-state optics

2.1 Maxwell equations in matter and complex optical quantities

In this chapter, we are considering light propagation in an uncharged, polarizable, magnetizable, and conducting material. This means the material is as general as possible. The only property that we are not considering is an extra electric charge. We assume that due to the conductivity any extra charge will dissipate within a short time. Because of the general property of the material, Maxwell's equations look much more complicated than in vacuum. The goal will therefore be to simplify Maxwell's equations to be able to solve the wave equation. The price we will have to pay for the simplification is the introduction of complex optical quantities. They will make the mathematical treatment much simpler at the cost of clearness.

We use in this chapter in parallel the SI system (***s**ystème **i**nternational d'unités*) and the Gaussian cgs system (**c**entimeter **g**ram **s**econd), common in atomic physics and still found in a lot of textbooks. The equations in cgs system will be put between curly brackets. Both can be defined by Coulomb's law:

$$\vec{F}_C = f^* \frac{q_1 q_2}{r^2} \frac{\vec{r}}{r}. \tag{2.1}$$

If we set $f^* = 1$, the Gaussian cgs system follows. Setting $f^* = 1/(4\pi\varepsilon_0)$ defines the SI system, where ε_0 is the electric constant or vacuum permittivity. In cgs system, the basic units of a force and length are, respectively, dyn and cm. An electric charge has the unit $\text{dyn}^{1/2}\ \text{cm} = \text{g}^{1/2}\ \text{cm}^{3/2}\text{s}^{-1}$ which is abbreviated as esu (**e**lectro**s**tatic **u**nit of charge).

In its differential form, the four Maxwell equations for an uncharged, polarizable, magnetizable, and conducting material are written as

$$\vec{\nabla} \times \vec{H} = \dot{\vec{D}} + \vec{j} \qquad \left\{\text{cgs: } \vec{\nabla} \times \vec{H} = \frac{1}{c}\dot{\vec{D}} + \frac{4\pi}{c}\vec{j}\right\}, \tag{2.2}$$

$$\vec{\nabla} \times \vec{E} = -\dot{\vec{B}} \qquad \left\{\text{cgs: } \vec{\nabla} \times \vec{E} = -\frac{1}{c}\dot{\vec{B}}\right\}, \tag{2.3}$$

$$\vec{\nabla} \cdot \vec{B} = 0 \qquad \{\text{cgs: } \vec{\nabla} \cdot \vec{B} = 0\}, \tag{2.4}$$

$$\vec{\nabla} \cdot \vec{D} = 0 \qquad \{\text{cgs: } \vec{\nabla} \cdot \vec{D} = 0\}, \tag{2.5}$$

where the nabla operator $\vec{\nabla}$ is defined as $\vec{\nabla} = \left(\frac{\partial}{\partial x}, \frac{\partial}{\partial y}, \frac{\partial}{\partial z}\right)$, c is the speed of light in vacuum. $\vec{E}$, $\vec{D}$, $\vec{B}$, and $\vec{H}$ are, respectively, the electric field, the electric displacement, the magnetic field (also called magnetic induction or magnetic flux density), and the magnetic field strength. The vector $\vec{j}$ is the current density.

https://doi.org/10.1515/9783110635300-002

In vacuum, the following relation holds in the SI system: $\vec{H} = \vec{B}_0/\mu_0$, where $\vec{B}_0$ describes the magnetic field in the absence of matter and μ_0 is the magnetic constant or vacuum permeability. In eq. (2.5), we assumed that no permanent charge density ρ can exist in a conducting material; that is, $\rho = 0$. The electric displacement $\vec{D}$ and the magnetic field $\vec{B}$ are associated with the electric field $\vec{E}$ and the magnetic field strength $\vec{H}$ by the following relations:

$$\vec{B} = \mu_0(\vec{H} + \vec{M}) \qquad \{\text{cgs}: \vec{B} = \vec{H} + 4\pi\vec{M}\}, \tag{2.6}$$

$$\vec{D} = \varepsilon_0\vec{E} + \vec{P} \qquad \{\text{cgs}: \vec{D} = \vec{E} + 4\pi\vec{P}\}. \tag{2.7}$$

$\vec{P}$ denotes the electric polarization, that is, the electric dipole moment per unit volume, and $\vec{M}$ the magnetization, that is, the magnetic moment per unit volume.

The interaction with light changes the state of the material and creates new properties which were not existent before. We say that the light induces these properties. The light induces a current density $\vec{j}$ due to the material's conductivity σ. We call it the ***optical*** conductivity because we are dealing with infrared, visible or ultraviolet light. Furthermore, the light will polarize the material and induce a polarization $\vec{P}$ via the polarizability α of the material. Finally, the light will also magnetize the material and induce a magnetization $\vec{M}$ due to the magnetic susceptibility χ_m of the material. If we restrict ourselves to linear effects, that is, to fields that are not too large, we can simplify Maxwell's equations by introducing these material properties. Deviations from linearity occur only at very large electric-field strengths, which can be produced, for example, with lasers. Also, in the vicinity of ferromagnetic or ferroelectric saturation, there is no linear relationship between induced effects and electric or magnetic field strength. In the aforementioned cases, the frequency dependence of the material properties must be considered to determine whether a linear approach is valid in the relevant frequency range. We will return to this later. In the linear case, we can write:

$$\vec{j} = \sigma\vec{E}, \qquad \vec{P} = \alpha\vec{E}, \qquad \vec{M} = \chi_\text{m}\vec{H}. \tag{2.8}$$

Because $\vec{E}$, $\vec{H}$, $\vec{j}$, $\vec{P}$, and $\vec{M}$ are vectors, σ, α, and χ_m are in general tensors $\overleftrightarrow{\sigma}$, $\overleftrightarrow{\alpha}$, and $\overleftrightarrow{\chi}_\text{m}$ of the following form:

$$\overleftrightarrow{\sigma} = \begin{pmatrix} \sigma_{xx} & \sigma_{xy} & \sigma_{xz} \\ \sigma_{yx} & \sigma_{yy} & \sigma_{yz} \\ \sigma_{zx} & \sigma_{zy} & \sigma_{zz} \end{pmatrix}. \tag{2.9}$$

In isotropic as well as in cubic materials, these tensors reduce to scalars (or, equivalently, to a diagonal tensor with identical elements). Of course, this applies only if no external field is present. We will see in the following chapters that in magneto-optics off-diagonal elements appear also in cubic or isotropic systems due to symmetry breaking by the applied magnetic field. By inserting the linear relations eqs. (2.6)–(2.8) into the Maxwell equations eqs. (2.2)–(2.5), we can express the Maxwell equations with only two fields, the electric field and the magnetic field:

$$\vec{\nabla}\times\frac{\vec{B}}{(1+\chi_m)\mu_0}=(\varepsilon_0+\alpha)\dot{\vec{E}}+\sigma\vec{E}\qquad\left\{\text{cgs}:\vec{\nabla}\times\frac{\vec{B}}{1+4\pi\chi_m}=\frac{1}{c}(1+4\pi\alpha)\dot{\vec{E}}+\frac{4\pi}{c}\sigma\vec{E}\right\},\tag{2.10}$$

$$\vec{\nabla}\times\vec{E}=-\dot{\vec{B}}\qquad\left\{\text{cgs}:\vec{\nabla}\times\vec{E}=-\frac{1}{c}\dot{\vec{B}}\right\},\tag{2.11}$$

$$\vec{\nabla}\cdot\vec{B}=0\qquad\{\text{cgs}:\vec{\nabla}\cdot\vec{B}=0\},\tag{2.12}$$

$$(\varepsilon_0+\alpha)\;\vec{\nabla}\cdot\vec{E}=0\qquad\{\text{cgs}:(1+4\pi\alpha)\vec{\nabla}\cdot\vec{E}=0\}.\tag{2.13}$$

If we finally introduce the dielectric function $\varepsilon(\omega)=1+\alpha/\varepsilon_0$ {cgs: $\varepsilon(\omega)=1+4\pi\alpha$} and the magnetic permeability $\mu=1+\chi_m$ {cgs: $\mu=1+4\pi\chi_m$}, we get $\vec{D}=\varepsilon\varepsilon_0\vec{E}$, $\vec{B}=\mu\mu_0\vec{H}$. By inserting these relations, the four Maxwell equations can be written as:

$$\vec{\nabla}\times\vec{B}=\mu\mu_0\varepsilon\varepsilon_0\dot{\vec{E}}+\mu\mu_0\sigma\vec{E}\qquad\left\{\text{cgs}:\vec{\nabla}\times\vec{B}=\frac{\mu\varepsilon}{c}\dot{\vec{E}}+\frac{4\pi\mu}{c}\sigma\vec{E}\right\},\tag{2.14}$$

$$\vec{\nabla}\times\vec{E}=-\dot{\vec{B}}\qquad\left\{\text{cgs}:\vec{\nabla}\times\vec{E}=-\frac{1}{c}\dot{\vec{B}}\right\},\tag{2.15}$$

$$\vec{\nabla}\cdot\vec{B}=0\qquad\{\text{cgs}:\vec{\nabla}\cdot\vec{B}=0\},\tag{2.16}$$

$$\varepsilon\varepsilon_0\vec{\nabla}\cdot\vec{E}=0\qquad\{\text{cgs}:\varepsilon\,\vec{\nabla}\cdot\vec{E}=0\}.\tag{2.17}$$

The equations still look rather complicated because they contain the first derivative of the electric and magnetic field. We can get rid of the time derivatives by assuming an explicit time dependence for the two fields. For that, we consider only periodic solutions as represented by harmonic plane waves. This constitutes no limitation because any periodic solution can be decomposed into a finite series of harmonic functions according to the Fourier theorem. This will even work for non-periodic solutions. We would have to consider then an infinite Fourier series. Let us further assume a complex representation of the waves and define the time dependence of the solutions as $e^{+i\omega t}$. This yields for the first time derivative $\dot{\vec{E}}=i\omega\vec{E}$ or $\vec{E}=\dot{\vec{E}}/(i\omega)$. If we substitute the latter expression in eq. (2.14), it can be expressed in an analogous manner as eq. (2.15):

$$\vec{\nabla}\times\vec{B}=\mu\mu_0\left(\varepsilon\varepsilon_0\dot{\vec{E}}-i\frac{\sigma}{\omega}\dot{\vec{E}}\right)=\mu\mu_0\tilde{\varepsilon}\varepsilon_0\dot{\vec{E}}\qquad\left\{\text{cgs}:\vec{\nabla}\times\vec{B}=\frac{\mu\varepsilon}{c}\dot{\vec{E}}-i\frac{4\pi\mu}{\omega c}\sigma\dot{\vec{E}}=\frac{\mu\tilde{\varepsilon}}{c}\dot{\vec{E}}\right\},\tag{2.18}$$

where

$$\tilde{\varepsilon}=\varepsilon-i\frac{\sigma}{\varepsilon_0\omega}\qquad\left\{\text{cgs}:\tilde{\varepsilon}=\varepsilon-i\frac{4\pi}{\omega}\sigma\right\}.\tag{2.19}$$

By defining a complex dielectric function,

$$\tilde{\varepsilon} = \varepsilon_1 - \mathrm{i}\varepsilon_2, \tag{2.20}$$

we can identify real and imaginary parts as

$$\varepsilon_1 = \varepsilon, \ \varepsilon_2 = \frac{\sigma}{\varepsilon_0 \omega} \qquad \left\{\text{cgs: } \varepsilon_1 = \varepsilon, \ \varepsilon_2 = \frac{4\pi}{\omega}\sigma\right\}. \tag{2.21}$$

So, the real part of $\tilde{\varepsilon}$ is the dielectric function of the material and the imaginary part is proportional to the optical conductivity σ. Note that the tilde symbol "~" is used throughout the book to denote complex quantities. Finally, we can solve eqs. (2.15)–(2.18) for one of the two fields. This will yield the wave equation. For that purpose, we apply $-\vec{\nabla} \times (\ldots)$ to eq. (2.15) and $\frac{\partial}{\partial t}(\ldots)$ $\{\text{cgs: } \frac{1}{c}\frac{\partial}{\partial t}(\ldots)\}$ to eq. (2.18) and sum up the two equations. The result is:

$$-\vec{\nabla} \times (\vec{\nabla} \times \vec{E}) = \frac{\mu\tilde{\varepsilon}}{c^2}\ddot{\vec{E}}, \tag{2.22}$$

where we used the identity $\mu_0\varepsilon_0 = 1/c^2$ in the SI system. Because the vector relations $\vec{\nabla} \times (\vec{\nabla} \times \vec{E}) = \vec{\nabla}(\vec{\nabla} \cdot \vec{E}) - \Delta\vec{E}$ and $\vec{\nabla} \cdot \vec{E} = 0$ hold, eq. (2.22) can be reduced to the wave equation for the electric field:

$$\Delta\vec{E} = \frac{\mu\tilde{\varepsilon}}{c^2}\ddot{\vec{E}}. \tag{2.23}$$

Analogously, we can derive a wave equation for the magnetic field:

$$\Delta\vec{B} = \frac{\mu\tilde{\varepsilon}}{c^2}\ddot{\vec{B}}. \tag{2.24}$$

Note that $\vec{\nabla} \cdot \vec{E} = 0$ holds only in the case of isotropic materials. For anisotropic materials, eq. (2.22) cannot be reduced to eq. (2.23). For further details, see the problem in Section 2.6.7. The solutions of the wave equations (2.23) and (2.24) are plane waves which have a time dependence as assumed above. They have the following form:

$$\vec{E} = \vec{E}_0 \, \mathrm{e}^{\mathrm{i}(\omega t - \tilde{\vec{g}} \cdot \vec{r})}, \tag{2.25}$$

$$\vec{B} = \vec{B}_0 \, \mathrm{e}^{\mathrm{i}(\omega t - \tilde{\vec{g}} \cdot \vec{r})}, \tag{2.26}$$

where $\tilde{\vec{g}}$ is the wave vector whose components might be complex. If the absolute value of the wave vector g is real, there exists no damping, that is, no absorption. So, we have $g = 2\pi/\lambda_{\text{mat}} = 2\pi n/\lambda_{\text{vac}}$, where n is the material's refractive index and λ_{mat} and λ_{vac} are the wavelength of the light in material and in vacuum, respectively. In general, the components and length of $\tilde{\vec{g}}$ are complex numbers and we can relate it to the complex refractive index $\tilde{n}$:

$$\sqrt{\vec{\tilde{g}}\cdot\vec{\tilde{g}}} = \left|\vec{\tilde{g}}\right| = \tilde{g} = g_1 - \mathrm{i}g_2 = \frac{2\pi}{\lambda_{\text{vac}}}(n - \mathrm{i}k) = \frac{\omega}{c}(n - \mathrm{i}k) = \frac{\omega}{c}\tilde{n}. \tag{2.27}$$

Here, we define a minus sign for the imaginary part of $\tilde{g}$ to be consistent with the definition of $\tilde{n}$, especially with the request that the absorption index must always have a positive value. Note that in the first term of eq. (2.27), the ordinary dot product for vectors in real space is used although the components of $\vec{\tilde{g}}$ might have complex values. Furthermore, g_1, $g_2 \in \mathbb{R}$. The real part n is called refractive index and the imaginary part k is called absorption index. In the problem in Section 2.6.2, we will have a closer look on the vector properties of the complex wave vector $\vec{\tilde{g}}$. The effect that $\vec{\tilde{g}}$ exerts on the electric field $\vec{E}$ is discussed in the problem in Section 2.6.3.

Before we continue the discussion, we should stop for a minute and reflect what the cost is we must pay for introducing complex optical quantities to bring Maxwell's equations into a simple form similar to the vacuum case. We started with assuming a complex time dependence of the solutions. As shown in eqs. (2.25) and (2.26), this leads to a complex representation of the plane waves that solve Maxwell's equations. However, this is a ***non-physical*** representation. To obtain the ***physical*** fields, we must always take the ***real*** part at the end. But we must be very careful. Although differentiation and integration are additive operations, it is ***not*** true that if a complex function solves Maxwell's equation (or any other complex equation), the real and the imaginary part will be a solution as well. Furthermore, for a complex wave vector $\vec{\tilde{g}}$ it is delicate to define the direction of propagation. Finally, the two field amplitude vectors $\vec{E}_0$ and $\vec{B}_0$ might have complex components as well. Then, although mathematical relations derived later look in form as in the vacuum case, the interpretation in terms of direction and relative orientation becomes intricate. We will address these problems in Section 2.3.2.

Now, let us continue with discussing the solutions of the wave equation. Inserting the complex representation of plane waves, eqs. (2.25) and (2.26), in the wave equations, eqs. (2.23) and (2.24), yields:

$$\vec{\tilde{g}}\cdot\vec{\tilde{g}} = \tilde{g}^2 = \frac{\mu\tilde{\varepsilon}}{c^2}\,\omega^2. \tag{2.28}$$

From comparison with eq. (2.27) follows a relation between the dielectric function and the refractive index:

$$\mu\tilde{\varepsilon} = \tilde{n}^2 = (n - \mathrm{i}k)^2. \tag{2.29}$$

In most materials, we can approximate $\mu = 1$. At optical frequencies (i.e., $\omega \geq 10^{11}\ \mathrm{s}^{-1}$), this holds even for ferromagnets according to the argumentation of Landau and Lifshitz (81) and Pershan (82). Thus, it follows:

$$\tilde{\varepsilon} = \tilde{n}^2 = (n - \mathrm{i}k)^2 = n^2 - k^2 - \mathrm{i}\,2nk. \tag{2.30}$$

By comparison with eq. (2.20), we get the dependence of the real and imaginary part of the dielectric function on the refractive index and on the absorption index:

$$\varepsilon_1 = n^2 - k^2, \tag{2.31}$$

$$\varepsilon_2 = 2nk. \tag{2.32}$$

This is not the only way to simplify the four Maxwell equations. Especially in metals, a complex optical conductivity $\tilde{\sigma}$ is introduced instead of a complex dielectric function $\tilde{\varepsilon}$. Substituting in eq. (2.14) $\dot{\vec{E}} = \mathrm{i}\omega\vec{E}$ instead of $\vec{E} = \dot{\vec{E}}/(\mathrm{i}\omega)$ yields the complex optical conductivity,

$$\tilde{\sigma} = \sigma_1 + \mathrm{i}\sigma_2. \tag{2.33}$$

Note the plus sign between real and imaginary part. The real part corresponds to the optical conductivity of the material and the imaginary part is proportional to the polarizability:

$$\tilde{\sigma} = \sigma + \mathrm{i}\;\omega\alpha, \quad \text{where } \sigma_1 = \sigma \text{ and } \sigma_2 = \omega\alpha. \tag{2.34}$$

Here, we defined the imaginary part of $\tilde{\sigma}$ using the polarizability α and not the dielectric function ε as we could have done as well. The reason for that is that with this definition, we obtain the vacuum equations when we set $\tilde{\sigma} = 0$. In eq. (2.21), we had already associated the real part of $\tilde{\sigma}$ with $\varepsilon_2 = \sigma/(\varepsilon_0\omega)$ {cgs: $\varepsilon_2 = 4\pi\sigma/\omega$}. Accordingly, we can derive from this a general relation between $\tilde{\sigma}$ and $\tilde{\varepsilon}$:

$$\begin{gathered}\tilde{\sigma} = \sigma_1 + \mathrm{i}\omega\alpha = \omega\varepsilon_0\varepsilon_2 + \mathrm{i}\omega\varepsilon_0(\varepsilon_1 - 1) = \mathrm{i}\omega\varepsilon_0(\tilde{\varepsilon} - 1) \\ \left\{\text{cgs: } \tilde{\sigma} = \sigma + \mathrm{i}\omega\alpha = \frac{\omega}{4\pi}\varepsilon_2 + \mathrm{i}\frac{\omega}{4\pi}(\varepsilon_1 - 1) = \mathrm{i}\frac{\omega}{4\pi}(\tilde{\varepsilon} - 1)\right\},\end{gathered} \tag{2.35}$$

or, solving for the dielectric function:

$$\tilde{\varepsilon} = 1 - \mathrm{i}\frac{\tilde{\sigma}}{\varepsilon_0\omega} \qquad \left\{\text{cgs: } \tilde{\varepsilon} = 1 - \mathrm{i}\frac{4\pi}{\omega}\tilde{\sigma}\right\}. \tag{2.36}$$

We now want to have a closer look at the solutions of the wave equation eq. (2.23). For this purpose, we first insert eqs. (2.25) and (2.26) into eqs. (2.16) and (2.17), respectively, and obtain:

$$\tilde{\vec{g}} \cdot \vec{E}_0 = 0 \text{ and } \tilde{\vec{g}} \cdot \vec{B}_0 = 0. \tag{2.37}$$

This is in form identical to the solutions in vacuum. Nevertheless, only if $\tilde{\vec{g}}$ has a well-defined direction of propagation, this will mean that the electric-field amplitude $\vec{E}_0$ and the magnetic-field amplitude $\vec{B}_0$ are perpendicular to the propagation direction. Now, let us consider a plane wave, propagating in z direction, which is linearly polarized along the x direction, that is, $E_y = E_z = B_z = 0$, $E_x = E_{0x}\,\mathrm{e}^{\mathrm{i}(\omega t - \tilde{g}z)}$ and

$\vec{B} = \vec{B}_0\, e^{i(\omega t - \tilde{g}z)}$. In this case, $\tilde{\vec{g}}$ has a well-defined direction: $\tilde{\vec{g}} = \tilde{g}\,\hat{e}_z$. From eq. (2.18), using in SI system again the identity $\mu_0\varepsilon_0 = 1/c^2$, we get:

$$(\vec{\nabla}\times\vec{B})_x = \frac{\partial B_z}{\partial y} - \frac{\partial B_y}{\partial z} = \frac{\mu\tilde{\varepsilon}}{c^2}\dot{E}_x = i\omega\frac{\mu\tilde{\varepsilon}}{c^2}E_x \quad \left\{\text{cgs: } (\vec{\nabla}\times\vec{B})_x = \frac{\partial B_z}{\partial y} - \frac{\partial B_y}{\partial z} = \frac{\mu\tilde{\varepsilon}}{c}\dot{E}_x = i\omega\frac{\mu\tilde{\varepsilon}}{c}E_x\right\}, \tag{2.38}$$

$$(\vec{\nabla}\times\vec{B})_y = \frac{\partial B_x}{\partial z} - \frac{\partial B_z}{\partial x} = \frac{\mu\tilde{\varepsilon}}{c^2}\dot{E}_y = 0 \qquad \left\{\text{cgs: } (\vec{\nabla}\times\vec{B})_y = \frac{\partial B_x}{\partial z} - \frac{\partial B_z}{\partial x} = \frac{\mu\tilde{\varepsilon}}{c}\dot{E}_y = 0\right\}, \tag{2.39}$$

$$(\vec{\nabla}\times\vec{B})_z = \frac{\partial B_y}{\partial x} - \frac{\partial B_x}{\partial y} = \frac{\mu\tilde{\varepsilon}}{c^2}\dot{E}_z = 0 \qquad \left\{\text{cgs: } (\vec{\nabla}\times\vec{B})_z = \frac{\partial B_y}{\partial x} - \frac{\partial B_x}{\partial y} = \frac{\mu\tilde{\varepsilon}}{c}\dot{E}_z = 0\right\}. \tag{2.40}$$

Performing the space derivative on the magnetic field and remembering that it only depends on the variable z and $B_z = 0$, we get:

$$B_y = -\frac{i\omega}{(-i\tilde{g})}\frac{\mu\tilde{\varepsilon}}{c^2}E_x = \frac{\mu\tilde{\varepsilon}}{c\tilde{n}}E_x = \frac{\tilde{n}}{c}E_x \qquad \left\{\text{cgs: } B_y = -\frac{i\omega\mu}{(-i\tilde{g})}\frac{\tilde{\varepsilon}}{c}E_x = \frac{\mu\tilde{\varepsilon}}{\tilde{n}}E_x = \tilde{n}E_x\right\}, \tag{2.41}$$

$$-i\tilde{g}B_x = 0 \quad \{\text{cgs: } -i\tilde{g}B_x = 0\}. \tag{2.42}$$

Equation (2.42) shows that the magnetic field has only a component in y direction. Thus, $\vec{B}$ and $\vec{E}$ are perpendicular to the wave vector $\tilde{\vec{g}}$, perpendicular to each other and the three vectors $\vec{E}$, $\vec{B}$, and $\tilde{\vec{g}}$ form a right-handed helix, exactly as in the vacuum case again. Finally, the amplitude of the magnetic field is proportional to the amplitude of the electric field and the factor is simply $\tilde{n}/c$. If absorption is present, that is, $k \neq 0$, the component B_y will be complex even if E_x is real. As only the real part of the field is physically relevant, this means that there is a phase shift between the magnetic and electric field. Consequently, the nodes of $\vec{B}$ and $\vec{E}$ are displaced along the propagation direction. For more details, see the problem in Section 2.6.4.

We have previously defined three pairs of variables to describe the optical properties of a material. However, there are many more options. For example, for the time dependence of the plane waves, we could have used $e^{-i\omega t}$ in place of $e^{+i\omega t}$. In all equations, we would then have to replace i by –i, for example, $\tilde{n} = n + ik$ and so forth. For the complex refractive index also the definition $\tilde{n} = n(1 - i\kappa)$ is found such that $\kappa = k/n$.

We finally need to introduce a quantity for absorption that is directly accessible to experiment. Usually, the absorption coefficient K (also called absorption constant) is used. K is equal to 1 m^{-1} {cgs: 1 cm^{-1}} if the intensity of a light beam is attenuated by a factor e (Euler's number) when traversing a material of 1 m length {cgs: 1 cm}. This can be expressed as Beer–Lambert law:

$$I = I_0\, e^{-Kd}, \tag{2.43}$$

where I_0 is the intensity of the incoming ray and I is the intensity of the light after having traversed a length d. The light intensity is related to the Poynting vector $\vec{S}$, which describes the energy flow of a wave:

$$\vec{S} = \mathrm{Re}\left\{\vec{E}\right\} \times \mathrm{Re}\left\{\vec{H}\right\} = \frac{1}{\mu\mu_0}\mathrm{Re}\left\{\vec{E}\right\} \times \mathrm{Re}\left\{\vec{B}\right\}$$

$$\left\{\text{cgs: } \vec{S} = \frac{c}{4\pi}\mathrm{Re}\left\{\vec{E}\right\} \times \mathrm{Re}\left\{\vec{H}\right\} = \frac{c}{4\pi\mu}\mathrm{Re}\left\{\vec{E}\right\} \times \mathrm{Re}\left\{\vec{B}\right\}\right\}. \tag{2.44}$$

Note that $\vec{S}$ is defined using the real part of the two field vectors. The reason for that is that the vector product involves multiplications of two complex field components and the real part of the product is not the product of the real parts of the multiplicands. With eqs. (2.25)–(2.27) and because $\vec{E}$ is perpendicular to $\vec{B}$, it follows (assuming again $\mu = 1$):

$$S = \frac{1}{c\mu_0}\sqrt{n^2+k^2}\,E_0^2\,\mathrm{e}^{-\frac{2\omega kd}{c}}\cos\left[\omega\left(t-\frac{nd}{c}\right)\right]\cos\left[\omega\left(t-\frac{nd}{c}\right)-\varphi\right], \quad \text{with } \varphi = \arctan\frac{k}{n}, \tag{2.45}$$

where we have used the polar representation for the complex refractive index. For a detailed calculation, see the problem in Section 2.6.5. Finally, the time-averaged absolute value of the Poynting vector $\langle S\rangle_t$ is equal to the intensity I:

$$I = \langle S\rangle_t = \frac{1}{2}cn\varepsilon_0|E_0|^2\mathrm{e}^{-\frac{2\omega kd}{c}} \qquad \left\{\text{cgs: } I = \langle S\rangle_t = \frac{1}{8\pi}\,n\,|E_0|^2\,\mathrm{e}^{-\frac{2\omega kd}{c}}\right\}. \tag{2.46}$$

Comparing Beer–Lambert law, eq. (2.43), with eq. (2.46) yields for the absorption coefficient K:

$$K = \frac{2\omega}{c}k = 4\pi\frac{k}{\lambda_{\mathrm{vac}}}. \tag{2.47}$$

Thus, the absorption coefficient K is directly proportional to the absorption index k times frequency of the light ω. The inverse of the absorption coefficient is a measure of the penetration depth of the light ray into a material. For a thickness $d = 1/K$, the intensity has decreased by a factor e. If the light is reflected, it will travel the thickness twice and the reduction is already more than a factor 7. Thus, $1/K$ is a good approximation for the penetration or interaction depth of light in material. From eq. (2.47), we see that $K^{-1} \cong \lambda_{\mathrm{vac}}/(12\,k)$. So, for $k = 1$, the penetration depth is only around 50 nm for green light. Metals have an absorption index which is usually much larger than one. Therefore, the penetration depth of light in metals is rarely much more than 10 nm in the visible and infrared.

Let us at the end discuss what the significance is of the different optical quantities. Assuming no absorption, that is, $k = 0$, it follows that the absorption coefficient, the imaginary part of the dielectric function, and the real part of the optical conductivity vanish: $K = 0$, $\varepsilon_2 = 0$, and $\sigma_1 = 0$. Consequently, the dielectric function $\tilde{\varepsilon}$ and the refractive index $\tilde{n}$ are real numbers, that is, $n = \sqrt{\varepsilon\mu} \cong \sqrt{\varepsilon}$. Therefore, the medium is transparent and no energy is absorbed. It is interesting to note that even for $k \neq 0$ no energy will be absorbed if $n = 0$. This is because the time-averaged absorbed power per unit volume P is given by microscopic Ohm's law:

$$P = \left\langle \mathrm{Re}\left\{\tilde{\vec{j}}\right\} \cdot \mathrm{Re}\left\{\vec{E}\right\}\right\rangle_t = \sigma_1 \left\langle (\mathrm{Re}\{E\})^2 \right\rangle_t, \tag{2.48}$$

and the real part of the optical conductivity $\sigma_1 = 2nk\omega\varepsilon_0$ {cgs: $\sigma_1 = nk\omega/2\pi$} is still zero. The Fresnel formulas, which will be derived in the next chapter, show that at normal incidence the reflectivity R is given by

$$R = \frac{(n-1)^2 + k^2}{(n+1)^2 + k^2}. \tag{2.49}$$

For $n = 0$, it follows that $R = 1$, that is, 100% of the light is reflected and nothing transmitted. We say the light is totally reflected.

2.2 Fresnel equations

So far, we have treated Maxwell's equations inside a material. However, to probe the optical properties of a material, the electromagnetic radiation must enter and exit the material through an interface. The Fresnel equations describe the relations that are valid for refraction and reflection at the interface. The Fresnel equations follow straightforwardly from Maxwell's equations considering the continuity conditions at the interface.

Without loss of generality, we choose the interface between two optically different materials 1 and 2, with refractive index n_1 and n_2, respectively, as the xz plane of a right-handed coordinate system. From half space $y > 0$, which is material 1, a linearly polarized plane wave impinges on the interface at an angle α with respect to the interface normal and enters material 2 as shown in Fig. 2.1. The plane of incidence is defined by the interface normal and the incoming light ray.

The restriction to linear polarization is no limitation. We will successively examine the special cases of polarization perpendicular and parallel to the plane of incidence. Any other polarization state can then be derived from the two cases. The incident wave will be refracted at an angle β and reflected at an angle α' with respect to the interface normal. The amplitude of the electric field of the incident, reflected, and refracted wave is E_{0i}, E_{0r}, and E_{0t}, respectively.

2.2.1 Electric field vector perpendicular to the plane of incidence

First, a plane wave of light polarized perpendicular to the plane of incidence will be considered. We call this type of polarization s-polarized (from the German word "senkrecht" for perpendicular). We use the ansatz of eqs. (2.25) and (2.26) but we disregard time dependence since it is the same for all waves. Thus, we obtain for the electric field of the incident, reflected and refracted wave (see Fig. 2.1):

$$E_{iz} = E_{0i}\, e^{-i\tilde{g}_1 (x \sin\alpha - y\cos\alpha)}, \tag{2.50}$$

$$E_{rz} = E_{0r}\, e^{-i\tilde{g}_1 (x \sin\alpha' + y\cos\alpha')}, \tag{2.51}$$

$$E_{tz} = E_{0t}\, e^{-i\tilde{g}_2 (x \sin\beta - y\cos\beta)}, \tag{2.52}$$

with the wave vectors $\vec{\tilde{g}}_i = \tilde{g}_1(\sin\alpha, -\cos\alpha, 0)$, $\vec{\tilde{g}}_r = \tilde{g}_1(\sin\alpha', \cos\alpha', 0)$, $\vec{\tilde{g}}_t = \tilde{g}_2(\sin\beta, -\cos\beta, 0)$ for the incoming, reflected, and refracted wave, respectively.

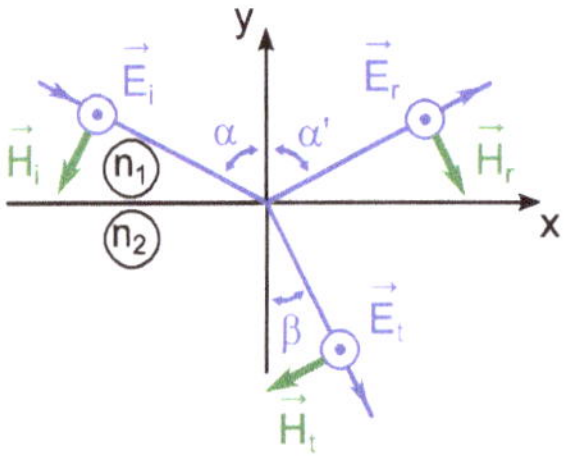

Fig. 2.1: Geometry for the derivation of the Fresnel equation for s-polarized light. $\vec{E}_i$ points out of the plane of incidence along the z direction. The interface (xz plane) is formed between material 1 ($y > 0$) with refractive index n_1 and material 2 ($y < 0$) with refractive index n_2.

At the interface, that is, at $y = 0$, the components which are parallel to the interface, $E_{||}$ and $H_{||}$, of the total electric field and of the total magnetic field strength, respectively, must be continuous everywhere on the interface, which means for all values of x and z. For the electric field, which is parallel to the interface due to its s-polarization, this requires that the sum of the electric fields in material 1 must be equal to the sum of the electric fields in material 2:

$$E_{0i}\, e^{-i\tilde{g}_1 x \sin\alpha} + E_{0r}\, e^{-i\tilde{g}_1 x \sin\alpha'} = E_{0t}\, e^{-i\tilde{g}_2 x \sin\beta}. \tag{2.53}$$

As said above, eq. (2.53) must apply simultaneously for all values of x. Therefore, for finite E_{0i}, it can be satisfied only if the exponential term is identical in all three partial waves:

$$\tilde{g}_1 \sin\alpha \equiv \tilde{g}_1 \sin\alpha' \equiv \tilde{g}_2 \sin\beta. \tag{2.54}$$

We note from eq. (2.54) that the component of the wave vector which is parallel to the interface is conserved. Furthermore, from the first identity, the ***law of reflection*** follows:

$$\alpha = \alpha' \tag{2.55}$$

and from the second, the ***law of refraction*** (Snell's law):

$$\tilde{g}_1 \sin\alpha = \tilde{g}_2 \sin\beta. \tag{2.56}$$

Using eq. (2.27), this can be rewritten in the familiar form:

$$\tilde{n}_1 \sin\alpha = \tilde{n}_2 \sin\beta. \tag{2.57}$$

After canceling the exponential terms in eq. (2.53), we find for the amplitudes:

$$E_{0\mathrm{i}} + E_{0\mathrm{r}} = E_{0\mathrm{t}}. \tag{2.58}$$

For determining the two unknown amplitudes, we must make a similar consideration for the magnetic field strength $\vec{H}$. According to eqs. (2.41), (2.6), and (2.8), the amplitude of $\vec{H}$ is calculated from the amplitude of $\vec{E}$ by multiplying with $\tilde{\varepsilon}/(c\mu_0\tilde{n}) = c\varepsilon_0\tilde{\varepsilon}/\tilde{n}\{\text{cgs: } \tilde{\varepsilon}/\tilde{n}\}$. We know that $\vec{E}$, $\vec{H}$, and $\vec{\tilde{g}}$ form a right-handed helix. With $\vec{E}$ perpendicular to the plane of incidence (see Fig. 2.1), $H_{0\mathrm{i}}$ and $H_{0\mathrm{t}}$ point, therefore, to the left and $H_{0\mathrm{r}}$ to the right. The boundary condition requires that the parallel component $H_{||}$ of total magnetic field strength must be continuous everywhere on the interface. For the chosen geometry, this is the x component of $\vec{H}$. To obtain the parallel component, $H_{0\mathrm{i}}$ and $H_{0\mathrm{t}}$ are therefore multiplied by $-\cos\alpha$ and $-\cos\beta$, respectively, and $H_{0\mathrm{r}}$ by $\cos\alpha' = \cos\alpha$. Thus, we get:

$$H_{\mathrm{i}x} = -c\,\varepsilon_0 \frac{\tilde{\varepsilon}_1}{\tilde{n}_1} \cos\alpha\, E_{0\mathrm{i}}\, \mathrm{e}^{-\mathrm{i}\tilde{g}_1 (x\sin\alpha - y\cos\alpha)}, \tag{2.59}$$

$$H_{\mathrm{r}x} = c\,\varepsilon_0 \frac{\tilde{\varepsilon}_1}{\tilde{n}_1} \cos\alpha\, E_{0\mathrm{r}}\, \mathrm{e}^{-\mathrm{i}\tilde{g}_1 (x\sin\alpha + y\cos\alpha)}, \tag{2.60}$$

$$H_{\mathrm{t}x} = -c\,\varepsilon_0 \frac{\tilde{\varepsilon}_2}{\tilde{n}_2} \cos\beta\, E_{0\mathrm{t}}\, \mathrm{e}^{-\mathrm{i}\tilde{g}_2 (x\sin\beta - y\cos\beta)}. \tag{2.61}$$

The continuity conditions at the interface, that is, at $y = 0$, requires $H_{\mathrm{i}x} + H_{\mathrm{r}x} = H_{\mathrm{t}x}$. Keeping in mind, eq. (2.54), this yields:

$$\frac{\tilde{\varepsilon}_1}{\tilde{n}_1} \cos\alpha\, (E_{0\mathrm{r}} - E_{0\mathrm{i}}) = -\frac{\tilde{\varepsilon}_2}{\tilde{n}_2} \cos\beta\, E_{0\mathrm{t}}, \tag{2.62}$$

or:

$$E_{0\mathrm{i}} - E_{0\mathrm{r}} = \frac{\tilde{\varepsilon}_2\tilde{n}_1}{\tilde{\varepsilon}_1\tilde{n}_2} \frac{\cos\beta}{\cos\alpha} E_{0\mathrm{t}}. \tag{2.63}$$

The expression

$$\tilde{\varepsilon}_2\tilde{n}_1/(\tilde{\varepsilon}_1\tilde{n}_2) = \mu_1\tilde{n}_2/(\mu_2\tilde{n}_1) = \sqrt{\mu_1\tilde{\varepsilon}_2/(\mu_2\tilde{\varepsilon}_1)} \tag{2.64}$$

is called the characteristic wave impedance. By adding or subtracting eq. (2.63) from eq. (2.58) one obtains:

$$2E_{0i} = \left(1 + \frac{\tilde{\varepsilon}_2\tilde{n}_1}{\tilde{\varepsilon}_1\tilde{n}_2}\frac{\cos\beta}{\cos\alpha}\right)E_{0t}, \tag{2.65}$$

$$2E_{0r} = \left(1 - \frac{\tilde{\varepsilon}_2\tilde{n}_1}{\tilde{\varepsilon}_1\tilde{n}_2}\frac{\cos\beta}{\cos\alpha}\right)E_{0t}. \tag{2.66}$$

In most cases, we can set $\mu = 1$ and therefore $\tilde{\varepsilon}_{1,2} = \tilde{n}_{1,2}^2$ (see eq. (2.29)). Using Snell's law, which is eq. (2.57), eqs. (2.65) and (2.66) can then be written as

$$2E_{0i} = \left(1 + \frac{\tilde{n}_2}{\tilde{n}_1}\frac{\cos\beta}{\cos\alpha}\right)E_{0t} = \left(1 + \frac{\sin\alpha\cos\beta}{\sin\beta\cos\alpha}\right)E_{0t}, \tag{2.67}$$

$$2E_{0r} = \left(1 - \frac{\tilde{n}_2}{\tilde{n}_1}\frac{\cos\beta}{\cos\alpha}\right)E_{0t} = \left(1 - \frac{\sin\alpha\cos\beta}{\sin\beta\cos\alpha}\right)E_{0t}. \tag{2.68}$$

By dividing eq. (2.68) by eq. (2.67), making of same denominator, and using the addition theorems for trigonometric functions on the one hand and solving eq. (2.67) for E_{0t} on the other hand, we can finally write the first set of Fresnel equations for s-polarized light:

$$E_{0r} = -\frac{\sin(\alpha-\beta)}{\sin(\alpha+\beta)}E_{0i}, \tag{2.69}$$

$$E_{0t} = \frac{2\sin\beta\cos\alpha}{\sin(\alpha+\beta)}E_{0i}. \tag{2.70}$$

It is noteworthy that upon reflecting from the optically denser material 2 with $\tilde{n}_2 > \tilde{n}_1$ (and, therefore, $\alpha > \beta$), E_{0r} has the opposite sign as E_{0i}, that is, a phase shift of 180° occurs when reflecting from the denser material.

2.2.2 Magnetic field vector perpendicular to the plane of incidence

Next, we will consider a plane wave of light polarized parallel to the plane of incidence. This type of polarization is called p-polarized. When $\vec{E}$ lies in the plane of incidence, $\vec{H}$ must be perpendicular to it, that is, pointing in the z direction: $\vec{H} = (0, 0, H_z)$, as shown in Fig. 2.2. Analogous to the case of s-polarized light, continuity of $\vec{H}$ at $y = 0$ requires for all values of x:

$$H_{0i}\,e^{-i\tilde{g}_1 x\sin\alpha} + H_{0r}\,e^{-i\tilde{g}_1 x\sin\alpha'} = H_{0t}\,e^{-i\tilde{g}_2 x\sin\beta}. \tag{2.71}$$

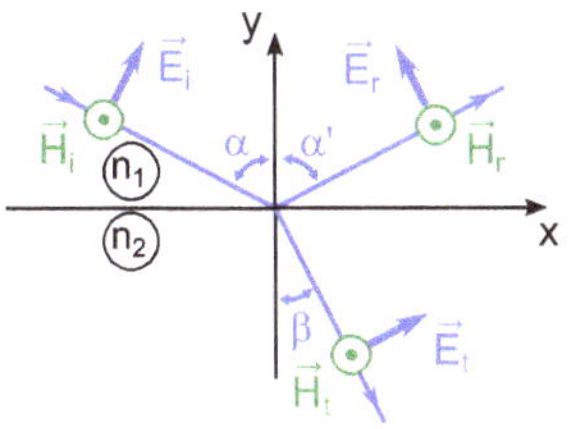

Fig. 2.2: Geometry for the derivation of the Fresnel equation for p-polarized light. $\vec{H}_i$ points out of the plane of incidence along the z direction. The interface (xz plane) is formed between material 1 ($y > 0$) with refractive index n_1 and material 2 ($y < 0$) with refractive index n_2.

We obtain the same law of reflection and refraction as in eqs. (2.55) and (2.56) and the following relation for the amplitudes:

$$H_{0i} + H_{0r} = H_{0t}. \tag{2.72}$$

Replacing $\vec{H}$ by the electric field according to eqs. (2.41), (2.6), and (2.8) yields:

$$\frac{\tilde{\varepsilon}_1}{\tilde{n}_1}(E_{0i} + E_{0r}) = \frac{\tilde{\varepsilon}_2}{\tilde{n}_2} E_{0t}. \tag{2.73}$$

A last condition is obtained from the continuity of E_x. The direction of the electric field of the incident, the refracted, and the reflected wave is again obtained from the rule that $\vec{E}$, $\vec{H}$, and $\vec{\tilde{g}}$ form a right-handed helix. In Fig. 2.2, the electric-field vectors are shown for the case that H_z is directed out of the plane of incidence. It follows that

$$\cos\alpha(E_{0i} - E_{0r}) = \cos\beta\, E_{0t}. \tag{2.74}$$

Combining eqs. (2.73) and (2.74) yields:

$$2E_{0i} = \left(\frac{\tilde{\varepsilon}_2\tilde{n}_1}{\tilde{\varepsilon}_1\tilde{n}_2} + \frac{\cos\beta}{\cos\alpha}\right) E_{0t}, \tag{2.75}$$

$$2E_{0r} = \left(\frac{\tilde{\varepsilon}_2\tilde{n}_1}{\tilde{\varepsilon}_1\tilde{n}_2} - \frac{\cos\beta}{\cos\alpha}\right) E_{0t}. \tag{2.76}$$

Setting again $\mu = 1$ and using Snell's law, which is eq. (2.57), we get:

$$2E_{0i} = \left(\frac{\sin\alpha}{\sin\beta} + \frac{\cos\beta}{\cos\alpha}\right) E_{0t} = \frac{\sin(\alpha+\beta)\,\cos(\alpha-\beta)}{\sin\beta\,\cos\alpha} E_{0t}, \tag{2.77}$$

$$2E_{0r} = \left(\frac{\sin\alpha}{\sin\beta} - \frac{\cos\beta}{\cos\alpha}\right) E_{0t} = \frac{\sin(\alpha-\beta)\,\cos(\alpha+\beta)}{\sin\beta\,\cos\alpha} E_{0t}. \tag{2.78}$$

For the rightmost term, we have used the addition theorems for trigonometric functions. By dividing eq. (2.78) by eq. (2.77) on the one hand and solving eq. (2.77) for E_{0t} on the other hand, the second set of Fresnel equations for p-polarized light can be derived:

$$E_{0r} = \frac{\tan(\alpha-\beta)}{\tan(\alpha+\beta)} E_{0i}, \tag{2.79}$$

$$E_{0t} = \frac{2\sin\beta\ \cos\alpha}{\sin(\alpha+\beta)\ \cos(\alpha-\beta)} E_{0i}. \tag{2.80}$$

Up to now, we have taken for granted that the polarization of the reflected and refracted wave points in the same direction as the polarization of the incoming beam. We just considered no rotation when deriving the Fresnel equations. Indeed, this is not self-evident. It must be proven in the same way as for deriving the Fresnel formulas. Namely, by applying the continuity conditions to the case that the reflected and diffracted rays have additional components perpendicular to the plane of incidence when the incoming beam is assumed to be polarized in the plane of incidence and vice versa. If one carefully performs this calculation, all the additional components will vanish. See the problem in Section 2.6.6 for a detailed treatment. As a result, it is correct to assume that for isotropic materials, purely s- or p-polarized light is reflected and refracted again as purely s- or p-polarized light, respectively.

2.3 Discussion of the Fresnel equations

The Fresnel equations are very powerful in the sense that they are valid for all combinations of isotropic materials at the interface and that they explain a wealth of physical phenomena. In the following chapter, we will jointly treat s- and p-polarization. However, we must distinguish between absorbing and non-absorbing materials and between the two cases of an incoming ray originating either from an optically less dense or from a denser material. For convenience, we introduce reflection coefficients as follows:

$$\rho_s = E_{0r}/E_{0i} \text{ for } \vec{E} \text{ perpendicular to the plane of incidence,} \tag{2.81}$$

$$\rho_p = E_{0r}/E_{0i} \text{ for } \vec{E} \text{ parallel to the plane of incidence,} \tag{2.82}$$

and transmission coefficients in the form of:

$$\sigma_s = E_{0t}/E_{0i} \text{ for } \vec{E} \text{ perpendicular to the plane of incidence,} \tag{2.83}$$

$$\sigma_p = E_{0t}/E_{0i} \text{ for } \vec{E} \text{ parallel to the plane of incidence.} \tag{2.84}$$

2.3.1 Non-absorbing materials

2.3.1.1 Incoming ray originates from optically less dense material

Let our discussion start with the simplest case: reflection and transmission at the interface of two non-absorbing materials. Non-absorbing means that the absorption index $k = 0$ and, consequently, the refractive index n is real. In addition, we assume that the refractive index n_1 of the material where the light originates from (material 1) is smaller than the refractive index n_2 of the material from which the light is reflected (material 2): $n_1 < n_2$. We say material 1 is optically less dense than material 2.

Let us introduce a relative refractive index as the ratio $n = n_2/n_1$. From the law of refraction for non-absorbing materials follows: $n = n_2/n_1 = \sin\alpha/\sin\beta$. Let us further consider a small incident angle $\alpha \cong 0$. Then, due to Snell's law, the refraction angle β must be small, too. So, we may approximate $\sin(\beta \pm \alpha) \cong \beta \pm \alpha \cong \tan(\beta \pm \alpha)$ and $n \cong \alpha/\beta$. Combining this with eqs. (2.69) and (2.79) yields:

$$\rho_s(\alpha \cong 0) = -\frac{\sin(\alpha-\beta)}{\sin(\alpha+\beta)} \cong -\frac{\alpha-\beta}{\alpha+\beta} = -\frac{\beta(\alpha/\beta-1)}{\beta(\alpha/\beta+1)} = -\frac{n-1}{n+1}, \tag{2.85}$$

$$\rho_p(\alpha \cong 0) = \frac{\tan(\alpha-\beta)}{\tan(\alpha+\beta)} \cong \frac{\alpha-\beta}{\alpha+\beta} = \frac{\beta(\alpha/\beta-1)}{\beta(\alpha/\beta+1)} = \frac{n-1}{n+1}. \tag{2.86}$$

The minus sign for s-polarized light indicates that there is a phase shift of 180° (or π) between reflected and incoming wave. The reason that there is no minus sign for p-polarized light can be found in the geometry (see Figs. 2.1 and 2.2). When we decrease the angle of incidence, reflected and incoming wave become more and more parallel. For s-polarized light, the electric field vector of reflected and incoming light point in the same direction but for p-polarized light, they point in the opposite direction. So, for p-polarized light, the sign change or 180° phase shift is already built in by definition.

From the reflection coefficients, the reflectivity can be derived, which is the ratio of the intensities of reflected and incoming light. Because the intensity is, according to eq. (2.46), proportional to the square of the absolute value of the electric-field amplitude, the reflectivity R is equal to the square of the absolute value of the reflection coefficient. Therefore, for small angles of incidence (normal incidence), R is equal for perpendicular and parallel polarization and amounts to:

$$R = \frac{(n-1)^2}{(n+1)^2}. \tag{2.87}$$

That is what we expect since we cannot distinguish any more between s- and p-polarization at normal incidence. As an example, we calculate the reflectivity of a glass window. For glass in the visible spectral region, we have $n \cong 1.5$. This yields a reflectivity of $R = (0.5/2.5)^2 = 0.04$. Hence, a glass window reflects only 4% of the incoming light at normal incidence. This is consistent with our everyday experience that it is inefficient to use a glass window as a mirror.

Next, we consider grazing incidence, where $\alpha = \pi/2$. Here, we can approximate sin =1, cos = 0 and tan $\alpha = \infty$. So we get:

$$\rho_s\left(\alpha = \frac{\pi}{2}\right) = -\frac{\sin(\alpha-\beta)}{\sin(\alpha+\beta)} = -\frac{\sin\left(\frac{\pi}{2}-\beta\right)}{\sin\left(\frac{\pi}{2}+\beta\right)} = -\frac{\cos\beta}{\cos\beta} = -1, \tag{2.88}$$

$$\rho_p\left(\alpha = \frac{\pi}{2}\right) = \frac{\tan(\alpha-\beta)}{\tan(\alpha+\beta)} = \frac{\sin\left(\frac{\pi}{2}-\beta\right)\cos\left(\frac{\pi}{2}+\beta\right)}{\cos\left(\frac{\pi}{2}-\beta\right)\sin\left(\frac{\pi}{2}+\beta\right)} = \frac{\cos\beta\,(-\sin\beta)}{\sin\beta\cos\beta} = -1. \tag{2.89}$$

We notice that while ρ_s is negative for all angles of incidence, ρ_p changes its sign between $\alpha = 0$ and $\alpha = \pi/2$. As noticed from eq. (2.89), this occurs when the denominator of ρ_p, $\tan(\alpha+\beta)$, jumps from $+\infty$ to $-\infty$. Then, the component which is p-polarized vanishes because $\rho_p = 0$. We call the corresponding angle of incidence the polarization angle or ***Brewster angle*** α_B. At this angle, the reflected light is completely s-polarized independent of the polarization of the incoming light. If the incoming light is p-polarized, no reflected beam exits. The Brewster angle α_B can be derived as follows:

$$\tan(\alpha_B+\beta) = \infty \Rightarrow \alpha_B + \beta = \pi/2;\ \text{therefore,}\ \sin\beta = \cos\alpha_B. \tag{2.90}$$

Note, eq. (2.90) implies that at the Brewster angle, the reflected and refracted waves propagate at a right angle. A simple physical explanation for this behavior is discussed in the problem in Section 2.6.9. On the other hand, Snell's law yields:

$$\sin\alpha_B = n\sin\beta, \tag{2.91}$$

combining with eq. (2.90) gives the relation between the Brewster angle and refractive index:

$$\tan\alpha_B = n. \tag{2.92}$$

We see that for $n > 1$, the Brewster angle α_B must always be larger than 45°. Equation (2.92) can be used to determine the refractive index of an unknown transparent material by simply measuring the Brewster angle.

The absolute values of ρ_s and ρ_p are plotted as a function of the angle of incidence α in Fig. 2.3 for values $n_1 = 1$ and $n_2 = 1.5$ of the refractive index, representing air and ordinary glass, respectively. The absolute values are plotted here for convenience, although this representation conceals the fact that ρ_s is indeed always negative corresponding to a phase shift of 180° or π that the incoming ray experiences when it is reflected. As we shall see later, the reflection coefficients may become complex and then plotting the absolute value will become favorable. At normal incidence ($\alpha = 0°$), ρ_s is negative and ρ_p is positive, and both have small values. This is known from everyday life. The reflection is quite small when looking straight through a glass window. With increasing angle of incidence, ρ_s decreases monotonically until

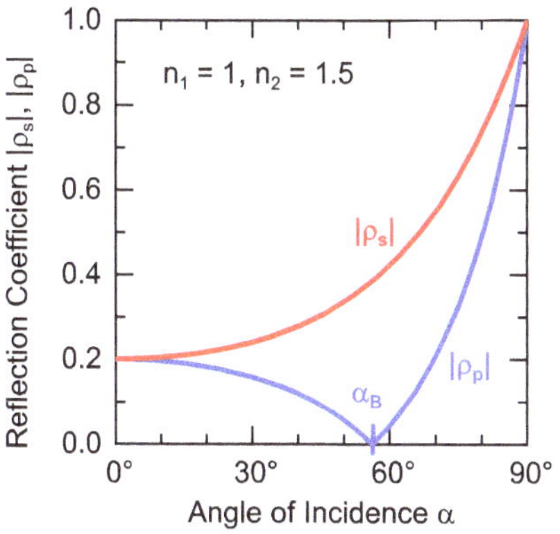

Fig. 2.3: Reflection coefficients ρ_s and ρ_p as a function of angle of incidence α for light polarized perpendicular (s) and parallel (p) to the plane of incidence for the case of reflection from an optically denser material. At the Brewster angle $\alpha_B = 56.3°$, the reflection coefficient ρ_p crosses zero and changes sign.

it reaches a value of −1 at grazing incidence ($\alpha = 90°$). Also ρ_p decreases monotonically but starting from a positive value. Hence, it crosses zero at $\alpha = \alpha_B$ and reaches a value of −1 at $\alpha = 90°$, too. This is perhaps less known from daily routine. When looking at grazing angle at a window, it becomes a perfect mirror.

Figure 2.4 shows, in addition, the phase difference between ρ_p and ρ_s as a function of angle of incidence α. Note the phase difference of 180° or π in the interval $0 \le \alpha < \alpha_B$ corresponding to the different sign of ρ_p and ρ_s. For $\alpha > \alpha_B$, the phase difference vanishes. ρ_p and ρ_s have now the same (negative) sign.

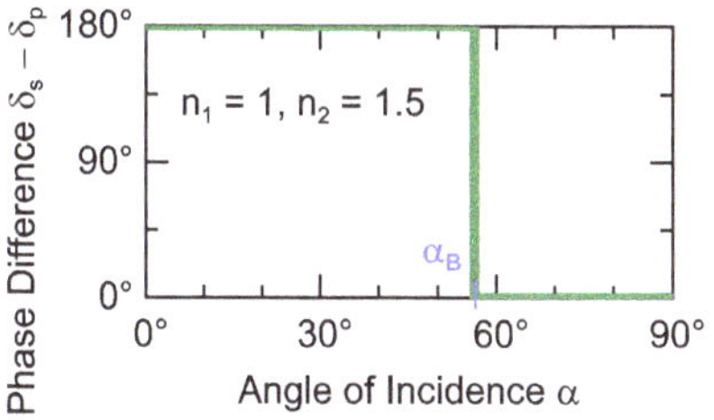

Fig. 2.4: Phase difference between ρ_p and ρ_s as a function of angle of incidence α for the case of reflection from an optically denser material. Note the step at the Brewster angle α_B indicating the sign change of ρ_p.

From eq. (2.58), it follows for σ_s:

$$\sigma_s = 1 + \rho_s, \tag{2.93}$$

and from eq. (2.73), assuming $\mu = 1$ and using the relative refractive index $n = n_2/n_1$, we derive for σ_p:

$$\sigma_p = (1 + \rho_p)/n. \tag{2.94}$$

These two relations are always valid independent of the value of n. The angular dependence of the transmission coefficients is shown for both polarizations in Fig. 2.5 using the same values for n_1 and n_2 as before. In contrast to the reflection coefficients, the transmission coefficients exhibit at normal incidence ($\alpha = 0°$) large values. This conforms again to everyday life. When looking straight through a glass window, the

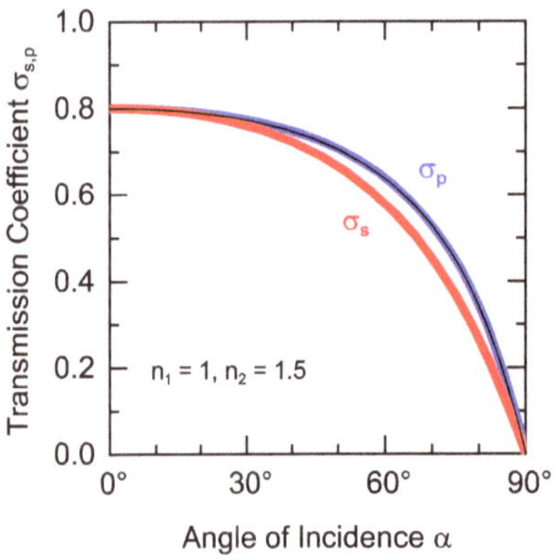

Fig. 2.5: Transmission coefficients $\sigma_{s,p}$ as a function of angle of incidence α for light polarized perpendicular (s) and parallel (p) to the plane of incidence when traveling from an optically less dense into a denser material.

transmission is very large. Both σ_s and σ_p decrease monotonically with increasing α and vanish at $\alpha = 90°$.

2.3.1.2 Incoming ray originates from optically denser material: Total internal reflection

For reflection from and refraction into an optically less dense material, exactly the same equations are valid as derived above because Snell's law is invariant upon swapping the two materials. However, because $n_2 < n_1$, we have $n < 1$ and $\sin\alpha < \sin\beta$ and this will lead to different behavior and new effects.

Starting again with normal incidence and looking at eq. (2.85), we see that because of $n < 1$, a phase shift no longer exists between incoming and reflected waves. Furthermore, as the angle of refraction β cannot be larger than 90°, we have $\sin\beta \le 1$. Because $\sin\alpha$ is always smaller than $\sin\beta$, there must exist a critical angle of incidence $\alpha = \alpha_T < 90°$ which leads to $\sin\beta = 1$. The critical angle α_T is defined by Snell's law:

$$\sin\alpha_T = n = n_2/n_1 \qquad \text{or} \qquad \alpha_T = \arcsin\,(n_2/n_1). \tag{2.95}$$

At the critical angle α_T, where β is 90°, the reflection coefficients both reach a value of 1, that is, 100% of the incoming light is reflected independent of polarization:

$$\rho_s\left(\alpha = \alpha_T, \beta = \frac{\pi}{2}\right) = \frac{\sin\left(\frac{\pi}{2} - \alpha_T\right)}{\sin\left(\alpha_T + \frac{\pi}{2}\right)} = \frac{\cos\alpha_T}{\cos\alpha_T} = 1, \tag{2.96}$$

$$\rho_p\left(\alpha = \alpha_T, \beta = \frac{\pi}{2}\right) = \frac{\tan\left(\alpha_T - \frac{\pi}{2}\right)}{\tan\left(\alpha_T + \frac{\pi}{2}\right)} = \frac{\sin\left(\alpha_T - \frac{\pi}{2}\right)\cos\left(\alpha_T + \frac{\pi}{2}\right)}{\cos\left(\alpha_T - \frac{\pi}{2}\right)\sin\left(\alpha_T + \frac{\pi}{2}\right)} = \frac{(-\cos\alpha_T)\,(-\sin\alpha_T)}{\sin\alpha_T\cos\alpha_T} = 1 \tag{2.97}$$

We call this phenomenon ***total internal reflection***. Figure 2.6 shows the reflection coefficients for both polarizations for values $n_1 = 1.5$ and $n_2 = 1$ as a function of angle of incidence α. In the interval $0 \le \alpha \le \alpha_T$, the plot looks like a compressed version of the case of light reflected from an optically denser material (Fig. 2.3) except for the

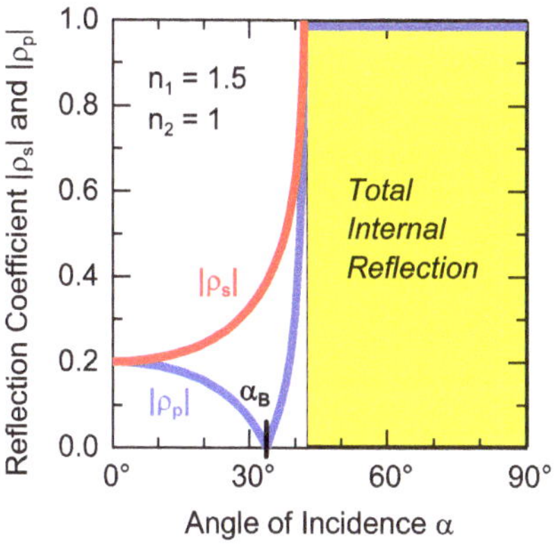

Fig. 2.6: Reflection coefficients ρ_s and ρ_p as a function of angle of incidence α for light polarized perpendicular (s) and parallel (p) to the plane of incidence for the case of reflection from an optically less dense material. At the Brewster angle $\alpha_B = 33.7°$, the reflection coefficient ρ_p vanishes. At the critical angle $\alpha_T = 41.8°$, the reflection reaches 100% for all polarization directions. This is called total internal reflection. Note that for α larger than α_T the reflection coefficients have complex values.

phase shift between incoming and reflected light which is now missing. The Brewster angle still exists but is below 45°, namely at $\alpha_B = \arctan(1/1.5) = 33.7°$.

Next, we would like to discuss the following question: What happens if we increase the angle of incidence above the critical angle α_T? As we can freely choose the angle of incidence between 0° and 90°, physics must give a well-defined answer. We know that Snell's law must still hold because it has been derived from continuity conditions, which are very strong. But, how to cope with $\sin\beta > 1$? Fresnel already found the solution to our problem two hundred years ago: We must consider not only real angles of refraction β but also imaginary ones. The condition that refraction must be continuous when crossing the critical angle, leads to the ansatz $\tilde{\beta} = \pi/2 + i\beta''$, where $\beta'' \in \mathbb{R}$. See the problem in Section 2.6.10 for a detailed treatment of this approach. The law of refraction is still valid in the case of total internal reflection with the additional condition that it must now be fulfilled simultaneously for the real and imaginary parts. Consequently, the coefficients ρ_s, ρ_p, σ_s and σ_p of the Fresnel equations turn to complex functions and can be obtained from eqs. (2.69) and (2.79) as follows:

$$\tilde{\rho}_s = -\frac{\sin\left(\alpha - \frac{\pi}{2} - i\beta''\right)}{\sin\left(\alpha + \frac{\pi}{2} + i\beta''\right)} = \frac{\cos\left(\alpha - i\beta''\right)}{\cos\left(\alpha + i\beta''\right)} = e^{i\delta_s}, \tag{2.98}$$

$$\tilde{\rho}_p = -\frac{\tan\left(\frac{\pi}{2} + i\beta'' - \alpha\right)}{\tan\left(\frac{\pi}{2} + i\beta'' + \alpha\right)} = \frac{\tan\left(\alpha + i\beta''\right)}{\tan\left(\alpha - i\beta''\right)} = e^{i\delta_p}, \tag{2.99}$$

where δ_s and δ_p have real values, since numerator and denominator are conjugate complex numbers. We notice that both reflection coefficients lie on the unit circle in the complex plane. This means that the incoming light is still 100% reflected for both polarizations. Hence, we still have total internal reflection. However, there exists an angle-dependent phase difference between the reflected s- and p-polarized waves. Dividing $\tilde{\rho}_s$ by $\tilde{\rho}_p$ yields the phase difference between the two polarization states:

$$e^{i(\delta_s - \delta_p)} = \frac{\sin(\alpha - i\beta'')}{\sin(\alpha + i\beta'')}. \tag{2.100}$$

There are two conditions where the phase difference $\delta_s - \delta_p$ between s- and p-polarized light is zero: for $\beta'' = 0$ and for $\alpha = \pi/2$, that is, there is no phase difference between $\tilde{\rho}_s$ and $\tilde{\rho}_p$ for $\alpha = \alpha_T$ and $\alpha = 90°$. In between these two angles, the phase shift is not zero and reaches a maximum defined by the following relation:

$$\tan\left(\frac{\delta_s - \delta_p}{2}\right)_{max} = \frac{n_1^2 - n_2^2}{2n_1n_2}. \tag{2.101}$$

It occurs at an angle of incidence α_{max}, which is defined by:

$$\sin^2\alpha_{max} = \frac{2n_2^2}{n_1^2 + n_2^2}. \tag{2.102}$$

Both relations are very tedious to derive. We present a detailed derivation of eq. (2.102) in the problem in Section 2.6.11 and of eq. (2.101) in the problem in Section 2.6.12. For the refraction of light from glass ($n_1 = 1.5$) to air ($n_2 = 1$), we obtain $(\delta_s - \delta_p)_{max} = 45.27°$. The angle of maximum phase difference is $\alpha_{max} = 51.67°$.

This result is shown in Fig. 2.7 where the phase difference $\delta_s - \delta_p$ is plotted as a function of angle of incidence α. For $0 \le \alpha \le \alpha_T$, the plot looks like a compressed version of the phase difference of light reflected from an optically denser material (Fig. 2.4). But for $\alpha > \alpha_T$, it is different. The phase difference increases until it reaches a shallow maximum at $\alpha_{max} = 51.67°$. For larger α, it decreases until it vanishes at $\alpha = 90°$. For the first time, we have generated a phase difference by reflection which depends on the

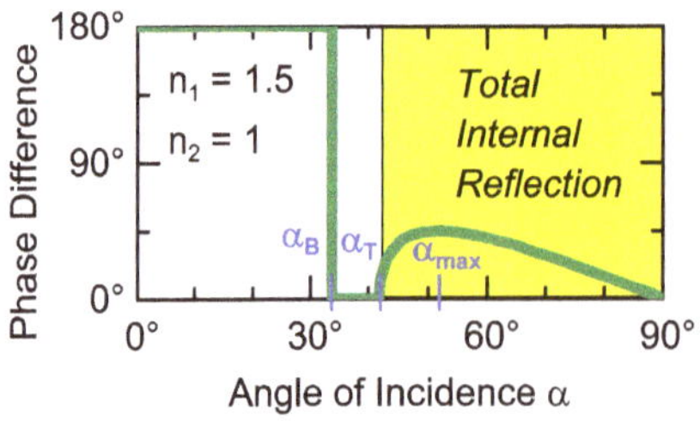

Fig. 2.7: Phase difference between $\tilde{\rho}_s$ and $\tilde{\rho}_p$ in the case of reflection from an optically less dense material. Note the step at the Brewster angle $\alpha_B = 33.7°$ indicating a sign change in $\tilde{\rho}_p$. At the critical angle $\alpha_T = 41.8°$, the phase difference increases again and reaches a maximum at $\alpha_{max} = 51.67°$.

refractive index of the material. Already Fresnel noticed two hundred years ago that this effect can be used to generate elliptically or circularly polarized light from linearly polarized light. In Fig. 2.8, the parallelepiped is shown that Fresnel designed to prove his famous formulas (83). It produces a phase difference of $\pi/2$ or 90° by a twofold total internal reflection in glass. Using his formulas, Fresnel calculated the angle needed to obtain a 45° phase shift for the glass he had available. Then he cut the glass and checked if the parallelepiped produced the desired result, what it did.

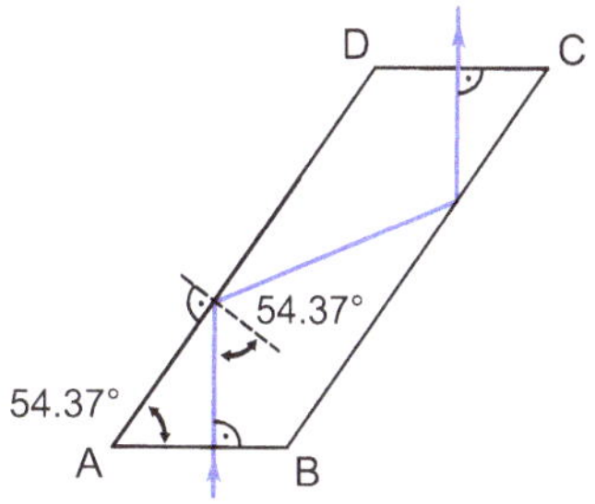

Fig. 2.8: Fresnel's parallelepiped, produces a phase shift of $\pi/2$ or 90° by a twofold total internal reflection. A 90° phase shift is needed to generate circularly polarized light from linearly polarized light.

We now want to briefly discuss if there exists a refracted wave in the case of total internal reflection. For that, we proceed from the ansatz in eq. (2.52) for a refracted wave polarized perpendicular to the plane of incidence, add the time dependence, and use a complex angle of refraction $\tilde{\beta} = \pi/2 + \mathrm{i}\beta''$ in accordance with the above-mentioned boundary conditions. Using the relations

$$\sin\tilde{\beta} = \sin\left(\frac{\pi}{2} + \mathrm{i}\beta''\right) = \cos(\mathrm{i}\beta'') = \cosh(\beta'') \quad \text{and}$$
$$\cos\tilde{\beta} = \cos\left(\frac{\pi}{2} + \mathrm{i}\beta''\right) = -\sin(\mathrm{i}\beta'') = -\mathrm{i}\sinh(\beta''), \tag{2.103}$$

we get for the refracted wave:

$$E_{tz} = E_{0t}\, \mathrm{e}^{g_2 y \sinh\beta''}\, \mathrm{e}^{\mathrm{i}\,(\omega t - g_2 x \cosh\beta'')}. \tag{2.104}$$

This represents an inhomogeneous wave. It spreads as a harmonic wave in the x direction, that is, along the interface while it is attenuated exponentially along the $-y$ direction, that is, going deeper into the second material. This is called a surface wave.

If a transmitted wave exists, then the transmission coefficients should be nonzero as well. We can check this by plotting the angular dependence of the transmission coefficients for s- and p-polarization in Fig. 2.9 using the same values for n_1 and n_2 as before. For $0 \le \alpha \le \alpha_T$, the plot looks completely different from the case of light transmitted through an optically denser material (Fig. 2.5). Both transmission coefficients start at normal incidence at a value greater than one indicating that the amplitude of the transmitted electric field is larger than that of the incoming one.

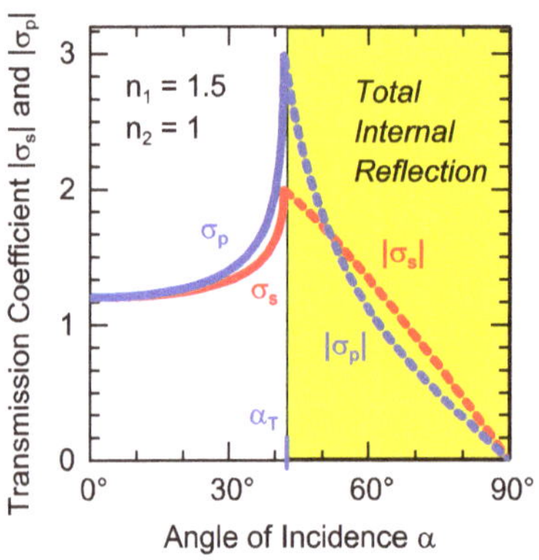

Fig. 2.9: Transmission coefficients σ_s and σ_p as a function of angle of incidence α for light polarized perpendicular (s) and parallel (p) to the plane of incidence when traveling form an optically dense into a less dense material. At the critical angle $\alpha_T = 41.8°$, the transmission reaches its maximum. Note that for both polarization directions, the maximum transmission is larger than one. Above α_T, the transmission coefficients have complex values and are nonzero despite the total internal reflection.

They increase until they reach a maximum at the critical angle α_T. For $\alpha > \alpha_T$, as expected, both transmission coefficients are nonzero. In addition, they assume complex values and the absolute value decreases monotonically reaching zero at $\alpha = 90°$. In this range, the transmission coefficients define the amplitude of the refracted surface wave. Nevertheless, the conservation of energy is not violated because there is no energy transported into the second material due to the exponential damping.

The fact that even in the case of total internal reflection a refracted wave exists, which intrudes into the less-dense material, is a consequence of the continuity conditions. Surprising, however, is that the time-averaged energy flow cancels out in y direction (see the problem in Section 2.6.10) when calculating the Poynting vector but not in x direction. Goos and Hähnchen have experimentally verified the energy transfer along the interface by observing that a beam displacement along the interface occurs during total internal reflection (84). The original measurement is shown in Fig. 2.10. The beam was collimated to a small slit and multiple times total internally reflected at the glass/air interface of a glass slab. The outgoing beam was recorded as a photograph. By moving the photographic film while chopping the beam, consecutive measurements were made. The middle section of the glass slab was covered with a silver film. Due to the high absorption of silver, the light encounters at the glass/silver interface a normal reflection and a displacement due to total internal reflection occurs only outside of the silver stripe. Therefore, the outside will be horizontally displaced with respect to the middle section of the beam as seen on the photograph.

Our rather simple treatment, using only the Fresnel equations, can explain that a surface wave exists. It cannot explain over which distance the surface wave will transport energy because our surface wave is not damped along its propagation

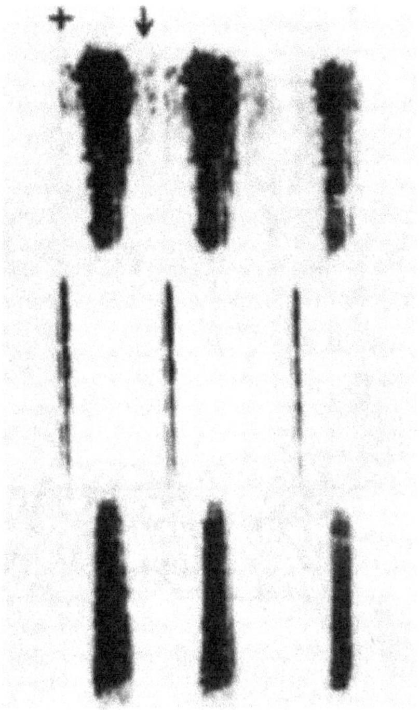

Fig. 2.10: Original photograph by Goos and Hähnchen of the total internally reflected light from a glass slab. The light was collimated to a slit and total internally reflected multiple times on a glass/air interface. On the slab was a thin silver film deposited in such a way that the middle section of the light beam was reflected from the silver film. Due to the high absorption, this reflection was a normal reflection. Therefore, the displacement of the light ray by total internal reflection occurs only on the outside and shows up as a horizontal displacement of the beam with respect to the middle section. Figure from Ref. (84).

direction, so it would go on forever. For a more realistic treatment, the final width of the light ray must be considered. If from the optically less dense side (usually air or vacuum) a third transparent material is placed close to the interface at a distance less than a couple of wavelengths, a kind of optical tunnel effect will occur. The total internal reflection diminishes while a light ray with finite intensity emanates from the interface and traverses the third material.

2.3.2 Absorbing materials

We now come to the most general case, which is also the normal case, as most materials are absorbing over a wide photon-energy range. Metals are typical examples of highly absorbing materials. In this context, we speak about metallic absorption and reflection. Furthermore, semiconductors and insulators are also absorbing at photon energies above the fundamental energy gap. The Fresnel equations apply for absorbing materials as well. However, $\tilde{n}_2$ and, hence, $\tilde{\beta}$ are now complex quantities. Considering the interface between a material 1 without absorption (n_1 real) and a material 2 with a complex refractive index $\tilde{n}_2 = n_2 - \mathrm{i}k_2$, we can define a ***relative*** complex refractive index as

$$\tilde{n} = \frac{\tilde{n}_2}{n_1} = n - \mathrm{i}k. \tag{2.105}$$

This definition works of course too if both materials are absorbing. The law of refraction is then expressed as

$$\frac{\sin\alpha}{\sin\tilde{\beta}} = \tilde{n}, \tag{2.106}$$

where $\tilde{\beta}$ must be complex because $\tilde{n}$ is complex. From eqs. (2.69) and (2.79), we obtain for the reflection coefficients, which are now complex as well:

$$\tilde{\rho}_\mathrm{s} = -\frac{\sin\left(\alpha - \tilde{\beta}\right)}{\sin\left(\alpha + \tilde{\beta}\right)} = \left|\tilde{\rho}_\mathrm{s}\right| \mathrm{e}^{\mathrm{i}\delta_\mathrm{s}}, \tag{2.107}$$

$$\tilde{\rho}_\mathrm{p} = \frac{\tan\left(\alpha - \tilde{\beta}\right)}{\tan\left(\alpha + \tilde{\beta}\right)} = \left|\tilde{\rho}_\mathrm{p}\right| \mathrm{e}^{\mathrm{i}\delta_\mathrm{p}}. \tag{2.108}$$

For small α and $\tilde{\beta}$, it follows from eq. (2.106) that $\alpha/\tilde{\beta} = \tilde{n}$. At normal incidence ($\alpha = 0$), we get similar to the case of nonabsorbing materials [eq. (2.85)]:

$$\tilde{\rho}_\mathrm{s} = -\tilde{\rho}_\mathrm{p} = -\frac{\tilde{n} - 1}{\tilde{n} + 1} = -\frac{n - \mathrm{i}k - 1}{n - \mathrm{i}k + 1}. \tag{2.109}$$

The reflectivity R at normal incidence follows by taking the square of the absolute value of eq. (2.109):

$$R = \left|\tilde{\rho}_\mathrm{p}\right|^2 = \left|\tilde{\rho}_\mathrm{s}\right|^2 = \frac{(n-1)^2 + k^2}{(n+1)^2 + k^2}. \tag{2.110}$$

It is again identical for both polarization states. The phases δ_s and δ_p for $\alpha = 0$ are obtained by expanding eq. (2.109) to make the denominator real:

$$\tilde{\rho}_\mathrm{s} = -\tilde{\rho}_\mathrm{p} = \frac{n^2 + k^2 - 1 - 2\mathrm{i}\,k}{(n+1)^2 + k^2}. \tag{2.111}$$

Then, the tangent of the phase is equal to the ratio of imaginary and real part:

$$\tan\delta_\mathrm{p} = \tan(\delta_\mathrm{s} + \pi) = -\frac{2k}{n^2 + k^2 - 1}. \tag{2.112}$$

We see from eqs. (2.107) and (2.108) that generally a phase difference $\delta_\mathrm{p} - \delta_\mathrm{s}$ exists between $\tilde{\rho}_\mathrm{s}$ and $\tilde{\rho}_\mathrm{p}$. But according to eq. (2.109), it has the extremal value π for $\alpha = 0$. For $\alpha = \pi/2$, eqs. (2.107) and (2.108) yield a difference $\delta_\mathrm{p} - \delta_\mathrm{s} = 0$. Thus, linearly polarized light will be elliptically polarized after reflection except for $\alpha = 0$ and $\alpha = \pi/2$.

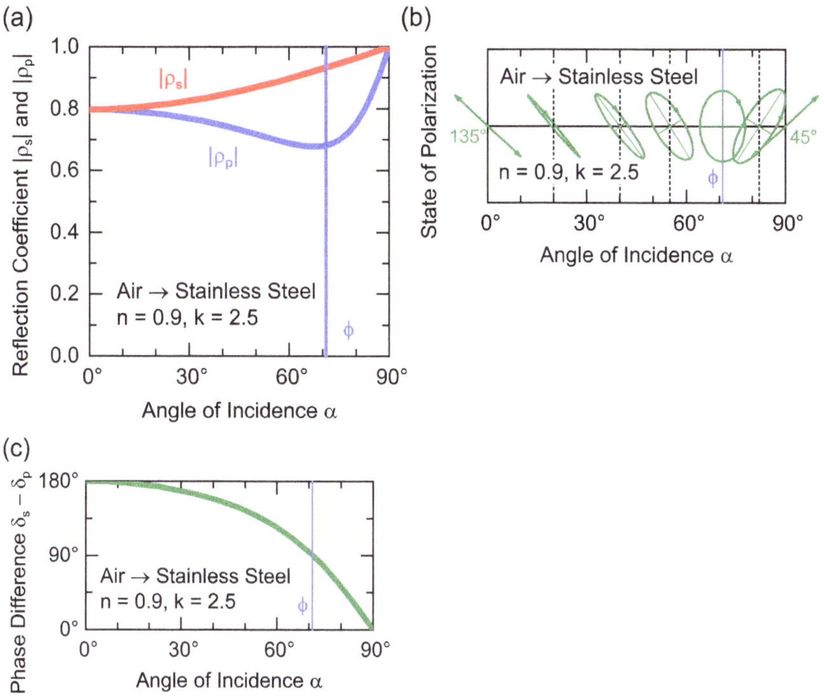

Fig. 2.11: Reflectivity of stainless steel. (a) Reflection coefficients ρ_s and ρ_p for light polarized perpendicular (s) and parallel (p) to the plane of incidence, (b) state of polarization of reflected light for incident light which is linearly polarized at 45°, (c) phase difference as a function of angle of incidence α. ϕ is the principal angle of incidence, where $\delta_s - \delta_p = 90°$.

Figure 2.11 shows the reflection coefficients $\tilde{\rho}_s$ and $\tilde{\rho}_p$ for stainless steel. The angle of incidence ϕ, for which the phase difference is $\delta_p - \delta_s = \pi/2$, is called ***principal angle of incidence***. Despite the phase difference of $\pi/2$, the reflected light is not circularly polarized, since $|\tilde{\rho}_p| \neq |\tilde{\rho}_s|$.

As said before, we write plane waves as complex quantities for the sake of mathematical simplicity. Hence, one should keep in mind that the ***real*** physical quantity or the physically ***measurable*** quantity is only the ***real part*** of this wave. Moreover, the situation gets really complicated when discussing propagation directions and relative orientation of the field vectors in absorbing materials. We now want to have a closer look at this problem. We write a plane wave propagating in a direction $\vec{r}$ as in eq. (2.25):

$$\vec{E} = \vec{E}_0 e^{i(\omega t - \tilde{\vec{g}} \cdot \vec{r})}, \tag{2.113}$$

where $\vec{\tilde{g}}$ is the wave vector and $\vec{r}$ the position vector. In an absorbing material, $\vec{\tilde{g}}$ is complex, that is, we have

$$\tilde{g}^2 = \vec{\tilde{g}} \cdot \vec{\tilde{g}} = \tilde{g}_x^2 + \tilde{g}_y^2 + \tilde{g}_z^2, \text{ with } \tilde{g}_i \in \mathbb{C} \text{ for } i = x, y, z. \tag{2.114}$$

This means, we apply the ordinary dot product of real space to a vector $\vec{\tilde{g}}$, which has complex projections $\tilde{g}_x, \tilde{g}_y, \tilde{g}_z \in \mathbb{C}$. A term such as $\tilde{g}_x^2$ involves a multiplication in complex plane and yields typically a complex number. Assuming the most general format for the wave vector, we get $\vec{\tilde{g}} = \vec{g}_1 - \mathrm{i}\vec{g}_2$, where $\vec{g}_1$ and $\vec{g}_2$ are vectors in real space with real components and the minus sign is defined in order to be consistent with eq. (2.27). If $\vec{g}_1$ and $\vec{g}_2$ are not collinear, it is impossible to relate $\vec{\tilde{g}}$ to a well-defined propagation direction because real and imaginary part point into different directions. Only with the restriction that $\vec{g}_1 || \vec{g}_2$, we can define a propagation direction $\hat{e}_{\vec{\tilde{g}}}$ as a real unit vector consistent with the definition in eq. (2.27) of the refractive index $\tilde{n}$ in an isotropic material:

$$\vec{\tilde{g}} = \tilde{g}\, \hat{e}_{\vec{\tilde{g}}} = \frac{\omega}{c} \tilde{n}\, \hat{e}_{\vec{\tilde{g}}}. \tag{2.115}$$

For plane waves as defined in eq. (2.25), we know that all the relations between electric and magnetic field vectors and propagation direction are valid in a form similar to the vacuum case. But what about the waves defined in eqs. (2.50) to (2.52) for reflection and refraction in the case of absorption? Here, we must be very careful with the interpretation. As already stated before, in the case of refraction in an absorbing material, the angle of refraction becomes complex: $\tilde{\beta} = \beta_1 + \mathrm{i}\beta_2$, with β_1, $\beta_2 \in \mathbb{R}$. So, we can ask the question: How does the propagation direction of the refracted wave depend on the complex angle of refraction $\tilde{\beta}$? One could assume that the propagation direction $\hat{e}_{\vec{\tilde{g}}\mathrm{t}}$ of the refracted wave is simply given by the real part β_1 of the complex angle of refraction. However, this is not true. To see that, we need a very general argument, how to define the direction of propagation for the case where it cannot be defined with the help of the wave vector because eq. (2.115) does not apply. For real plane waves, the propagation direction can also be defined as the direction perpendicular to the planes of constant amplitude of the field vectors or to the planes of constant phase. In absorbing materials, the first definition does not make sense because we know that the amplitudes of the field vectors are exponentially damped. Therefore, we can define the direction of propagation in a very general way as the direction perpendicular to the planes of constant phase.

Next, let us determine the direction of propagation for the refracted wave in an absorbing material with respect to the electric field. For a detailed treatment, see the problem in Section 2.6.13. First, we insert the complex angle of refraction $\tilde{\beta} = \beta_1 + \mathrm{i}\beta_2$ into the refracted wave as defined in eq. (2.52). By explicitly writing $\tilde{n}$ and the sine and cosine functions with real and imaginary parts, we can separate the phase term from the amplitude. As a first result, we notice that the amplitude is exponentially damped in $-y$ direction. This means that the planes of constant amplitude are parallel to the interface (= xz plane) and decline exponentially while penetrating deeper into the absorbing material. For the planes of constant phase, we obtain:

$$(n\sin\beta_1\cosh\beta_2 + k\cos\beta_1\sinh\beta_2)\,x - (n\cos\beta_1\cosh\beta_2 - k\sin\beta_1\sinh\beta_2)\,y = \text{const.} \tag{2.116}$$

Now, we can define the direction of propagation $\hat{e}_{\tilde{g}t}$ perpendicular to the planes of constant phase (the last equation follows from Snell's law):

$$\begin{aligned}\hat{e}_{\tilde{g}t} &\propto \left(\sin\beta_1 + \frac{k}{n}\cos\beta_1\tanh\beta_2\right)\hat{e}_x - \left(\cos\beta_1 - \frac{k}{n}\sin\beta_1\tanh\beta_2\right)\hat{e}_y\\ &= (\sin\beta_1\,\hat{e}_x - \cos\beta_1\,\hat{e}_y) + \frac{k}{n}\tanh\beta_2\,(\cos\beta_1\,\hat{e}_x + \sin\beta_1\,\hat{e}_y)\\ &= (\sin\beta_1\,\hat{e}_x - \cos\beta_1\,\hat{e}_y) + \left(\frac{k}{n}\right)^2\tan\beta_1(\cos\beta_1\,\hat{e}_x + \sin\beta_1\,\hat{e}_y).\end{aligned} \tag{2.117}$$

We immediately conclude from eq. (2.117) that to the direction $(\sin\beta_1\,\hat{e}_x - \cos\beta_1\,\hat{e}_y)$, defined by the real part β_1, a vector with positive length $(k/n)^2\tan\beta_1(\cos\beta_1\,\hat{e}_x + \sin\beta_1\,\hat{e}_y)$ is added. Therefore, the direction of propagation of the refracted wave is clearly not defined by the real part β_1. It will in fact be slightly tilted toward the interface. Finally, we can perform the same calculation as outlined above for the magnetic field. As a result, we get the same direction of propagation as for the electric field. In conclusion, the refracted wave is still a propagating wave but no longer a plane wave because both field amplitudes are exponentially damped in the direction perpendicular to the interface. Furthermore, there is no simple relation anymore between the refractive angle and the direction of propagation.

In anisotropic materials the situation gets even more complicated. There are for each direction of propagation exactly two waves that can emanate. These waves are called normal waves. They are elliptically polarized (for a definition see next chapter) and orthogonal with different complex refractive indices.

2.4 Polarization and sign convention

2.4.1 Definition of polarization states

In the previous chapter, we have used linearly polarized light without explaining in detail what polarization is and how a polarized state can be described mathematically. This will now be done. We always assume plane waves, where the various types of polarization are well defined. When we talk of a certain polarization state, we always refer to the polarization of the electric field. Of course, the magnetic field must possess a similar polarization state due to its relation to the electric field defined by Maxwell's equations.

In this book, we restrict ourselves to materials that are isotropic in the absence of a magnetic field. Then, owing to Maxwell's equations, the electric field vector lies always within a plane perpendicular to the wave vector (defining the propagation direction). Just as a side remark: In anisotropic materials, this would be true only for the electric displacement vector. Without loss of generality, we can assume that the wave propagates in z direction. Consequently, the electric (and the magnetic) field vector lies in the xy plane and all polarization states are defined within that plane.

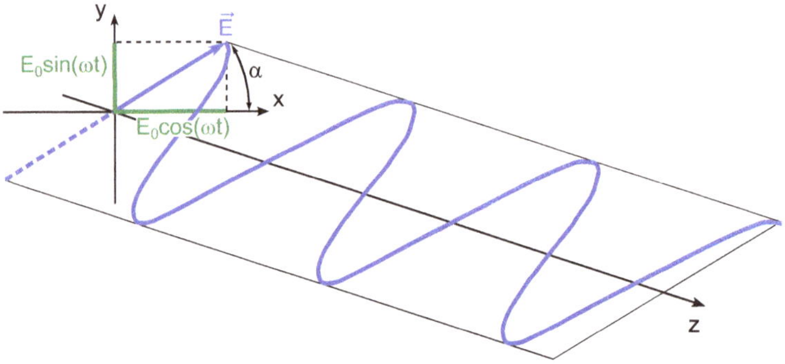

Fig. 2.12: Definition of linearly polarized light. Electric-field vector $\vec{E}$ of a linearly polarized wave traveling along the positive z axis.

The simplest polarization state is ***linearly polarized***. We say a plane wave is (linearly) polarized in x direction when the electric field oscillates along the x axis. The most general case of linearly polarized light is shown in Fig. 2.12. The electric-field vector oscillates at a fixed position in space along the polarization direction at an angle α with respect to the x axis. The angle α is called the polarization angle. The oscillation is within a plane (here the xy plane) perpendicular to the propagation direction (here the z axis). In space, the wave propagates within the polarization plane which is defined by the propagation direction and the fixed polarization direction. Mathematically, a linearly polarized plane wave traveling in z direction can, hence, be written as:

$$\vec{E} = \vec{E}_0 e^{i(\omega t - \tilde{g}z)} \quad \text{with} \quad \vec{E}_0 = E_0(\cos\alpha, \sin\alpha, 0) = E_0\left(\cos\alpha\, \hat{e}_x + \sin\alpha\, \hat{e}_y\right), \tag{2.118}$$

where the polarization vector $\vec{E}_0$ defines the linear polarization. E_0 is the amplitude and α the polarization angle, measured with respect to the x axis. $\hat{e}_x$ and $\hat{e}_y$ are the unit vectors (length 1) pointing along the x and y axis, respectively.

It is advantageous in magneto-optics to consider not only linearly polarized but also ***circularly polarized*** light. As we will discuss later, the reason for that is found in quantum mechanics, more specific, in the interaction of light with atoms. Selection rules for optical transitions in an applied magnetic field involve circularly polarized states. In addition, the eigenstates of a photon, that is, a light quantum, correspond classically to circularly polarized states.

Let us now define this peculiar polarization state. For circularly polarized light, the electric field vector $\vec{E}$ has a constant norm $|\vec{E}| = E_0$ and rotates in a plane perpendicular to the direction of propagation. There are two rotation directions possible, clockwise and counter-clockwise, defining two different states of circular polarization. To obtain an unambiguous definition, we assume that the observer looks ***into*** the light source. Then, in the case of ***right circularly polarized*** (**rcp**) light, the electric field vector $\vec{E}$ rotates in a ***clockwise*** direction as a function of time at a fixed position in space. This is depicted in Fig. 2.13. For ***left circularly polarized*** (**lcp**) light, $\vec{E}$ rotates ***counterclockwise*** as a function of time at a fixed position in space.

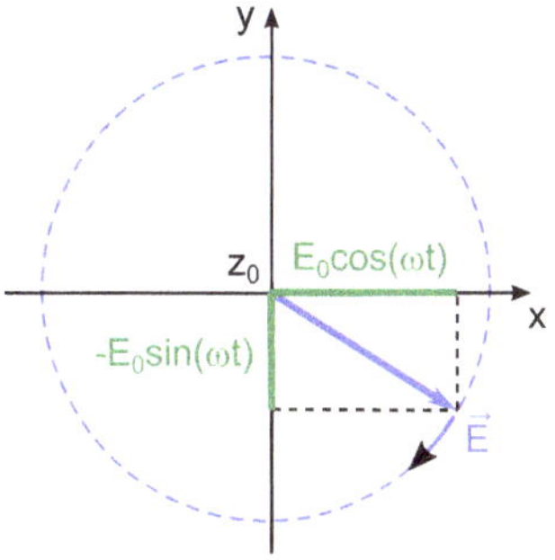

Fig. 2.13: Definition of right circularly polarized light. The light source is located below the paper plane. z_0 is a fixed point on the propagation axis z of the light wave.

Using Euler's formula $e^{i\omega t} = \cos(\omega t) + i\sin(\omega t)$ to express the exponential function as a combination of cosine and sine functions, the time dependence at a fixed location z_0 of the components of a plane wave $\vec{E}$ from eq. (2.113) can be written as:

$$\mathrm{Re}\{E_{0x}\,e^{i\omega t}\} = \mathrm{Re}\{E_{0x}[\cos(\omega t) + i\sin(\omega t)]\} = E_{0x}\cos(\omega t), \tag{2.119}$$

$$\mathrm{Re}\{i\,E_{0y}\,e^{i\omega t}\} = \mathrm{Re}\{E_{0y}[-\sin(\omega t) + i\cos(\omega t)]\} = -E_{0y}\sin(\omega t). \tag{2.120}$$

This describes a circle as a function of time. Assuming a propagation direction along the z axis, we can thus represent **rcp** (or +) light as:

$$\vec{E}_+ = \frac{1}{\sqrt{2}}E_0\left(\hat{e}_x + i\,\hat{e}_y\right)e^{i(\omega t - \tilde{g}z)}, \tag{2.121}$$

and **lcp** (or −) light accordingly as:

$$\vec{E}_{-} = \frac{1}{\sqrt{2}} E_0 \left(\hat{e}_x - \mathrm{i}\hat{e}_y\right) \mathrm{e}^{\mathrm{i}(\omega t - \tilde{g} z)} . \tag{2.122}$$

The prefactor is due to the normalizing of the amplitude. Let us now consider how an rcp light ray looks like in space at a fixed time. Considering the real part of eq. (2.121), an rcp wave appears for an observer in space at a fixed time as a right-handed helix (see Fig. 2.14). Lcp light would thus look like a left-handed helix. A reversal of the two definitions would occur if we had written $\mathrm{e}^{-\mathrm{i}\omega t}$ in place of $\mathrm{e}^{\mathrm{i}\omega t}$ in eq. (2.121).

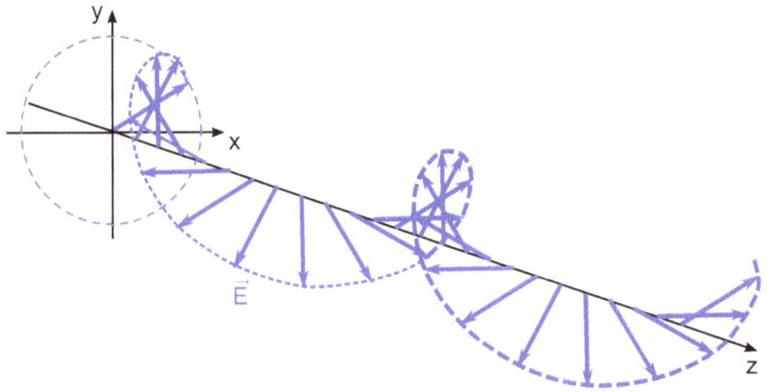

Fig. 2.14: Electric-field vector $\vec{E}$ of an rcp wave traveling along the positive z axis. The direction of rotation of the helix would be reversed for an lcp wave.

For ***elliptically polarized*** light, $\vec{E}$ describes an ellipse at a fixed point in space (see Fig. 2.15). The ratio of the semiminor to the semimajor axis is defined as the ellipticity. In many cases, the ellipticity is sloppily defined as the angle whose tangent is equivalent to the ratio of the semiaxes. For small angles, both definitions are identical. We say the ellipticity is ***positive*** (***negative***), when $\vec{E}$ rotates along the ellipse in ***clockwise*** (***counterclockwise***) direction as seen by an observer who looks into the light source (similar to circular polarization). Consequently, elliptically polarized light looks in space at a fixed time for positive (negative) ellipticity like a deformed right-handed (left-handed) helix with an elliptical cross section.

Circular and linear polarizations are special cases of elliptical polarization. Assuming again propagation direction parallel to the z axis and denoting the phase difference between the x and y component of $\vec{E}$ as δ, the expression

$$\vec{E} = \frac{1}{\sqrt{2}} E_0 \left(\hat{e}_x + \mathrm{e}^{\mathrm{i}\delta} \hat{e}_y\right) \mathrm{e}^{\mathrm{i}(\omega t - \tilde{g} z)} \tag{2.123}$$

describes for $\delta = 0$ linearly polarized light with the direction of oscillation along the bisector of the first and third quadrant. For $0 < \delta < \pi/2$, the light is elliptically polarized with the semimajor axis along the same bisector and with a positive ellipticity

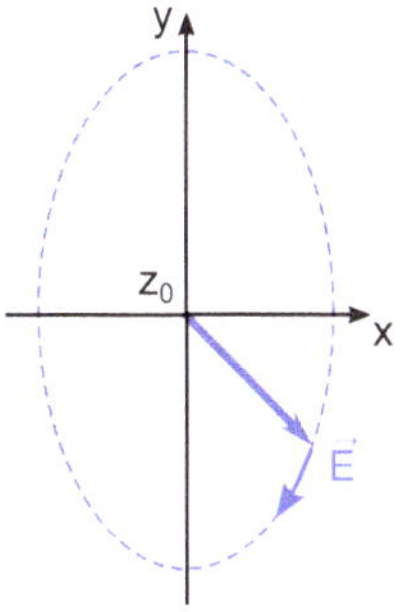

Fig. 2.15: Elliptically polarized light with positive ellipticity. The light source is located below the paper plane. z_0 is a fixed point on the propagation axis z of the light wave.

amounting to $\tan(\delta/2)$. This can be illustrated by separating the vector term in eq. (2.123) into its real and imaginary part:

$$\hat{e}_x + e^{i\delta}\,\hat{e}_y = e^{i\frac{\delta}{2}}\left(e^{-i\frac{\delta}{2}}\,\hat{e}_x + e^{i\frac{\delta}{2}}\,\hat{e}_y\right) = e^{i\frac{\delta}{2}}\left[\cos\frac{\delta}{2}\,(\hat{e}_x + \hat{e}_y) + i\sin\frac{\delta}{2}\,(-\hat{e}_x + \hat{e}_y)\right]. \tag{2.124}$$

The first term on the right side, $\cos(\delta/2)\;(\hat{e}_x + \hat{e}_y)$, describes the semimajor axis of the ellipse. It points along the bisector of the first and third quadrant and has a length of $\cos(\delta/2)$. The second term, $\sin(\delta/2)\;(-\hat{e}_x + \hat{e}_y)$, describes the semiminor axis which points along the bisector of the second and fourth quadrant. Its length is $\sin(\delta/2)$, and it is phase shifted by $\pi/2$ with respect to the semimajor axis due to the prefactor i. Keep in mind that only the real part of eq. (2.123) is the physically relevant one. A prefactor i will hence swap real and imaginary part of the exponential function containing the complex phase of the wave. According to Euler's formula, stating that an exponential function with complex argument can be written as a combination of cosine function as the real part and sine function as the imaginary part, this will swap the cosine with the sine function. Thus, the ratio of the two semiaxes, that is, the ellipticity, is $\tan(\delta/2)$, as claimed before. If $\delta = \pi/2$, eq. (2.123) transforms into eq. (2.121); that is, we obtain an rcp wave. For $\pi/2 < \delta < \pi$, the light is again elliptically polarized but the semimajor axis is now along the bisector of the second and fourth quadrant. The ellipticity amounts to $\tan^{-1}(\delta/2)$, which is still positive. At $\delta = \pi$, we obtain again linearly polarized light with the direction of oscillation along the same bisector. A phase $\pi < \delta < 3\pi/2$ yields once more elliptically polarized light. The semimajor axis lies still along the bisector of the second and fourth quadrant but the ellipticity is now negative. At $\delta = 3\pi/2$, eq. (2.123) transforms into eq. (2.122), that is, we obtain an lcp wave. Finally, for $3\pi/2 < \delta < 2\pi$, we have again elliptically polarized light. The semimajor axis lies as at the beginning along the bisector of the first and third quadrants but the ellipticity is still negative. A phase of $\delta = 2\pi$ is equivalent to the case with $\delta = 0$. Obviously, eq. (2.123) does not describe the most general elliptically polarized state as the semimajor axis cannot enclose an arbitrary angle with the x axis. We discuss in the problem in Section 2.6.14 how the most general elliptically polarized state is described mathematically.

The three polarizations – linear, circular, and elliptical – form each a basis system in the plane perpendicular to the propagation direction. Consequently, we can express any polarization state by a set of two orthogonal polarization states in any of the three polarizations. In addition, there exists a well-defined basis transformation between the three polarizations and we can apply methods known from linear algebra when switching between the various polarizations. Just as an example, from eqs. (2.118), (2.121), and (2.122) it is easy to see that light, linearly polarized in x direction, can be decomposed into rcp and lcp light as follows: $\vec{E}_x = \frac{1}{\sqrt{2}}\left(\vec{E}_+ + \vec{E}_-\right)$. See the problem in Section 2.6.8 for details.

2.4.2 Phenomenological model of the Faraday rotation

In the following section, we want to propose a phenomenological model to explain the occurrence of a Faraday rotation. We simply assume that under the influence of an applied magnetic field an optical anisotropy is induced in the material. However, we do not assume a linear anisotropy, that is, an anisotropy between two orthogonal directions. Rather, an anisotropy affecting circularly polarized states is supposed. We will discuss later in Section 3.2 why this must happen. Consequently, the refractive index will now be different for rcp and lcp light. For rcp light, the refractive index is n_+ and for lcp light n_-. Neglecting absorption, the wavelength of rcp and lcp light then amounts to λ_{vac}/n_+ and λ_{vac}/n_-, respectively, where λ_{vac} is the wavelength of light in vacuum. We consider Faraday geometry as depicted in Fig. 1.3. Let us suppose that a plane wave, polarized linearly along the x axis, is propagating along the z direction. The transparent material with length d is positioned along the z axis, starting at $z = 0$ and ending at $z = d$. A constant magnetic field is applied to the material along the z direction. At $z = 0$, the light enters the material. Outside of the material, the refractive index is isotropic and equal to 1 for vacuum (or approximately for air). So, we describe the plane wave for $z \leq 0$ as

$$\vec{E} = \vec{E}_0 e^{i(\omega t - gz)} = \frac{1}{2}E_0 \left[\underbrace{\left(\hat{e}_x + i\hat{e}_y\right)}_{\text{rcp}} + \underbrace{\left(\hat{e}_x - i\hat{e}_y\right)}_{\text{lcp}} \right] e^{i(\omega t - gz)}, z < 0. \tag{2.125}$$

In the material, the refractive index will be different for rcp and lcp light. Hence, in the material we must write the plane wave in the circular (polarization) basis as:

$$\vec{E} = \frac{1}{2}E_0 \left[\left(\hat{e}_x + i\hat{e}_y\right) e^{i\left(\omega t - \frac{2\pi n_+ z}{\lambda_{vak}}\right)} + \left(\hat{e}_x - i\hat{e}_y\right) e^{i\left(\omega t - \frac{2\pi n_- z}{\lambda_{vak}}\right)} \right]. \tag{2.126}$$

Next, we define an average refractive index of the material in magnetic field,

$$\bar{n} = \frac{1}{2}(n_+ + n_-), \tag{2.127}$$

which reduces without magnetic field to n, the index of refraction of the material without magnetic field. Furthermore, we define a phase difference,

$$\delta = 2\pi\,(n_+ - n_-)\,z/\lambda_{\text{vac}}, \tag{2.128}$$

proportional to the splitting of the refractive index in the circular basis. We call the splitting a ***magnetic circular birefringence*** because it is present only when a magnetic field is applied. Substituting in eq. (2.126) n_+ and n_- by the two new quantities and taking the common factor to the front of the square bracket yields:

$$\vec{E} = \frac{1}{2}E_0 \mathrm{e}^{\mathrm{i}\left(\omega t - \frac{2\pi\bar{n}}{\lambda_{\text{vak}}}z\right)} \left[(\hat{e}_x + \mathrm{i}\hat{e}_y)\,\mathrm{e}^{-\mathrm{i}\frac{\delta(z)}{2}} + (\hat{e}_x - \mathrm{i}\hat{e}_y)\,\mathrm{e}^{\mathrm{i}\frac{\delta(z)}{2}}\right], 0 < z < d. \tag{2.129}$$

Bundling terms belonging to the same unit vector results in:

$$\vec{E} = \frac{1}{2}E_0 \mathrm{e}^{\mathrm{i}\left(\omega t - \frac{2\pi\bar{n}}{\lambda_{\text{vak}}}z\right)} \left[\hat{e}_x \left(\mathrm{e}^{\mathrm{i}\frac{\delta(z)}{2}} + \mathrm{e}^{-\mathrm{i}\frac{\delta(z)}{2}}\right) - \mathrm{i}\hat{e}_y \left(\mathrm{e}^{\mathrm{i}\frac{\delta(z)}{2}} - \mathrm{e}^{-\mathrm{i}\frac{\delta(z)}{2}}\right)\right], 0 < z < d. \tag{2.130}$$

Using Euler's formula, we end up with the following representation of the electric field:

$$\vec{E} = E_0\left(\cos\frac{\delta}{2}\hat{e}_x + \sin\frac{\delta}{2}\hat{e}_y\right)\mathrm{e}^{\mathrm{i}\left(\omega t - \frac{2\pi\bar{n}z}{\lambda_{\text{vak}}}\right)}. \tag{2.131}$$

As one can see, eq. (2.131) describes again linearly polarized light but the polarization direction is now rotated by an angle $\delta/2$ from the x axis toward the y axis. For positive phase difference δ, that is, $n_+ > n_-$, we get a counterclockwise rotation of the plane of oscillation after passing the material when the observer is looking into the light source (see Fig. 2.16). We will call this rotation ***positive*** if it is generated by a magnetic field applied in the same direction as the propagation direction of the light. In this case, the rotation is equal to the rotation that a positive current in a coil would need to generate a magnetic field of the given direction. This definition agrees with Faraday's original definition of the sign.

For absorbing materials, the real refractive index must be substituted in eqs. (2.127) and (2.128) by the complex one,

$$\tilde{n}_\pm = n_\pm - \mathrm{i}k_\pm, \tag{2.132}$$

where $k_\pm$ is the absorption index which is related to the absorption coefficient K (see Section 2.1):

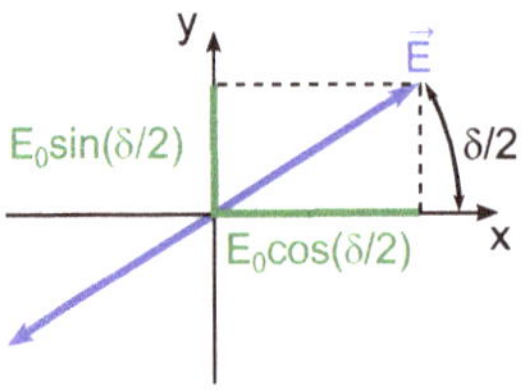

Fig. 2.16: Sketch to visualize eq. (2.131).

$$K_{\pm} = \frac{4\pi k_{\pm}}{\lambda_{\text{vak}}}. \tag{2.133}$$

This leads to a phase difference $\tilde{\delta}$ which is now complex. Consequently, after passing a distance d through the material, light is elliptically polarized in the presence of absorption. To investigate the change in polarization state, we consider the vector part $\cos\left(\tilde{\delta}/2\right)\hat{e}_x + \sin\left(\tilde{\delta}/2\right)\hat{e}_y$ of eq. (2.131) in the case of a complex refractive index $\tilde{n}_{\pm}$ and, therefore, complex angle $\tilde{\delta} = \delta_1 + \mathrm{i}\,\delta_2$:

$$\begin{aligned}\cos\frac{\tilde{\delta}}{2}\hat{e}_x + \sin\frac{\tilde{\delta}}{2}\hat{e}_y &= \cos\left(\frac{\delta_1}{2} + \mathrm{i}\frac{\delta_2}{2}\right)\hat{e}_x + \sin\left(\frac{\delta_1}{2} + \mathrm{i}\frac{\delta_2}{2}\right)\hat{e}_y \\ &= \cosh\frac{\delta_2}{2}\left(\cos\frac{\delta_1}{2}\hat{e}_x + \sin\frac{\delta_1}{2}\hat{e}_y\right) + \mathrm{i}\sinh\frac{\delta_2}{2}\left(-\sin\frac{\delta_1}{2}\hat{e}_x + \cos\frac{\delta_1}{2}\hat{e}_y\right).\end{aligned} \tag{2.134}$$

The first term on the right side of eq. (2.134) describes the semimajor axis of the polarization ellipse, which is tilted by an angle $\delta_1/2$ with respect to the x axis. The second term corresponds to the semiminor axis which is perpendicular to the semimajor axis, phase-shifted by $\pi/2$ and rotated by an angle $\delta_1/2$ with respect to the y axis. The ratio of the two semiaxes, that is, the ellipticity, is given by $-\tanh(\delta_2/2)$ where the minus sign is in accordance with the sign convention of the Faraday effect. For the rotation of the semimajor axis, we thus obtain from eq. (2.128):

$$\theta_{\mathrm{F}} = \mathrm{Re}\left\{\frac{\tilde{\delta}}{2}\right\} = \frac{\pi d}{\lambda_{\text{vak}}}(n_+ - n_-) = \frac{\omega d}{2c}(n_+ - n_-). \tag{2.135}$$

The ellipticity $\tan\eta_{\mathrm{F}}$ is given by the imaginary part δ_2 of $\tilde{\delta}$:

$$\tan\eta_{\mathrm{F}} = -\tanh\left(\mathrm{Im}\left\{\frac{\tilde{\delta}}{2}\right\}\right) = \tanh\left[\frac{\pi d}{\lambda_{\text{vak}}}(k_+ - k_-)\right] = \tanh\left[\frac{\omega d}{2c}(k_+ - k_-)\right]. \tag{2.136}$$

For small angles η_{F}, we have:

$$\eta_{\mathrm{F}} = \frac{\omega d}{2c}(k_+ - k_-). \tag{2.137}$$

Thus, if the circular birefringence is caused by a magnetic field, eqs. (2.135) and (2.137) describe the Faraday rotation θ_{F} and ellipticity η_{F}, respectively. Empirically, the following law, proposed by Wiedemann and Verdet, is valid:

$$\theta_F = V B d \cos\alpha, \tag{2.138}$$

where B is the applied magnetic field, d is the distance traveled in the material, α the angle between the direction of light propagation and magnetic field, and V is the Verdet constant, which is a material constant and in general wavelength and temperature dependent. For diamagnetic and paramagnetic materials, V is independent of magnetic field at moderate magnetic-field strength. From the sign of the Verdet constant V one can generally not conclude on the nature of magnetism. This is a consequence of the fact that magnetism is a property of the occupied ground state of the electrons at a given temperature and magnetic field, whereas magneto-optics also depends crucially on the final states of the optically excited electrons as we shall see later.

So far, we have not yet explained why an external magnetic field causes a circular birefringence along the propagation direction of the light ray. In the next chapter, we will derive a dispersion relation for the dielectric function using a classical model where a bound electron is excited by an electromagnetic wave. We will solve the equation of motion considering an external magnetic field. It will be evident already from this simple classical approach that in Faraday geometry a circular magnetic birefringence must be induced by a static magnetic field.

2.5 Comprehension questions

2.5.1 Basic magneto-optic effects

(a) State the complete definition of the Faraday and the polar Kerr effect using a drawing.
(b) Explain the principle of the Rayleigh light trap.

2.5.2 Maxwell equations

(a) Explain how the introduction of complex quantities for the optical functions such as $\tilde{\varepsilon}$, $\tilde{\sigma}$, or $\tilde{n}$ relates to the Maxwell equations. What are the preconditions for being able to introduce these complex quantities?
(b) How are $\tilde{\varepsilon}$, $\tilde{\sigma}$, $\tilde{\vec{g}}$, and $\tilde{n}$ interrelated? Discuss the physical meaning of the real and the imaginary part of $\tilde{\varepsilon}$, $\tilde{\sigma}$, $\tilde{\vec{g}}$ and $\tilde{n}$.
(c) Why is the absorption index k called absorption index? How is the absorption coefficient K defined?
(d) What are the limitations of the wave vector $\tilde{\vec{g}}$ with respect to direction in space and to the magnitude as opposed to a general complex vector in two-dimensional complex vector space?

(e) Discuss the vector relations between the vectors $\tilde{\vec{E}}$, $\tilde{\vec{D}}$, and $\tilde{\vec{g}}$ in the case of isotropic materials.
(f) Which relation exists between $\tilde{\vec{E}}$ and $\tilde{\vec{B}}$? How can it be derived?
(g) How is the Poynting vector $\vec{S}$ defined? What is its physical meaning?
(h) Which physical quantity measures a detector that is illuminated by light? How is it related to the Poynting vector?

2.5.3 Fresnel equations

(a) What are the boundary conditions for deriving the Fresnel formulas?
(b) What is a plane of incidence?
(c) Why do we have to consider only two polarization states (s- and p-polarized light) when describing reflection and transmission through an interface?
(d) Write down the formula for the electric field vector of the incident, the reflected, and the refracted electromagnetic wave (s- and p-polarized) and for the corresponding wave vectors.
(e) How is the law of reflection (i.e., angle of incidence is equal to angle of reflection) and Snell's law derived?
(f) Give the definition of the reflection and transmission coefficient and the relative index of refraction. How are reflection and transmission coefficient related to each other?
(g) Assume absorption index $k = 0$. For the case of light incident onto (1) an optically denser material and (2) an optically less dense material, plot the typical behavior of the reflection and transmission coefficient considering the two cases, polarized perpendicular and parallel to the plane of incidence. In addition, plot the phase difference between the two polarization states.
(h) Knowing the initial state of a light wave, how could you find out whether light has been reflected form a surface or not?
(i) How is the Brewster angle defined? How does it change when the light is reflected from an optically less dense material instead of from an optically denser material? Does a Brewster angle, or something similar, exist in absorbing materials?
(j) Give a microscopic, physical explanation why a Brewster angle must exist when light impinges on an interface.
(k) What is total internal reflection? When does it happen?
(l) Does a refracted wave exist in the case of total internal reflection? If yes, how does it look like? Does it influence the totally reflected beam? If yes, how?
(m) In the case of absorbing materials, is the direction of propagation of the refracted wave given by the real part β_1 of the complex angle of refraction $\tilde{\beta}$? If not, can you give a general statement in which direction the refracted wave travels with respect to β_1?

(n) Explain the principle of the Fresnel parallelepiped that is used to generate circularly polarized light.
(o) How are the reflectivity R and the absorption coefficient K defined?
(p) In a metal, the absorption is usually very large in the visible energy range, that is, $k > 10$ and $k > n$. What follows from that for the reflectivity R at normal incidence (i.e., $\alpha = 0°$)?
(q) Glass is usually transparent when looking at normal incidence, with a small reflectivity of approx. $R = 4\%$. How large is the reflectivity when looking at grazing incidence (i.e., $\alpha \cong 90°$)?

2.5.4 Polarization states

(a) What are the three typical polarization states of light? How do they look like at (1) a fixed position in space and (2) at a fixed time? Write down a mathematical representation for each of them. What do they have in common?
(b) Give a simple phenomenological model to explain the rotation of linearly polarized light when traveling through a transparent material in an applied magnetic field (Faraday effect).
(c) Can the model be extended to include absorbing materials? How? How does absorption affect the state of polarization of the light after having traveled through the material?
(d) What distinguishes the Faraday rotation from optical activity that is observed, for example, in a transparent sugar solution?
(e) How is a positive Faraday rotation defined (sign convention)?

2.6 Problems

2.6.1 Kerr geometries

Discuss the magnetic-field dependence of the Kerr effect for polar and longitudinal Kerr geometry in the limit of normal (angle of incidence $\alpha = 0°$) and grazing ($\alpha = 90°$) incidence. Refer to the definition of the various Kerr effects in Fig. 1.5. Use the simple argument that the Kerr effect is proportional to the magnetic field when the propagation direction of the light is parallel to the magnetic-field direction and that there is a quadratic dependence if the magnetic-field direction is perpendicular to the light-propagation direction.

2.6.2 Complex wave vectors

The complex wave vector $\tilde{\vec{g}}$ can be defined in an absorbing, isotropic material in two different ways:

$$\tilde{\vec{g}} = \vec{g}_1 - \mathrm{i}\vec{g}_2, \quad \text{where } g_{1j},\ g_{2j} \in \mathbb{R}, \qquad j = x,\ y,\ z, \tag{2.139}$$

or

$$\tilde{\vec{g}} = \tilde{g}'\hat{e}_{\tilde{\vec{g}}}, \text{ where } \left|\hat{e}_{\tilde{\vec{g}}}\right| = 1 \text{ with components } \hat{e}_{\tilde{\vec{g}}j} \in \mathbb{R},\ j = x,y,z, \text{ and } \tilde{g}' = g_1' - \mathrm{i}\, g_2' \text{ with } g_1',\ g_2' \in \mathbb{R}. \tag{2.140}$$

(a) Discuss if both representations are identical.

(b) If not, what are the constraints that must be made to make the two representations identical?

(c) Can these constraints be reconciled with the Maxwell equations?

Hint: Discuss the relation between the two subvectors $\mathrm{Re}\left\{\tilde{\vec{g}}\right\} = \vec{g}_1$ *and* $\mathrm{Im}\left\{\tilde{\vec{g}}\right\} = \vec{g}_2$

2.6.3 Absorption

Assume a linearly polarized electromagnetic plane wave traveling in z direction. The electric field is expressed as the real part of

$$\vec{E} = \begin{pmatrix} E_x \\ E_y \\ 0 \end{pmatrix} \mathrm{e}^{\mathrm{i}(\omega t - gz)}, \tag{2.141}$$

where g is the wave vector and ω the angular frequency. Within a material, the wave vector g depends on the index of refraction n of the material:

$$g_{\mathrm{mat}} = \frac{2\pi}{\lambda_{\mathrm{mat}}} = n\,\frac{2\pi}{\lambda} = ng, \tag{2.142}$$

where g, λ and g_{mat}, λ_{mat} are the wave vector and wavelength in vacuum and in the material, respectively.

(a) Discuss how the electromagnetic wave is changing when the material is absorbing. Then, the refractive index contains an imaginary part $k > 0$, that is, it converts to a complex refractive index: $n \to \tilde{n} = n - \mathrm{i}k$.

(b) Discuss how the electromagnetic wave is changing when it is multiplied by a complex number $\tilde{a} = a_1 - \mathrm{i}a_2$.

2.6.4 Phase difference in absorbing materials

Calculate the phase difference between the electric field $\vec{E}$ and the magnetic field $\vec{B}$ of a plane wave in an absorbing, isotropic material (i.e. absorption index $k \neq 0$) as a function of refractive index n and absorption index k.

Hint: Assume a magnetic permeability $\mu = 1$. Use the complex index of refraction $\tilde{n} = n - \mathrm{i}k$ and the Maxwell equations to derive a relation between $\vec{E}$ and $\vec{B}$. The phase difference is the difference between the arguments of the (complex) plane-wave representation for $\vec{E}$ and $\vec{B}$.

2.6.5 Poynting vector

Write down the space and time dependence of a plane electromagnetic wave traveling in an absorbing material (i.e., it includes a complex refractive index $\tilde{n} = n - \mathrm{i}k$). Calculate the time-averaged Poynting vector $\langle \vec{S} \rangle_t$, where

$$\vec{S} = \mathrm{Re}\left\{\vec{E}\right\} \times \mathrm{Re}\left\{\vec{H}\right\} = \frac{1}{\mu\mu_0} \mathrm{Re}\left\{\vec{E}\right\} \times \mathrm{Re}\left\{\vec{B}\right\}. \tag{2.143}$$

Hint: Use SI units and assume a magnetic permeability $\mu = 1$. The time average is calculated by integrating in time over a full period of the plane wave and dividing by the period.

2.6.6 Rotation of polarization by reflection from an isotropic absorbing material

Show that for incident light which is exactly s- or p-polarized, no rotation of the polarization appears when it is reflected from an isotropic and absorbing material. Assume a complex index of refraction $\tilde{n}$ and set the magnetic permeability $\mu = 1$. There is no applied magnetic field.

Hint: Discuss for both cases, s- and p-polarized incident light. Assume for the reflected and refracted fields $\vec{E}_{\mathrm{r,t}}$ and $\vec{H}_{\mathrm{r,t}}$ a component perpendicular to the chosen incident polarization. Using the boundary conditions, show that this component must then vanish at the interface.

2.6.7 Dielectric tensor in an absorbing anisotropic material

In isotropic materials, the electric displacement $\vec{D}$ is always collinear to the electric field $\vec{E}$. Therefore, the fourth Maxwell equation is also valid for $\vec{E}$. In an anisotropic material, this is no longer true because the dielectric function $\overleftrightarrow{\varepsilon}$ is now a tensor of rank two with the components $\tilde{\varepsilon}_{jk} = \varepsilon_{1jk} - \mathrm{i}\,\varepsilon_{2jk}$ and $\varepsilon_{1jk},\ \varepsilon_{2jk} \in \mathbb{R}$. This yields for the complex electric displacement:

$$D_j = \sum_k \tilde{\varepsilon}_{jk} E_k, \qquad j,\ k = x,\ y,\ z. \tag{2.144}$$

(a) Due to the complexification of Maxwell's equations, $\vec{D}$ represents no longer just the static electric displacement defined by the product of physical dielectric tensor and the physical electric field. $\vec{D}$ has now a more general meaning. Discuss the fourth Maxwell equation for the real part of the complex electric displacement $\vec{\nabla} \cdot \mathrm{Re}\{\vec{D}\} = 0$ considering plane electromagnetic waves. How can the equation be interpreted in the context of induced charge and current densities?

Hint: Write the differential equation explicitly as a sum of real and imaginary part and discuss the two terms by, first, replacing them with corresponding physical quantities know from electrodynamics and, second, by taking the time derivative.

(b) Derive the wave equation in an absorbing anisotropic material from Maxwell's equations. Considering plane electromagnetic waves, this will yield a relation between $\tilde{\vec{g}}$, $\tilde{\overleftrightarrow{\varepsilon}}$ and $\vec{E}$.

(c) Using the wave equation, find out how $\vec{D}$, $\vec{E}$ and $\tilde{\vec{g}}$ are correlated.

Hint: Discuss the result of problem 2.6.7(b) in terms of collinearity and/or coplanarity by looking at the direction of the three vectors $\vec{D}$, $\vec{E}$ and $\tilde{\vec{g}}$.

2.6.8 Polarization states

Show that linearly polarized light can be expressed as a linear combination of rcp (+) and lcp (–) light.

2.6.9 Brewster angle

Give a microscopic explanation for the appearance of the Brewster angle (no calculation, just argumentation).

Hint: Treat atoms at the interface as Hertzian dipoles which are excited by the refracted electromagnetic wave. Discuss their emission direction with respect to the polarization of the refracted wave.

2.6.10 Complex angle of refraction in a nonabsorbing material

Discuss Snell's law for arbitrary complex angle of refraction $\tilde{\beta} = \beta_1 + i\beta_2$ for the refraction from a nonabsorbing material with $n_1 > 1$ into vacuum ($n_2 = 1$).

(a) Show that in this way the case of refraction below the critical angle ($\alpha < \alpha_T$) as well as the case of total internal reflection ($\alpha > \alpha_T$) can be derived.

Hint: Snell's law must be fulfilled both, for its real as well as for its imaginary part.

(b) Considering total internal reflection, show that the energy does not flow perpendicular to the surface but parallel to it.

Hint: Consider the Poynting vector and discuss the components parallel and perpendicular to the interface.

2.6.11 Angle of maximum phase difference by total internal reflection

Considering total internal reflection in a nonabsorbing isotropic medium with $n_1 > n_2$ (where n_2 is the index of refraction of the outer space), show that for angle of incidence α larger than the critical angle α_T, the maximum phase difference occurs at an angle α_{max}, where α_{max} is given by

$$\sin^2(\alpha_{max}) = \frac{2n_2^2}{n_1^2 + n_2^2}. \tag{2.145}$$

Hint: The phase difference can be easily described by looking at the ratio $\tilde{\rho}_s / \tilde{\rho}_p$. *Using Snell's law, the angle* β *can be replaced in this formula by the relative index of refraction* $n = n_2/n_1$. *Finally, the maximum is found by determining the zero crossings of the first derivative* $\partial(\delta_s - \delta_p)/\partial\alpha = 0$.

2.6.12 Maximum of phase difference by total internal reflection

Show that the maximum of the phase difference by total internal reflection in a nonabsorbing isotropic material is given by:

$$\tan\left(\frac{\delta_s - \delta_p}{2}\right)_{max} = \frac{n_1^2 - n_2^2}{2n_1 n_2}. \tag{2.146}$$

Hint: Consider, as in the problem in Section 2.6.11, the ratio $\tilde{\rho}_s/\tilde{\rho}_p$ *and insert the result for the angle* α_{max} *of the maximum phase difference. Remember that the tangent of half the phase difference can be obtained from the ratio between imaginary and real part of the square root of* $\tilde{\rho}_s/\tilde{\rho}_p$.

2.6.13 Planes of constant phase and constant amplitude in an absorbing material

Determine the planes of constant phase and constant amplitude as a function of angle of incidence α in the case of refraction from vacuum into an absorbing material ($\tilde{n}$ complex). Find the direction of propagation $\hat{e}_{\vec{g}}$. It is defined as the direction perpendicular to the planes of constant phase.

Hint: Consider the complex law of refraction with $\tilde{\beta} = \beta_1 + \mathrm{i}\beta_2$ *and derive therefrom* $\sin\beta_1$ *and* $\sin\beta_2$ *as a function of* $\sin\alpha$. *Then, consider the refracted wave and divide it up into phase and amplitude part. These two fractions can be simplified according to the first part of the problem.*

2.6.14 Elliptically polarized light

(a) Show that elliptically polarized light,

$$\vec{E} = \frac{1}{\sqrt{2}} E_0 \left(\hat{e}_x + \mathrm{e}^{\mathrm{i}\delta} \hat{e}_y\right) \mathrm{e}^{\mathrm{i}(\omega t - \vec{g}\cdot\vec{r})}, \tag{2.147}$$

can be described as a linear combination of circularly polarized light.

Hint: First, express *linearly polarized light in x and y directions by rcp and lcp light. Enter your findings in the formula for elliptically polarized light above and bundle the terms for rcp and lcp light, respectively.*

(b) Show that the following equation describes elliptically polarized light in the most general way:

$$\vec{E} = \left(E_x \mathrm{e}^{\mathrm{i}\delta_x} \hat{e}_x + E_y \mathrm{e}^{\mathrm{i}\delta_y} \hat{e}_y\right) \mathrm{e}^{\mathrm{i}(\omega t - \vec{g}\cdot\vec{r})}. \tag{2.148}$$

Hint: First, consider a local x'y' coordinate system, wherein the mathematical representation of an ellipse is the simplest. In this local coordinate system, the ellipticity $\tan\eta$ *is easily defined. Then, rotate the ellipse from the local coordinate system back into a general xy coordinate system where it adopts its general form above (this yields the rotation angle* θ *of the major half axis). Now, verify if the coefficients* $E_{x,y}$ *and* $\delta_{x,y}$ *can be unambiguously determined by the two parameters, rotation* θ *and the ellipticity* η, *that define an ellipse in the most general representation as well.*

2.7 Solutions

2.7.1 Kerr geometries

For normal incidence $\alpha = 0°$, the polar Kerr effect is proportional to the magnetic field B because the magnetic field direction is parallel to the propagation direction of the light. The transverse and longitudinal Kerr effect is no longer distinguishable because the electric field vector is in both cases parallel to the sample surface and perpendicular to the propagation direction. Therefore, both Kerr effects show a quadratic dependence on B because B is perpendicular to the propagation direction of the light.

For grazing incidence $\alpha = 90°$, the longitudinal Kerr effect is proportional to the magnetic field B because the magnetic field is parallel to the propagation direction. The polar Kerr effect shows, on the other hand, a quadratic dependence on B because B is now perpendicular to the propagation direction of the light.

For an angle α between normal and grazing incidence, a mixture of linear and quadratic effects is measured but the quadratic component is usually much smaller and can be neglected.

2.7.2 Complex wave vectors

(a) Let us write explicitly the components of the wave vectors in the two representations. For eq. (2.139), we get:

$$\tilde{\vec{g}} = \vec{g}_1 - \mathrm{i}\,\vec{g}_2 = \begin{pmatrix} g_{1x} - \mathrm{i}\,g_{2x} \\ g_{1y} - \mathrm{i}\,g_{2y} \\ g_{1z} - \mathrm{i}\,g_{2z} \end{pmatrix}. \tag{2.149}$$

For eq. (2.140), we get:

$$\tilde{\vec{g}}' = \tilde{g}'\,\hat{e}_{\tilde{\vec{g}}'} = (g_1' - \mathrm{i}\,g_2')\begin{pmatrix} \alpha \\ \beta \\ \gamma \end{pmatrix}, \quad \text{with } \alpha^2 + \beta^2 + \gamma^2 = 1. \tag{2.150}$$

Now, we can write eq. (2.150) in a similar way as eq. (2.149):

$$\tilde{\vec{g}}' = g_1' \begin{pmatrix} \alpha \\ \beta \\ \gamma \end{pmatrix} - \mathrm{i}g_2' \begin{pmatrix} \alpha \\ \beta \\ \gamma \end{pmatrix} = \vec{g}_1' - \mathrm{i}\,\vec{g}_2'. \tag{2.151}$$

From eq. (2.151) it follows that the two subvectors, representing the real and imaginary part, are collinear: $\vec{g}_1' || \vec{g}_2'$. But for the first representation $\tilde{\vec{g}} = \vec{g}_1 - \mathrm{i}\,\vec{g}_2$, the vector $\vec{g}_1$, representing the real part, is not necessarily parallel to the vector $\vec{g}_2$, representing the imaginary part. Therefore, the two representations eqs. (2.139) and (2.140) are not identical. Equation (2.139) is more general.

(b) For the two representations to be identical, the vectors representing the real and imaginary part of eq. (2.139) must be collinear: $\vec{g}_1 || \vec{g}_2$. So, we can write:

$$\tilde{\vec{g}} = \vec{g}_1 - \mathrm{i}\vec{g}_2 = \vec{g}_1\,(1 - \mathrm{i}a), a \in \mathbb{R}. \tag{2.152}$$

Then, eq. (2.139) can be written in the form of eq. (2.150) as

$$\tilde{\vec{g}} = \tilde{g}\,\hat{e}_{\tilde{g}}, \text{ with } \hat{e}_{\tilde{g}} = \frac{\vec{g}_1}{|\vec{g}_1|} \text{ and } \tilde{g} = (1 - \mathrm{i}a)\,|\vec{g}_1|. \tag{2.153}$$

(c) The collinearity of the real part $\vec{g}_1$ and the imaginary part $\vec{g}_2$ of $\tilde{\vec{g}}$ is a direct consequence of the second and fourth Maxwell equations in isotropic materials, eqs. (2.3) and (2.5), respectively. Assuming the most general representation according to eq. (2.139), it follows from the fourth Maxwell equation $\vec{\nabla} \cdot \vec{D} = \varepsilon \vec{\nabla} \cdot \vec{E} = 0$ that $\tilde{\vec{g}} \cdot \vec{E} = 0$. From the second Maxwell equation $\vec{\nabla} \times \vec{E} = -\dot{\vec{B}}$, it follows that $\tilde{\vec{g}} \times \vec{E} = \omega \vec{B}$. From the fourth Maxwell equation, we see that $\vec{g}_1$ and $\vec{g}_2$ are in a plane perpendicular to $\vec{E}$. From the second Maxwell equation, it results that $\vec{g}_1$ and $\vec{g}_2$ are in the same plane as $\vec{E}$. The two requirements can be fulfilled only if $\vec{g}_1$ is collinear to $\vec{g}_2$ and, thus, $\tilde{\vec{g}}$ can be written as $\tilde{\vec{g}} = \tilde{g}\,\hat{e}_{\tilde{g}}$. Note that this is true only if the amplitude $\vec{E}_0$ of the electric field vector is real. If $\vec{E}_0$ is complex, as, for example, for the refracted wave in an absorbing material, the two vectors $\vec{g}_1$ and $\vec{g}_2$ are no longer collinear and the wave vector must be represented in its most general form according to eq. (2.139).

2.7.3 Absorption

Remember that the genuine electric field is always the real part of the complex electric field $\tilde{\vec{E}}$.

(a) If the refractive index is $n \to \tilde{n} = n - \mathrm{i}k$ (and $E_x, E_y \in \mathbb{R}$) then eq. (2.142) changes to eq. (2.27). Therefore, we can write the electric field according to eq. (2.141) as:

$$\mathrm{Re}\left\{\tilde{\vec{E}}\right\} = \begin{pmatrix} E_x \\ E_y \\ 0 \end{pmatrix} \mathrm{Re}\left\{\mathrm{e}^{\mathrm{i}[\omega t - (n - \mathrm{i}k)gz]}\right\} = \mathrm{e}^{-kgz} \begin{pmatrix} E_x \\ E_y \\ 0 \end{pmatrix} \mathrm{Re}\left\{\mathrm{e}^{\mathrm{i}(\omega t - ngz)}\right\}$$
$$= \mathrm{e}^{-kgz} \begin{pmatrix} E_x \\ E_y \\ 0 \end{pmatrix} \cos(\omega t - ngz). \tag{2.154}$$

As a result, we see that the amplitude is exponentially damped due to absorption. The damping constant is merely proportional to the imaginary part k of the complex refractive index. If the amplitude is damped, the intensity will be exponentially damped, too. Furthermore, the space dependence of the phase depends only on the real part n of the complex refractive index.

(b) Let the complex number be $\tilde{a} = a_1 - \mathrm{i}a_2$, then the product between the electric field and $\tilde{a}$ is:

$$\mathrm{Re}\left\{\tilde{\vec{E}}\right\} = \mathrm{Re}\left\{(a_1 - \mathrm{i}a_2)\begin{pmatrix} E_x \\ E_y \\ 0 \end{pmatrix} \mathrm{e}^{\mathrm{i}(\omega t - gz)}\right\} = \mathrm{Re}\left\{\underbrace{\sqrt{a_1^2 + a_2^2}\,\mathrm{e}^{-\mathrm{i}\varphi}}_{\text{polar representation}}\begin{pmatrix} E_x \\ E_y \\ 0 \end{pmatrix} \mathrm{e}^{\mathrm{i}(\omega t - gz)}\right\}$$
$$= \sqrt{a_1^2 + a_2^2}\begin{pmatrix} E_x \\ E_y \\ 0 \end{pmatrix} \mathrm{Re}\left\{\mathrm{e}^{\mathrm{i}(\omega t - gz - \varphi)}\right\}$$
$$= \sqrt{a_1^2 + a_2^2}\begin{pmatrix} E_x \\ E_y \\ 0 \end{pmatrix} \cos(\omega t - gz - \varphi)\,, \tag{2.155}$$

with $\varphi = \arctan(a_2/a_1)$. Note that we have used the polar representation for $\tilde{a}$ because the phase will then simply add to the phase of the wave. As a result, the amplitude of the electromagnetic wave is multiplied by the absolute value of $\tilde{a}$ and the wave exhibits a phase shift $-\varphi$ according to the phase of $\tilde{a}$.

2.7.4 Phase difference in absorbing materials

Assume a plane electromagnetic wave traveling along the z direction and linearly polarized in x direction:

$$E_x = E_0\ \mathrm{e}^{\mathrm{i}(\omega t - \tilde{g}_z z)}. \tag{2.156}$$

The physically relevant real part is given by eq. (2.154) of the solution of part (a) of the problem in Section 2.6.3 with $E_x = E_0$ and $E_y = 0$. The amplitude E_0 is real. According to eq. (2.41), the following relation holds between the magnetic and electric field amplitude:

$$B_y = \frac{\tilde{n}}{c} E_x. \tag{2.157}$$

Because the refractive index $\tilde{n}$ is a complex number, we immediately see that according to eq. (2.155) of the solution of part (b) of the problem in Section 2.6.3, the magnetic-field amplitude B_y must exhibit a phase difference with respect to E_x. To derive the phase difference, let us write $\tilde{n}$ in polar representation:

$$\tilde{n} = n - \mathrm{i}k = \sqrt{n^2 + k^2}\,\mathrm{e}^{-\mathrm{i}\arctan\left(\frac{k}{n}\right)} = |\tilde{n}|\,\mathrm{e}^{\mathrm{i}\varphi}. \tag{2.158}$$

where the phase of $\tilde{n}$ is $\varphi = -\arctan(k/n)$. Inserting eq. (2.158) and the electric field from eq. (2.156) into eq. (2.157), the magnetic field can be written as:

$$B_y = B_0\ \mathrm{e}^{\mathrm{i}(\omega t - \tilde{g}_z z + \varphi)}, \tag{2.159}$$

The magnetic-field amplitude is then:

$$B_0 = \frac{\sqrt{n^2 + k^2}}{c}\,E_0, \tag{2.160}$$

and the phase difference with respect to E_x is given by φ. It amounts to:

$$\varphi = -\arctan\left(\frac{k}{n}\right). \tag{2.161}$$

The phase difference implies that the nodes of the magnetic field are displaced with respect to the nodes of the electric field along the propagation direction of the wave. For a transparent material with $k = 0$, it follows that the magnetic field amplitude is $B_0 = (n/c)\,E_0$ and the phase difference vanishes, that is, $\varphi = 0$.

2.7.5 Poynting vector

Assume a plane electromagnetic wave traveling in z direction which is linearly polarized in x direction as in eq. (2.156): $E_x = E_0\ \mathrm{e}^{\mathrm{i}(\omega t - \tilde{g}_z z)}$. Because it is traveling in an absorbing material, the wave vector is complex:

$$\tilde{g}_z = \frac{\omega}{c}(n - \mathrm{i}\,k) = \frac{\omega}{c}\tilde{n}. \tag{2.162}$$

Inserting this into the expression of the plane electromagnetic wave, we get:

$$E_x = E_0\ \mathrm{e}^{\mathrm{i}(\omega t - \tilde{g}_z z)} = E_0\ \mathrm{e}^{-\frac{\omega k z}{c}}\ \mathrm{e}^{\mathrm{i}\left(\omega t - \frac{\omega n z}{c}\right)}. \tag{2.163}$$

For the Poynting vector, we need the real part of the electric field:

$$\mathrm{Re}\{E_x\} = E_0\ \mathrm{e}^{-\frac{\omega k z}{c}}\ \cos\left(\omega t - \frac{\omega n z}{c}\right). \tag{2.164}$$

Because $\vec{E}$, $\vec{B}$, and $\vec{g}$ form a right-handed helix as shown in Section 2.1, the magnetic field vector must point along the y axis. From the relation between the magnetic and electric field amplitude, eq. (2.41), and the relation between the magnetic field and the magnetic field strength $\vec{B} = \mu\mu_0\vec{H}$ (see Section 2.1), we get the magnetic field strength of the plane electromagnetic wave. We write the complex refractive index again in polar form according to eq. (2.158) to have an explicit representation of its phase: $\tilde{n} = \sqrt{n^2 + k^2}\ e^{i\varphi}$ with phase $\varphi = -\arctan(k/n)$:

$$H_y = H_0\ e^{i(\omega t - \tilde{g}_z z)} = \frac{\tilde{n}}{\mu\mu_0 c} E_x \overset{\mu=1}{=} \frac{\tilde{n}}{\mu_0 c} E_x = \frac{\sqrt{n^2+k^2}}{\mu_0 c} E_0\ e^{-\frac{\omega k z}{c}}\ e^{i\left(\omega t - \frac{\omega n z}{c} + \varphi\right)}. \quad (2.165)$$

The real part of H_y is then:

$$\mathrm{Re}\{H_y\} = \frac{\sqrt{n^2+k^2}}{\mu_0 c} E_0\ e^{-\frac{\omega k z}{c}} \cos\left(\omega t - \frac{\omega n z}{c} + \varphi\right). \quad (2.166)$$

Due to the vector product between $\mathrm{Re}\{\vec{E}\}$ and $\mathrm{Re}\{\vec{H}\}$, the Poynting vector has only a z component S_z. The time average is calculated by integrating S_z in time over a full period of the plane wave and dividing by the period. This yields:

$$\begin{aligned}
\left\langle\left|\vec{S}\right|\right\rangle_t &= \langle S_z\rangle_t = \left\langle \mathrm{Re}\{E_x\}\cdot\mathrm{Re}\{H_y\}\right\rangle_t \\
&= \frac{\sqrt{n^2+k^2}}{\mu_0 c} E_0^2\ e^{-\frac{2\omega k z}{c}} \left\langle \cos\left(\omega t - \frac{\omega n z}{c}\right) \cos\left(\omega t - \frac{\omega n z}{c} + \varphi\right)\right\rangle_t \\
&= \frac{\sqrt{n^2+k^2}}{\mu_0 c} E_0^2\ e^{-\frac{2\omega k z}{c}} \left\langle \cos\left(\omega t - \frac{\omega n z}{c}\right) \left[\cos\left(\omega t - \frac{\omega n z}{c}\right)\cos\varphi - \sin\left(\omega t - \frac{\omega n z}{c}\right)\sin\varphi\right]\right\rangle_t \\
&= \frac{\sqrt{n^2+k^2}}{\mu_0 c} E_0^2\ e^{-\frac{2\omega k z}{c}} \left[\frac{\cos\varphi}{T}\int_0^T \cos^2\left(\omega t - \frac{\omega n z}{c}\right) dt - \frac{\sin\varphi}{T}\int_0^T \cos\left(\omega t - \frac{\omega n z}{c}\right)\sin\left(\omega t - \frac{\omega n z}{c}\right) dt\right] \\
&= \frac{\sqrt{n^2+k^2}}{\mu_0 c} E_0^2\ e^{-\frac{2\omega k z}{c}} \left[\frac{\cos\varphi}{2T}\int_0^T \left[1 + \cos\left[2\left(\omega t - \frac{\omega n z}{c}\right)\right]\right] dt - \frac{\sin\varphi}{2T}\int_0^T \sin\left[2\left(\omega t - \frac{\omega n z}{c}\right)\right] dt\right] \\
&= \frac{\sqrt{n^2+k^2}}{\mu_0 c} E_0^2\ e^{-\frac{2\omega k z}{c}} \left[\frac{\cos\varphi}{2T}\underbrace{\left[t + \frac{1}{2\omega}\sin\left[2\left(\omega t - \frac{\omega n z}{c}\right)\right]\right]_0^T}_{T} + \frac{\sin\varphi}{2T}\underbrace{\left[\frac{1}{2\omega}\cos\left[2\left(\omega t - \frac{\omega n z}{c}\right)\right]\right]_0^T}_{0}\right].
\end{aligned} \quad (2.167)$$

Finally, the time average $\langle S_z\rangle_t$ of the Poynting vector amounts to:

$$\left\langle\left|\vec{S}\right|\right\rangle_t = \langle S_z\rangle_t = \frac{1}{2}\cos\varphi \frac{\sqrt{n^2+k^2}}{\mu_0 c} E_0^2\ e^{-\frac{2\omega k z}{c}}. \quad (2.168)$$

We can get rid of the cosine factor by expressing it by the tangent of φ and using eq. (2.161):

$$\cos\varphi = \frac{1}{\sqrt{1+\tan^2\varphi}} = \frac{n}{\sqrt{n^2+k^2}}. \tag{2.169}$$

Then, we get as a result for the time average of the Poynting vector:

$$\left\langle \left|\vec{S}\right| \right\rangle_t = \frac{1}{2}\frac{n}{\mu_0 c} E_0^2\, \mathrm{e}^{-\frac{2\omega kz}{c}} = \frac{1}{2} cn\varepsilon_0\, E_0^2\, \mathrm{e}^{-Kz}. \tag{2.170}$$

The Poynting vector points along the direction of the wave vector. As expected from the connection to the intensity, see eq. (2.46), $\langle S_z \rangle_t$ is exponentially damped along the propagation direction of the plane wave. The damping constant is the absorption coefficient K. Furthermore, $\langle S_z \rangle_t$ is proportional to the square of the electric-field amplitude E_0 and to the real part n of the complex refractive index, that is, the ordinary refractive index of the material.

2.7.6 Rotation of polarization by reflection from an isotropic absorbing material

To prove the hypothesis, we will make a proof by contradiction. Therefore, we assume a rotation of polarization for the reflected and the refracted wave. By rotating, the field vectors generate an additional component ***perpendicular*** to the incoming polarization. In addition, eq. (2.41) holds for the magnitude of the field vectors. At the interface, the following boundary condition must apply: The components $\vec{E}_{||}$ and $\vec{H}_{||}$ of the electric field and of the magnetic field strength, respectively, which are parallel to the interface, are continuous. The boundary conditions must then be fulfilled for both components separately.

We start with a plane electromagnetic wave which is s-polarized: $\vec{E}_\mathrm{i}||\hat{e}_z$. The geometry is given in Fig. 2.17.

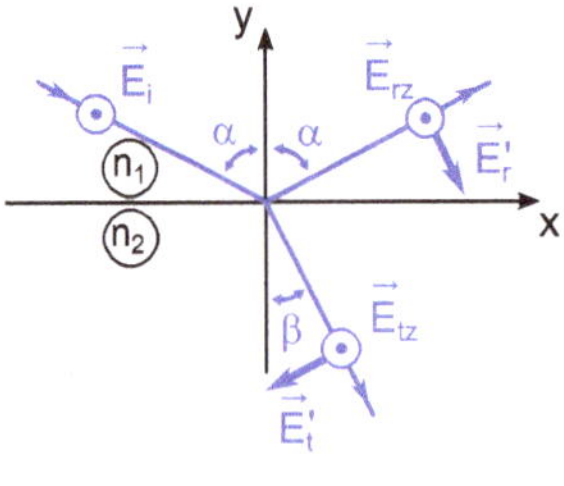

Fig. 2.17: Geometry of the electric-field vectors including a rotated, reflected, and refracted beam. $\vec{E}_\mathrm{i}$ points out of the plane of incidence.

We assume that the reflected wave has an additional component $\vec{E}_\mathrm{r}'$ in the xy plane. The entire reflected electric field is then:

$$\vec{E}_\mathrm{r}'' = \vec{E}_{\mathrm{r}z} + \vec{E}_\mathrm{r}', \tag{2.171}$$

with

$$\vec{E}'_{\mathrm{r}} = \vec{E}'_{0\mathrm{r}}\ \mathrm{e}^{-\mathrm{i}\tilde{g}_1 x \sin\alpha}, \tag{2.172}$$

where $E'_{0\mathrm{r}}$ is the length of the additional, rotated component of the reflected electric-field vector. The non-rotated component $E_{\mathrm{r}z}$ is given by eq. (2.51).

The refracted wave has an additional component $\vec{E}'_{\mathrm{t}}$ in xy plane, too. The entire refracted electric field is:

$$\vec{E}''_{\mathrm{t}} = \vec{E}_{\mathrm{t}z} + \vec{E}'_{\mathrm{t}}, \tag{2.173}$$

with

$$\vec{E}'_{\mathrm{t}} = \vec{E}'_{0\mathrm{t}}\ \mathrm{e}^{-\mathrm{i}\tilde{g}_2 x \sin\tilde{\beta}}, \tag{2.174}$$

where $E'_{0\mathrm{t}}$ is the length of the additional, rotated component of the refracted electric-field vector. The non-rotated component $E_{\mathrm{t}z}$ is given by eq. (2.52).

From the boundary condition that $\vec{\mathrm{E}}_{||}$ must be continuous follows for the x component of the electric field according to the geometry defined in Fig. 2.17:

$$E'_{\mathrm{r}x} \underbrace{+\ 0}_{\text{incoming wave}} = -E'_{\mathrm{t}x}, \tag{2.175}$$

with

$$E'_{\mathrm{r}x} = \cos\alpha\ E'_{0\mathrm{r}}\ \mathrm{e}^{-\mathrm{i}\tilde{g}_1 x \sin\alpha}, \tag{2.176}$$

and

$$E'_{\mathrm{t}x} = \cos\tilde{\beta}\ E'_{0\mathrm{t}}\ \mathrm{e}^{-\mathrm{i}\tilde{g}_2 x \sin\tilde{\beta}}. \tag{2.177}$$

The exponential terms cancel because of Snell's law as for the non-rotated components. This yields the simplified boundary condition:

$$\cos\alpha\ E'_{0\mathrm{r}} = -\cos\tilde{\beta}\ E'_{0\mathrm{t}}. \tag{2.178}$$

Solving for the transmitted component gives:

$$E'_{0\mathrm{t}} = -\frac{\cos\alpha}{\cos\tilde{\beta}}\ E'_{0\mathrm{r}}. \tag{2.179}$$

To determine the two unknown components, we need a second equation. That can be derived in an analogous way with the help of the boundary condition for the magnetic-field strength. The geometry is given in Fig. 2.18.

We assume again that the reflected wave $\vec{H}_{\mathrm{r}}$ has an additional component $\vec{H}'_{\mathrm{r}z}$ in $-z$ direction. The entire reflected magnetic-field-strength vector is then:

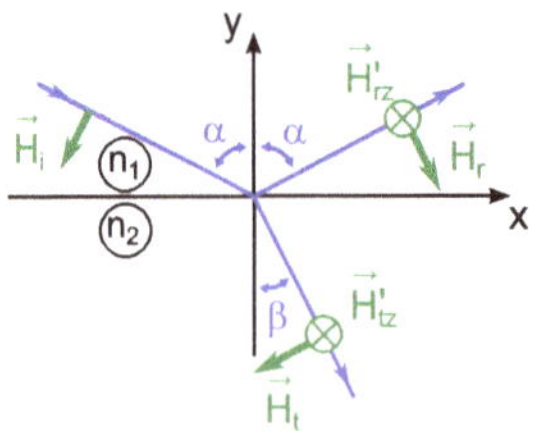

Fig. 2.18: Geometry of the magnetic-field-strength vectors including rotated, reflected, and refracted beams. $\vec{H}_i$ lies in the plane of incidence.

$$\vec{H}''_r = \vec{H}_r + \vec{H}'_{rz}, \tag{2.180}$$

with

$$\vec{H}'_{rz} = \vec{H}'_{0r}\, e^{-i\tilde{g}_1 x \sin\alpha}, \tag{2.181}$$

where H'_{0r} is the length of the additional, rotated component of the reflected magnetic-field-strength vector. The continuous, non-rotated component H_{rx} is given by eq. (2.60).

The refracted wave $\vec{H}_t$ has an additional component $\vec{H}'_{tz}$ in $-z$ direction, as well. The entire refracted magnetic field strength is:

$$\vec{H}''_t = \vec{H}_t + \vec{H}'_{tz}, \tag{2.182}$$

with

$$\vec{H}'_{tz} = \vec{H}'_{0t}\, e^{-i\tilde{g}_2 x \sin\tilde{\beta}}, \tag{2.183}$$

where H'_{0t} is the length of the additional, rotated component of the refracted magnetic-field-strength vector. The continuous, non-rotated component H_{tx} is given by eq. (2.61).

From the boundary condition that $\vec{H}_{||}$ must be continuous follows for the z component of the magnetic field strength according to the geometry defined in Fig. 2.18:

$$-H'_{rz} + \underbrace{0}_{\text{incoming wave}} = -H'_{tz}. \tag{2.184}$$

Using eq. (2.41) and noticing that a proportionality between the field-vector magnitudes implies the same proportionality between corresponding vector components, we get (assuming again $\mu = 1$)

$$H'_{0r,t} = c\varepsilon_0 \frac{\tilde{\varepsilon}}{\tilde{n}}\, E'_{0r,t} \overset{\mu=1}{=} c\varepsilon_0 \tilde{n}\, E'_{0r,t}. \tag{2.185}$$

Remembering that the exponential terms in eq. (2.184) cancel again because of Snell's law as for the non-rotated components and inserting eq. (2.185) into eq. (2.184) yields the simplified boundary condition:

$$\tilde{n}_1\, E'_{0r} = \tilde{n}_2\, E'_{0t}. \tag{2.186}$$

Solving for the transmitted component and combining with eq. (2.179) results in:

$$E'_{0t} = \frac{\tilde{n}_1}{\tilde{n}_2}\, E'_{0r} = -\frac{\cos\alpha}{\cos\tilde{\beta}}\, E'_{0r}. \tag{2.187}$$

From this and using Snell's law, we get

$$\frac{\tilde{n}_1}{\tilde{n}_2} = \frac{\sin\tilde{\beta}}{\sin\alpha} = -\frac{\cos\alpha}{\cos\tilde{\beta}}, \tag{2.188}$$

and further:

$$\sin\tilde{\beta}\cos\tilde{\beta} = -\sin\alpha\cos\alpha, \tag{2.189}$$

and finally

$$\sin 2\tilde{\beta} = -\sin 2\alpha. \tag{2.190}$$

This equation has only one solution within the allowed range of angles of incidence, that is, for $0 \le \alpha \le 90°$:

$$\tilde{\beta} = -\alpha. \tag{2.191}$$

This condition leads to a contradiction because it can be fulfilled only with $\alpha = 0$ but the angle of incidence can be freely chosen. Therefore, rotated components cannot exist and $E'_{0r} = E'_{0t} = 0$, or in other words: Purely s-polarized light is reflected and refracted as purely s-polarized light.

For p-polarized light, the procedure is completely analogous. We, therefore, just state the major steps and leave out the detailed calculation. For the incoming plane wave, we have $\vec{H}_i || \hat{e}_z$. The geometry is analogous to Fig. 2.17 but the electric field $\vec{E}$ is replaced by the magnetic field strength $\vec{H}$. We assume again that the reflected wave has an additional magnetic-field-strength component in the *xy* plane. The entire reflected magnetic field strength is then:

$$\vec{H}''_r = \vec{H}_{rz} + \vec{H}'_r. \tag{2.192}$$

The entire refracted magnetic field strength is

$$\vec{H}''_t = \vec{H}_{tz} + \vec{H}'_t. \tag{2.193}$$

From the boundary condition that $\vec{H}_{||}$ must be continuous follows:

$$-H'_{rx} + \underbrace{0}_{\text{incoming wave}} = H'_{tx}. \tag{2.194}$$

Using again the relation between the electric and magnetic field amplitudes, we get

$$-\cos\alpha\, \tilde{n}_1\, E'_{0r} = \cos\tilde{\beta}\, \tilde{n}_2\, E'_{0t}. \tag{2.195}$$

A second equation we get again by considering the continuity of the electric field. The geometry is analogous to Fig. 2.18 but the magnetic field strength $\vec{H}$ is replaced by the electric field $\vec{E}$. The reflected electric field has an additional component in z direction. The entire reflected electric field vector is then:

$$\vec{E}''_r = \vec{E}_r + \vec{E}'_{rz}, \tag{2.196}$$

and the entire refracted electric field is

$$\vec{E}''_t = \vec{E}_t + \vec{E}'_{tz}. \tag{2.197}$$

From the boundary condition that $\vec{E}_{||}$ must be continuous follows:

$$-\vec{E}'_{rz} + \underbrace{0}_{\text{incoming wave}} = -\vec{E}'_{tz}, \tag{2.198}$$

which results in:

$$E'_{0r} = E'_{0t}. \tag{2.199}$$

Combining eq. (2.195) with eq. (2.199), we get

$$-\cos\alpha\, \tilde{n}_1\, E'_{0r} = \cos\tilde{\beta}\, \tilde{n}_2\, E'_{0r}, \tag{2.200}$$

and further:

$$-\cos\alpha\, \tilde{n}_1 = \cos\tilde{\beta}\, \tilde{n}_2, \tag{2.201}$$

and then

$$-\frac{\cos\alpha}{\cos\tilde{\beta}} = \frac{\tilde{n}_2}{\tilde{n}_1} = \frac{\sin\alpha}{\sin\tilde{\beta}}. \tag{2.202}$$

This reduces to

$$-\tan\tilde{\beta} = \tan\alpha, \tag{2.203}$$

and, finally, to

$$\tilde{\beta} = -\alpha. \tag{2.204}$$

This condition leads again to a contradiction because it can be fulfilled only with $\alpha = 0$. Therefore, rotated components cannot exist and $E'_{0r} = E'_{0t} = 0$. This completes the proof. In conclusion, we can state that purely s- or p-polarized light is reflected and refracted again as purely s- or p-polarized light, respectively.

2.7.7 Dielectric tensor in absorbing anisotropic material

(a) Let us first collect all the relations connecting electric field, electric displacement, current density, and charge. We write the complex dielectric tensor as the sum of two tensors, representing the real and imaginary part:

$$\overleftrightarrow{\tilde{\varepsilon}} = \overleftrightarrow{\varepsilon}_1 - \mathrm{i}\overleftrightarrow{\varepsilon}_2 = \overleftrightarrow{\varepsilon} - \mathrm{i}\,\frac{1}{\varepsilon_0\omega}\overleftrightarrow{\sigma}. \tag{2.205}$$

This can be expressed in terms of the tensor components as

$$\tilde{\varepsilon}_{jk} = \varepsilon_{1jk} - \mathrm{i}\,\varepsilon_{2jk} = \varepsilon_{jk} - \mathrm{i}\frac{1}{\varepsilon_0\omega}\sigma_{jk}. \tag{2.206}$$

The complex electric displacement is then expressed as the product of complex dielectric tensor and complex electric field:

$$\vec{\tilde{D}} = \varepsilon_0 \overleftrightarrow{\tilde{\varepsilon}}\, \vec{\tilde{E}}. \tag{2.207}$$

The components of the complex electric displacement are

$$\tilde{D}_j = \varepsilon_0 \sum_k \tilde{\varepsilon}_{jk}\tilde{E}_k \quad \text{for } j,\ k = x, y, z. \tag{2.208}$$

The physical current density is connected to the physical electric field by the optical conductivity tensor:

$$\vec{j} = \overleftrightarrow{\sigma}\, \vec{E}. \tag{2.209}$$

Finally, the physical charge density follows from the fourth Maxwell equation, eq. (2.5), with the physical electric field:

$$\vec{\nabla}\cdot\vec{D} = \vec{\nabla}\cdot\left(\varepsilon_0 \overleftrightarrow{\varepsilon}\vec{E}\right) = \rho. \tag{2.210}$$

Here, $\vec{D}$ is the static electric displacement and $\vec{E}$ is the physical electric field. Note that the physical electric field is just the real part of the complex electric field.

Now, we are ready to discuss the complex representation of the fourth Maxwell equation:

$$\vec{\nabla}\cdot\mathrm{Re}\left\{\vec{D}\right\}=\vec{\nabla}\cdot\mathrm{Re}\left\{\varepsilon_0\overleftrightarrow{\varepsilon}\vec{E}\right\}=\vec{\nabla}\cdot\mathrm{Re}\left\{\varepsilon_0\left[\overleftrightarrow{\varepsilon}_1-\mathrm{i}\overleftrightarrow{\varepsilon}_2\right]\vec{E}\right\}$$
$$=\underbrace{\vec{\nabla}\cdot\left(\varepsilon_0\overleftrightarrow{\varepsilon}\ \mathrm{Re}\left\{\vec{E}\right\}\right)}_{\rho}-\mathrm{Re}\left\{\mathrm{i}\frac{\varepsilon_0}{\varepsilon_0\omega}\vec{\nabla}\cdot\left(\overleftrightarrow{\sigma}\vec{E}\right)\right\}=0. \quad (2.211)$$

Assuming harmonic waves and taking the time derivative of the above equation yields:

$$\dot{\rho}-\mathrm{Re}\left\{\mathrm{i}\frac{\varepsilon_0}{\varepsilon_0\omega}\vec{\nabla}\cdot\left(\overleftrightarrow{\sigma}\dot{\vec{E}}\right)\right\}=\dot{\rho}-\mathrm{Re}\left\{\mathrm{i}\frac{1}{\omega}\vec{\nabla}\cdot\left(\overleftrightarrow{\sigma}\,\mathrm{i}\omega\vec{E}\right)\right\}=\dot{\rho}+\vec{\nabla}\cdot\left(\underbrace{\overleftrightarrow{\sigma}\,\mathrm{Re}\left\{\vec{E}\right\}}_{\vec{j}}\right)=0. \quad (2.212)$$

From this follows:

$$\dot{\rho}+\vec{\nabla}\cdot\vec{j}=0. \quad (2.213)$$

This is the continuity equation! So, the complex electric displacement not only contains information on the charge density but includes the current density as well. The fourth Maxwell equation is equivalent to the integrated continuity relation.

(b) We start with the first Maxwell equation, eq. (2.18),

$$\vec{\nabla}\times\vec{B}=\mu\mu_0\left(\varepsilon\varepsilon_0\dot{\vec{E}}-\mathrm{i}\frac{\sigma}{\omega}\dot{\vec{E}}\right)=\mu\mu_0\tilde{\varepsilon}\varepsilon_0\dot{\vec{E}}, \quad (2.214)$$

and take the first time derivative:

$$\frac{\partial}{\partial t}\left(\vec{\nabla}\times\vec{B}\right)=\vec{\nabla}\times\dot{\vec{B}}=\mu\mu_0\tilde{\varepsilon}\varepsilon_0\ddot{\vec{E}}. \quad (2.215)$$

Next, we take the 2nd Maxwell equation,

$$\vec{\nabla}\times\vec{E}=-\dot{\vec{B}}, \quad (2.216)$$

and take the rotation:

$$\underbrace{\vec{\nabla}\times\vec{\nabla}\times\vec{E}}_{(-\mathrm{i}\vec{g})\times\left[(-\mathrm{i}\vec{g})\times\vec{E}\right]}=-\vec{\nabla}\times\dot{\vec{B}}=-\mu\mu_0\overleftrightarrow{\tilde{\varepsilon}}\varepsilon_0\ddot{\vec{E}}\overset{\mu=1}{=}\frac{1}{\varepsilon_0c^2}\ddot{\vec{D}}. \quad (2.217)$$

Assuming harmonic waves, we can execute the time and space derivative:

$$\left(\tilde{\vec{g}}\right)^2 \vec{E} - \left(\tilde{\vec{g}} \cdot \vec{E}\right) \tilde{\vec{g}} = \frac{\omega^2}{c^2} \overset{\leftrightarrow}{\varepsilon} \vec{E} = \frac{\omega^2}{\varepsilon_0 c^2} \vec{D}. \tag{2.218}$$

Note that $\tilde{\vec{g}} \cdot \vec{E}$ is not zero because of the dielectric function being a tensor.

(c) From eq. (2.218) follows. that $\vec{D}$, $\vec{E}$, and $\tilde{\vec{g}}$ lie in the same plane and from the fourth Maxwell equation $\vec{\nabla} \cdot \vec{D} = 0$ follows $\vec{D} \perp \tilde{\vec{g}}$.

2.7.8 Polarization states

According to eqs. (2.121) and (2.122), we write rcp (+) and lcp (–) light as:

$$\vec{E}_+ = \frac{1}{\sqrt{2}} \begin{pmatrix} 1 \\ \mathrm{i} \\ 0 \end{pmatrix} E_{0+}\, \mathrm{e}^{\mathrm{i}(\omega t - gz)}, \quad \vec{E}_- = \frac{1}{\sqrt{2}} \begin{pmatrix} 1 \\ -\mathrm{i} \\ 0 \end{pmatrix} E_{0-}\, \mathrm{e}^{\mathrm{i}(\omega t - gz)} \tag{2.219}$$

and, according to eq. (2.118), light linearly polarized along the x or y axis as

$$\vec{E}_{x,y} = E_{0x,y} \hat{e}_{x,y} \mathrm{e}^{\mathrm{i}(\omega t - gz)}. \tag{2.220}$$

Then, we can write x-polarized light as the sum of rcp and lcp light:

$$\vec{E}_x = \frac{1}{\sqrt{2}} \left(\vec{E}_+ + \vec{E}_-\right) = \frac{1}{2} \begin{pmatrix} E_{0+} + E_{0-} \\ \mathrm{i}E_{0+} - \mathrm{i}E_{0-} \\ 0 \end{pmatrix} \mathrm{e}^{\mathrm{i}(\omega t - gz)} = \frac{1}{2} \begin{pmatrix} 2E_{0x} \\ 0 \\ 0 \end{pmatrix} \mathrm{e}^{\mathrm{i}(\omega t - gz)}, \tag{2.221}$$

with $E_{0+} = E_{0-} = E_{0x}$. Similarly, y-polarized light can be written as the difference between rcp and lcp light:

$$\vec{E}_y = \frac{1}{\mathrm{i}\sqrt{2}} \left(\vec{E}_+ - \vec{E}_-\right). \tag{2.222}$$

Any other linear polarization can be generated by a linear combination of $\vec{E}_x$ and $\vec{E}_y$.

2.7.9 Brewster angle

At the Brewster angle, reflected and refracted beam are exactly ***perpendicular*** to each other. At the interface, the electric field of the light excites the electrons of material 2. Consequently, they behave like Hertzian dipoles; that is, they oscillate and emit electromagnetic radiation in every direction except along the excitation axis which is parallel to the electric field vector of the refracted wave as shown in Fig. 2.19. The electric field vector is perpendicular to the propagation direction of the refracted ray. Therefore, for p-polarized light at the Brewster angle, it is

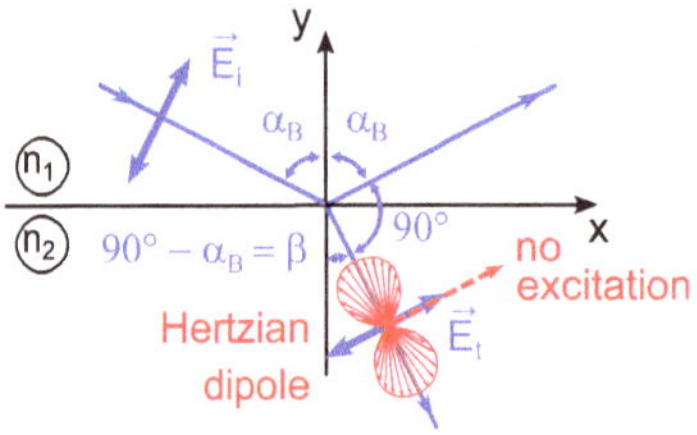

Fig. 2.19: Visualization of the reflection at the Brewster angle.

parallel to the direction of the reflected wave. Hence, no reflected beam can emerge. This simple model supposes that at the interface the atoms of material 2 generate the refracted wave as well as the reflected wave.

For s-polarized light, the excitation axis is perpendicular to the plane of incidence Hence, a reflected beam will exist because its direction is not parallel to the excitation axis.

2.7.10 Complex angle of refraction in a nonabsorbing material

(a) We write Snell's law with a complex angle of refraction:

$$\sin\alpha = n\ \sin(\beta_1 + \mathrm{i}\beta_2) = n\ \sin\beta_1 \cosh\beta_2 + \mathrm{i}\ n\cos\beta_1 \sinh\beta_2. \tag{2.223}$$

Using the identities between trigonometric and hyperbolic functions as shown in eq. (2.103), the real part becomes:

$$\sin\alpha = n\sin\beta_1 \cosh\beta_2, \tag{2.224}$$

and the imaginary part:

$$0 = n\cos\beta_1 \sinh\beta_2. \tag{2.225}$$

The only two solutions for the imaginary part, eq. (2.225), are (1) $\beta_2 = 0$ and (2) $\beta_1 = \pi/2$.

From case (1), it follows that

$$\sin\alpha = n\sin\beta_1. \tag{2.226}$$

This is simply Snell's law for normal refraction valid at angles of incidence $\alpha \le \alpha_T$. From case (2), it follows that

$$\sin\alpha = n\ \cosh\beta_2. \tag{2.227}$$

This is Snell's law at angles of incidence $\alpha \ge \alpha_T$, where total internal reflection occurs. Consequently, the angle of refraction is here complex with the special form:

$$\tilde{\beta} = \frac{\pi}{2} + \mathrm{i}\beta_2. \tag{2.228}$$

For the limiting case $\alpha = \alpha_\mathrm{T}$, we get

$$\sin\alpha_\mathrm{T} = n = n\sin\beta_1\cosh\beta_2. \tag{2.229}$$

From this equation, we conclude that $\beta_1 = \pi/2$ and $\beta_2 = 0$, which is consistent with case (1) and (2).

(b) The refracted wave in the case of total internal reflection is defined in eq. (2.104):

$$E_{\mathrm{t}z} = E_{0\mathrm{t}}\,\mathrm{e}^{\frac{\omega}{c}y\sinh\beta''}\,\mathrm{e}^{\mathrm{i}\left(\omega t - \frac{\omega}{c}x\cosh\beta''\right)} = A\,\mathrm{e}^{\mathrm{i}\left(\omega t - \vec{g}\cdot\vec{r}\right)}. \tag{2.230}$$

Note that we have set $n_2 = 1$ for vacuum. The amplitude of the refracted wave is

$$A = E_{0\mathrm{t}}\,\mathrm{e}^{\frac{\omega}{c}y\sinh\beta''} \tag{2.231}$$

and the wave vector (in vacuum) is

$$\vec{g} = \left(\tfrac{\omega}{c}\cosh\beta'', 0, 0\right). \tag{2.232}$$

We see that the refracted wave propagates in x direction. Because $\vec{E}$, $\vec{H}$ and $\vec{g}$ are perpendicular to each other, we have $H_{\mathrm{t}y} = \varepsilon_0 c E_{\mathrm{t}z}$ (because $n_2 = 1$). Now, we can calculate the Poynting vector according to eq. (2.44):

$$\vec{S} = \mathrm{Re}\left\{\vec{E}\right\} \times \mathrm{Re}\left\{\vec{H}\right\} = \frac{1}{\mu\mu_0}\mathrm{Re}\left\{\vec{E}\right\} \times \mathrm{Re}\left\{\vec{B}\right\}. \tag{2.233}$$

The Poynting vector has only an x component. It is aligned parallel to the wave vector, $\vec{S}||\vec{g}$. The time average of the x component amounts to:

$$\langle S_x\rangle_t = c\varepsilon_0 E_{0\mathrm{t}}^2\,\mathrm{e}^{2\frac{\omega}{c}y\sinh\beta''}\underbrace{\left\langle\cos^2\left(\omega t - \frac{\omega}{c}x\cosh\beta''\right)\right\rangle_t}_{1/2} = \frac{1}{2}c\varepsilon_0 E_{0\mathrm{t}}^2\,\mathrm{e}^{2\frac{\omega}{c}y\sinh\beta''}, \tag{2.234}$$

This implies that the energy flows only in x direction and its magnitude is exponentially decreasing with distance from the interface (i.e., in $-y$ direction). As the Poynting vector has no component along the y direction, there is no energy flow perpendicular to the surface or, in other words, out of the material into the vacuum.

2.7.11 Angle of maximum phase difference by total internal reflection

In this problem, we want to calculate the angle where the maximum phase difference occurs for $\alpha > \alpha_\mathrm{T}$, that is in the regime of total internal reflection. The angle of refraction is then complex and written as

$$\tilde{\beta} = \frac{\pi}{2} + \mathrm{i}\beta''. \tag{2.235}$$

From this we can express the imaginary part as follows:

$$\mathrm{i}\beta'' = \tilde{\beta} - \frac{\pi}{2}, \quad -\mathrm{i}\beta'' = \frac{\pi}{2} - \tilde{\beta}. \tag{2.236}$$

Considering eqs. (2.98) and (2.99), the phase difference between s- and p-polarized light can be derived by inserting the Fresnel equations into the ratio of the complex reflection coefficients:

$$\mathrm{e}^{\mathrm{i}(\delta_\mathrm{s} - \delta_\mathrm{p})} = \frac{\mathrm{e}^{\mathrm{i}\delta_\mathrm{s}}}{\mathrm{e}^{\mathrm{i}\delta_\mathrm{p}}} = \frac{\tilde{\rho}_\mathrm{s}}{\tilde{\rho}_\mathrm{p}} = -\frac{\sin\left(\alpha - \tilde{\beta}\right)\tan\left(\alpha + \tilde{\beta}\right)}{\sin\left(\alpha + \tilde{\beta}\right)\tan\left(\alpha - \tilde{\beta}\right)} = -\frac{\cos\left(\alpha - \tilde{\beta}\right)}{\cos\left(\alpha + \tilde{\beta}\right)} = -\frac{\cos\alpha\cos\tilde{\beta} + \sin\alpha\sin\tilde{\beta}}{\cos\alpha\cos\tilde{\beta} - \sin\alpha\sin\tilde{\beta}}. \tag{2.237}$$

With Snell's law,

$$\sin\alpha = \frac{n_2}{n_1}\sin\tilde{\beta} = n\,\sin\tilde{\beta}, \tag{2.238}$$

and

$$\cos\tilde{\beta} = \sqrt{1 - \sin^2\tilde{\beta}}, \tag{2.239}$$

it follows:

$$\mathrm{e}^{\mathrm{i}\,(\delta_\mathrm{s} - \delta_\mathrm{p})} = -\frac{\cos\alpha\sqrt{1 - \frac{1}{n^2}\sin^2\alpha} + \frac{1}{n}\sin^2\alpha}{\cos\alpha\sqrt{1 - \frac{1}{n^2}\sin^2\alpha} - \frac{1}{n}\sin^2\alpha}. \tag{2.240}$$

Here, we have used the relative refractive index $n = n_2/n_1$. Taking the natural logarithm on both sides yields:

$$\begin{aligned}\mathrm{i}\left(\delta_\mathrm{s} - \delta_\mathrm{p}\right) &= \ln\left[\frac{\cos\alpha\sqrt{1 - \frac{1}{n^2}\sin^2\alpha} + \frac{1}{n}\sin^2\alpha}{-\cos\alpha\sqrt{1 - \frac{1}{n^2}\sin^2\alpha} + \frac{1}{n}\sin^2\alpha}\right] \\ &= \ln\left[\cos\alpha\sqrt{1 - \frac{1}{n^2}\sin^2\alpha} + \frac{1}{n}\sin^2\alpha\right] - \ln\left[\frac{1}{n}\sin^2\alpha - \cos\alpha\sqrt{1 - \frac{1}{n^2}\sin^2\alpha}\right].\end{aligned} \tag{2.241}$$

The maximum phase difference α_max is obtained by taking the first derivative of eq. (2.241) and setting it equal zero:

$$\frac{\partial(\delta_s - \delta_p)}{\partial\alpha}\Big|_{\alpha = \alpha_{max}} = 0. \tag{2.242}$$

Performing the first derivative results in a lengthy expression:

$$0 = \frac{1}{\cos\alpha\sqrt{1-\frac{1}{n^2}\sin^2\alpha}+\frac{1}{n}\sin^2\alpha}\left[-\sin\alpha\sqrt{1-\frac{1}{n^2}\sin^2\alpha}-\frac{\cos\alpha\,\frac{1}{n^2}\sin\alpha\cos\alpha}{\sqrt{1-\frac{1}{n^2}\sin^2\alpha}}+2\frac{1}{n}\sin\alpha\cos\alpha\right]$$
$$-\frac{1}{\frac{1}{n}\sin^2\alpha-\cos\alpha\sqrt{1-\frac{1}{n^2}\sin^2\alpha}}\left[2\frac{1}{n}\sin\alpha\cos\alpha+\sin\alpha\sqrt{1-\frac{1}{n^2}\sin^2\alpha}+\frac{\cos\alpha\,\frac{1}{n^2}\sin\alpha\cos\alpha}{\sqrt{1-\frac{1}{n^2}\sin^2\alpha}}\right]. \tag{2.243}$$

We can reformulate the equation by dividing it by the product of the two prefactors in front of the square brackets:

$$0 = \left[\frac{1}{n}\sin^2\alpha-\cos\alpha\sqrt{1-\frac{1}{n^2}\sin^2\alpha}\right]\left[\frac{2}{n}\sin\alpha\cos\alpha-\sin\alpha\sqrt{1-\frac{1}{n^2}\sin^2\alpha}-\frac{\frac{1}{n^2}\sin\alpha\cos^2\alpha}{\sqrt{1-\frac{1}{n^2}\sin^2\alpha}}\right]$$
$$-\left[\frac{1}{n}\sin^2\alpha+\cos\alpha\sqrt{1-\frac{1}{n^2}\sin^2\alpha}\right]\left[\frac{2}{n}\sin\alpha\cos\alpha+\sin\alpha\sqrt{1-\frac{1}{n^2}\sin^2\alpha}+\frac{\frac{1}{n^2}\sin\alpha\cos^2\alpha}{\sqrt{1-\frac{1}{n^2}\sin^2\alpha}}\right]. \tag{2.244}$$

By expanding the brackets, a lot of terms cancel, and we get a much more simplified equation:

$$0 = -\frac{2}{n}\sin^3\alpha\sqrt{1-\frac{1}{n^2}\sin^2\alpha}-\frac{\frac{2}{n^3}\sin^3\alpha\cos^2\alpha}{\sqrt{1-\frac{1}{n^2}\sin^2\alpha}}-\frac{4}{n}\sin\alpha\cos^2\alpha\sqrt{1-\frac{1}{n^2}\sin^2\alpha}. \tag{2.245}$$

This can be further reduced by dividing by the factor $(2/n)\sin\alpha$:

$$0 = \sin^2\alpha\sqrt{1-\frac{1}{n^2}\sin^2\alpha}+\frac{\frac{1}{n^2}\sin^2\alpha\cos^2\alpha}{\sqrt{1-\frac{1}{n^2}\sin^2\alpha}}+2\cos^2\alpha\sqrt{1-\frac{1}{n^2}\sin^2\alpha}. \tag{2.246}$$

Further simplification is obtained by expanding with the factor $\sqrt{1-(1/n^2)\sin^2\alpha}$:

$$0 = \sin^2\alpha\left(1-\frac{1}{n^2}\sin^2\alpha\right)+\frac{1}{n^2}\sin^2\alpha\cos^2\alpha+2\cos^2\alpha\left(1-\frac{1}{n^2}\sin^2\alpha\right). \tag{2.247}$$

Now, we bundle the terms and get:

$$0=\underbrace{(\sin^2\alpha+\cos^2\alpha)}_{=1}\left(1-\frac{1}{n^2}\sin^2\alpha\right)+\frac{1}{n^2}\sin^2\alpha\cos^2\alpha+\cos^2\alpha\left(1-\frac{1}{n^2}\sin^2\alpha\right). \tag{2.248}$$

This simplifies to:

$$0=\left(1-\frac{1}{n^2}\sin^2\alpha\right)+\underbrace{\cos^2\alpha}_{1-\sin^2\alpha}=2-\sin^2\alpha\left(1+\frac{1}{n^2}\right). \tag{2.249}$$

By solving for $\sin^2\alpha_{\max}$, we obtain the desired expression for the maximum phase shift:

$$\sin^2\alpha_{\max}=\frac{2}{1+\frac{1}{n^2}}=\frac{2}{1+\frac{n_1^2}{n_2^2}}=\frac{2n_2^2}{n_1^2+n_2^2}. \tag{2.250}$$

2.7.12 Maximum of phase difference by total internal reflection

Now, we want to calculate the maximum phase difference that occurs in the regime of total internal reflection at the angle $\alpha_{\max}>\alpha_T$, which we have calculated in the problem in Section 2.6.11. The angle of refraction is complex and written as in eq. (2.235). To describe the phase difference between s- and p-polarized light, we consider again, as in the problem in Section 2.6.11, the ratio $\tilde{\rho}_s/\tilde{\rho}_p$ of the complex reflection coefficients and insert the Fresnel equations as in eq. (2.237):

$$e^{i(\delta_s-\delta_p)}=\frac{e^{i\delta_s}}{e^{i\delta_p}}=\frac{\tilde{\rho}_s}{\tilde{\rho}_p}=-\frac{\cos\alpha\cos\tilde{\beta}+\sin\alpha\sin\tilde{\beta}}{\cos\alpha\cos\tilde{\beta}-\sin\alpha\sin\tilde{\beta}}. \tag{2.251}$$

Using eq. (2.239), we can replace $\cos\tilde{\beta}$ by $\sin\tilde{\beta}$. In the next step, using Snell's law, eq. (2.238), $\sin\tilde{\beta}$ is replaced by $\sin\alpha$ which results again in eq. (2.240) but with $\alpha_{\max}$ instead of α. Finally, replacing the refractive index n by the ratio n_2/n_1 and entering the result for $\sin\alpha_{\max}$ from eq. (2.250), the phase difference at $\alpha=\alpha_{\max}$ is expressed as:

$$\begin{aligned}e^{i(\delta_s-\delta_p)}&=-\frac{\cos\alpha_{\max}\sqrt{1-\frac{n_1^2}{n_2^2}\frac{2n_2^2}{n_1^2+n_2^2}}+\frac{2n_2^2}{n_1^2+n_2^2}\frac{n_1}{n_2}}{\cos\alpha_{\max}\sqrt{1-\frac{n_1^2}{n_2^2}\frac{2n_2^2}{n_1^2+n_2^2}}-\frac{2n_2^2}{n_1^2+n_2^2}\frac{n_1}{n_2}}=-\frac{i\cos\alpha_{\max}\sqrt{\frac{n_1^2-n_2^2}{n_1^2+n_2^2}}+\frac{2n_1n_2}{n_1^2+n_2^2}}{i\cos\alpha_{\max}\sqrt{\frac{n_1^2-n_2^2}{n_1^2+n_2^2}}-\frac{2n_1n_2}{n_1^2+n_2^2}}\\&=\frac{\left(i\cos\alpha_{\max}\sqrt{\frac{n_1^2-n_2^2}{n_1^2+n_2^2}}+\frac{2n_1n_2}{n_1^2+n_2^2}\right)^2}{\cos^2\alpha_{\max}\frac{n_1^2-n_2^2}{n_1^2+n_2^2}+\left(\frac{2n_1n_2}{n_1^2+n_2^2}\right)^2}\end{aligned} \tag{2.252}$$

Note that we have expanded the right part of the equation with the conjugate complex expression of the denominator to make the latter real. Next, we take the square root to obtain half the phase difference in the exponent:

$$e^{i\left(\frac{\delta_s-\delta_p}{2}\right)}=\sqrt{e^{i(\delta_s-\delta_p)}}=\frac{\left(i\cos\alpha_{max}\sqrt{\frac{n_1^2-n_2^2}{n_1^2+n_2^2}}+\frac{2n_1n_2}{n_1^2+n_2^2}\right)}{\sqrt{\cos^2\alpha_{max}\frac{n_1^2-n_2^2}{n_1^2+n_2^2}+\left(\frac{2n_1n_2}{n_1^2+n_2^2}\right)^2}} \tag{2.253}$$

Finally, by taking the ratio between the imaginary and real part and replacing $\cos\alpha_{max}$ by $(1-\sin^2\alpha_{max})^{1/2}$ the sought-after expression for half the phase difference is obtained:

$$\tan\left(\frac{\delta_s-\delta_p}{2}\right)=\frac{\mathrm{Im}\left\{e^{i\left(\frac{\delta_s-\delta_p}{2}\right)}\right\}}{\mathrm{Re}\left\{e^{i\left(\frac{\delta_s-\delta_p}{2}\right)}\right\}}=\frac{\cos\alpha_{max}\sqrt{\frac{n_1^2-n_2^2}{n_1^2+n_2^2}}}{\frac{2n_1n_2}{n_1^2+n_2^2}}=\frac{\sqrt{1-\frac{2n_2^2}{n_1^2+n_2^2}}\sqrt{\frac{n_1^2-n_2^2}{n_1^2+n_2^2}}}{\frac{2n_1n_2}{n_1^2+n_2^2}}=\frac{\frac{n_1^2-n_2^2}{n_1^2+n_2^2}}{\frac{2n_1n_2}{n_1^2+n_2^2}}=\frac{n_1^2-n_2^2}{2n_1n_2}. \tag{2.254}$$

2.7.13 Planes of constant phase and constant amplitude in an absorbing material

We start with writing Snell's law with complex angle of refraction $\tilde{\beta}=\beta_1+i\beta_2$:

$$\sin\alpha=\tilde{n}\sin\tilde{\beta}=(n-ik)\ \sin(\beta_1+i\beta_2). \tag{2.255}$$

Using the sum and difference formulas for trigonometric functions and the identities between trigonometric and hyperbolic functions as shown in eq. (2.103), this yields two equations, one for the real and one for the imaginary part. The real part becomes:

$$\sin\alpha=n\sin\beta_1\cosh\beta_2+k\ \cos\beta_1\sinh\beta_2, \tag{2.256}$$

and the imaginary part:

$$0=n\ \cos\beta_1\sinh\beta_2-k\sin\beta_1\cosh\beta_2. \tag{2.257}$$

We can separate the two angles in eq. (2.257) yielding:

$$\frac{k}{n}\tan\beta_1=\tanh\beta_2. \tag{2.258}$$

By replacing $\sinh\beta_2$ in eq. (2.256) using eq. (2.258), we get:

$$\sin\alpha = n\sin\beta_1\cosh\beta_2 + k\frac{k}{n}\sin\beta_1\cosh\beta_2 = \frac{n^2+k^2}{n}\sin\beta_1\cosh\beta_2. \tag{2.259}$$

Squaring eq. (2.259) and then inserting eq. (2.258) by using the expression $\cosh^2\beta_2 = 1/(1-\tanh^2\beta_2)$ yields:

$$\sin^2\alpha = \left(\frac{n^2+k^2}{n}\right)^2\sin^2\beta_1\cosh^2\beta_2 = \left(n^2+k^2\right)^2\frac{\sin^2\beta_1}{n^2-k^2\tan^2\beta_1}. \tag{2.260}$$

Replacing $\cos\beta_1$ by $\sin\beta_1$ in eq. (2.260) yields:

$$\sin^2\alpha = \frac{\left(n^2+k^2\right)^2\sin^2\beta_1\left(1-\sin^2\beta_1\right)}{n^2\left(1-\sin^2\beta_1\right)-k^2\sin^2\beta_1}. \tag{2.261}$$

This can be solved for $\sin\beta_1$ resulting in a biquadratic equation:

$$\sin^4\beta_1 - \left[1+\frac{\sin^2\alpha}{n^2+k^2}\right]\sin^2\beta_1 + \frac{n^2}{\left(n^2+k^2\right)^2}\sin^2\alpha = 0. \tag{2.262}$$

The solutions for the real part β_1 of the complex angle of refraction are:

$$\sin^2\beta_1 = \frac{1}{2}\left[\left[1+\frac{\sin^2\alpha}{n^2+k^2}\right] - \sqrt{1+\frac{\sin^2\alpha}{\left(n^2+k^2\right)^2}\left[2\left(k^2-n^2\right)+\sin^2\alpha\right]}\right]. \tag{2.263}$$

Note that the solution with the plus sign is discarded because the right part of the equation would be greater than one in this case. Finally, from eq. (2.258), the imaginary part β_2 can be determined.

Now, we are ready to have a look at the refracted wave to determine the planes of constant phase and constant amplitude as a function of angle of incidence. Taking eq. (2.52), adding the time dependence, replacing $\tilde{g}_2$ by $\tilde{\vec{g}}$, and β by the complex angle of refraction $\tilde{\beta} = \beta_1 + i\beta_2$, the refracted wave is written as:

$$E_t = E_{0t}\, e^{i\left(\omega t - \tilde{\vec{g}}\cdot\vec{r}\right)} = E_{0t}\, e^{i\left[\omega t - \frac{\omega}{c}\tilde{n}\left(x\sin\tilde{\beta} - y\cos\tilde{\beta}\right)\right]}. \tag{2.264}$$

Using the sum and difference formulas for trigonometric functions and the identities between trigonometric and hyperbolic functions as shown in eq. (2.103), yields:

$$\begin{aligned} E_t = E_{0t}e^{i\omega t}\, & e^{-i\frac{\omega}{c}\underbrace{\left(n\sin\beta_1\cosh\beta_2 + k\cos\beta_1\sinh\beta_2\right)}_{=\sin\alpha,\ \text{due to eq.}\ (2.256)}x}\, e^{i\frac{\omega}{c}\left(n\cos\beta_1\cosh\beta_2 - k\sin\beta_1\sinh\beta_2\right)y} \\ & \cdot e^{\frac{\omega}{c}\underbrace{\left(n\cos\beta_1\sinh\beta_2 - k\sin\beta_1\cosh\beta_2\right)}_{=0,\ \text{due to eq.}\ (2.257)}x}\, e^{\frac{\omega}{c}\left(n\sin\beta_1\sinh\beta_2 + k\cos\beta_1\cosh\beta_2\right)y}. \end{aligned} \tag{2.265}$$

By bundling all real and imaginary arguments, we get:

$$E_t = E_{0t}\, e^{\frac{\omega}{c}\left(n\sin\beta_1\sinh\beta_2 + k\cos\beta_1\cosh\beta_2\right) y} e^{i\left[\omega t - \frac{\omega}{c}\left(\sin\alpha\, x - \left(n\cos\beta_1\cosh\beta_2 - k\sin\beta_1\sinh\beta_2\right) y\right)\right]}. \tag{2.266}$$

The first two factors on the right side are real. They form the amplitude of the refracted wave:

$$E_{0t}\, e^{\frac{\omega}{c}\left(n\sin\beta_1\sinh\beta_2 + k\cos\beta_1\cosh\beta_2\right) y}. \tag{2.267}$$

We see that the amplitude is exponentially damped in $-y$ direction. From this we can derive the planes of constant amplitude. They must be perpendicular to the y direction that is parallel to the xz plane which is the interface.

The phase is defined in eq. (2.266) by the purely imaginary argument of the last factor on the right side. The planes of constant phase can be defined by setting at a fixed time the phase equal to a constant:

$$-\frac{\omega}{c}\left[\sin\alpha\, x - \left(n\cos\beta_1\cosh\beta_2 - k\sin\beta_1\sinh\beta_2\right) y\right] = \hat{e}_{\tilde{g}} \cdot \vec{r} = \text{const.} \tag{2.268}$$

This must be equivalent to the space part of the phase which is the dot product between wave vector direction $\hat{e}_{\tilde{g}}$ and position vector $\vec{r}$. The direction of wave propagation $\hat{e}_{\tilde{g}}$ is defined to be perpendicular to these planes. From vector geometry, we know that the components of $\hat{e}_{\tilde{g}}$ are proportional to the prefactors of the components x and y of $\vec{r}$ in eq. (2.268). Using eq. (2.256), we can express $\sin\alpha$ by the real and imaginary part of the complex angle of refraction $\tilde{\beta}$ and separate the vector part into a vector defined by the real part of $\tilde{\beta}$ and the rest:

$$\begin{aligned}
\hat{e}_{\tilde{g}} \propto \begin{pmatrix} \sin\alpha \\ -n\cos\beta_1\cosh\beta_2 + k\sin\beta_1\sinh\beta_2 \\ 0 \end{pmatrix} &= \begin{pmatrix} n\sin\beta_1\cosh\beta_2 + k\cos\beta_1\sinh\beta_2 \\ -n\cos\beta_1\cosh\beta_2 + k\sin\beta_1\sinh\beta_2 \\ 0 \end{pmatrix} \\
&= n\cosh\beta_2 \begin{pmatrix} \sin\beta_1 \\ -\cos\beta_1 \\ 0 \end{pmatrix} + k\sinh\beta_2 \begin{pmatrix} \cos\beta_1 \\ \sin\beta_1 \\ 0 \end{pmatrix} \neq n\cosh\beta_2 \begin{pmatrix} \sin\beta_1 \\ -\cos\beta_1 \\ 0 \end{pmatrix}.
\end{aligned} \tag{2.269}$$

We notice from eq. (2.269) that the direction of propagation $\hat{e}_{\tilde{g}}$ of the refracted wave is, not given by the real part β_1 of $\tilde{\beta}$. This would just be the first vector. The second vector vanishes for $k = 0$ and is perpendicular to the first one. It describes a deviation of the direction of propagation away from the direction defined by β_1 toward the interface.

2.7.14 Elliptically polarized light

(a) We know that circularly polarized light forms a basis for polarization states. Therefore, elliptically polarized light can be described as a linear combination of rcp and lcp light which are written as:

$$\vec{E}_{\pm} = \frac{1}{\sqrt{2}} E_0 \left(\hat{e}_x \pm \mathrm{i}\, \hat{e}_y\right) \mathrm{e}^{\mathrm{i}(\omega t - \vec{g}\cdot\vec{r})}. \tag{2.270}$$

With this, we can express linearly polarized light in x and y directions by, respectively, building the sum and the difference between rcp and lcp light:

$$\sqrt{2} E_0\, \hat{e}_x\, \mathrm{e}^{\mathrm{i}(\omega t - \vec{g}\cdot\vec{r})} = \vec{E}_+ + \vec{E}_-, \tag{2.271}$$

$$\mathrm{i}\sqrt{2} E_0\, \hat{e}_y\, \mathrm{e}^{\mathrm{i}(\omega t - \vec{g}\cdot\vec{r})} = \vec{E}_+ - \vec{E}_-. \tag{2.272}$$

Now, we can substitute in eq. (2.147) the unit vectors for the x and y directions with the help of eqs. (2.271) and (2.272). This yields the decomposition of elliptically polarized light into rcp and lcp light:

$$\vec{E} = \frac{1}{\sqrt{2}} E_0 \left(\hat{e}_x + \mathrm{e}^{\mathrm{i}\delta} \hat{e}_y\right) \mathrm{e}^{\mathrm{i}(\omega t - \vec{g}\cdot\vec{r})} = \frac{1}{2}\left(\vec{E}_+ + \vec{E}_-\right) + \frac{\mathrm{e}^{\mathrm{i}\delta}}{2\mathrm{i}}\left(\vec{E}_+ - \vec{E}_-\right) = \frac{1}{2}\left(1 - \mathrm{i}\mathrm{e}^{\mathrm{i}\delta}\right) \vec{E}_+ + \frac{1}{2}\left(1 + \mathrm{i}\mathrm{e}^{\mathrm{i}\delta}\right) \vec{E}_-. \tag{2.273}$$

(b) We want to prove that every elliptically polarized state can be described by the following equation:

$$\vec{E} = \left(E_x \mathrm{e}^{\mathrm{i}\delta_x} \hat{e}_x + E_y \mathrm{e}^{\mathrm{i}\delta_y} \hat{e}_y\right) \mathrm{e}^{\mathrm{i}(\omega t - \vec{g}\cdot\vec{r})}. \tag{2.274}$$

To prove this, we will construct the most general elliptically polarized state from a highly symmetric elliptically polarized state, defined in a local $x'y'$ coordinate system, which is mathematically easy to describe. By rotating the local coordinate system, we will get the most general elliptically polarized state. This procedure is sketched in Fig. 2.20. In such a highly symmetric state, the semiaxes are parallel to the local coordinate axes. In the local $x'y'$ coordinate system, the ellipse can, therefore, be represented as follows:

$$\vec{E}' = \left(E_{x'}\, \hat{e}_{x'} + \mathrm{i}\, E_{y'}\, \hat{e}_{y'}\right) \mathrm{e}^{\mathrm{i}(\omega t - \vec{g}\cdot\vec{r})}. \tag{2.275}$$

Here, $E_{x'}$ and $E_{y'}$ is the length of the semimajor and semiminor axis, respectively, and the factor "i" represents the 90° phase shift to create a rotation of the electric-field vector, visible only in the real part of $\vec{E}'$. We can easily define the ellipticity as the ratio of the length of the two semiaxes:

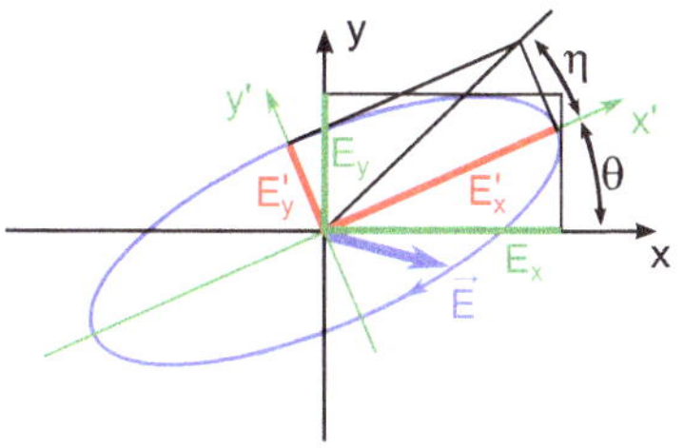

Fig. 2.20: Sketch of the construction of the most general elliptically polarized state from a highly symmetric elliptically polarized state, defined in a local *x'y'* coordinate system, by rotation by an angle *θ*.

$$\tan\eta = \frac{E_{y'}}{E_{x'}}. \tag{2.276}$$

Rotating the $x'y'$ coordinate system with respect to the xy coordinate system by an angle θ yields the most general orientation of the ellipse and hence of elliptically polarized light. Using the rotation matrix,

$$R_\theta = \begin{pmatrix} \cos\theta & -\sin\theta \\ \sin\theta & \cos\theta \end{pmatrix}, \tag{2.277}$$

we can rotate the electric field vector $\vec{E}'$ describing the elliptically polarized state according to eq. (2.275) in the local $x'y'$ coordinate system. This yields:

$$\vec{E} = R_\theta \vec{E}' = \left[\left(\cos\theta\, E_{x'} - \mathrm{i}\sin\theta\, E_{y'}\right)\hat{e}_x + \left(\sin\theta\, E_{x'} + \mathrm{i}\cos\theta\, E_{y'}\right)\hat{e}_y\right] \mathrm{e}^{\mathrm{i}(\omega t - \vec{g}\cdot\vec{r})}. \tag{2.278}$$

For this equation to describe the most general elliptically polarized state, the x and y component must be equal to the x and y component of eq. (2.274). For the x component, we get:

$$E_x \mathrm{e}^{\mathrm{i}\delta_x} = \cos\theta\, E_{x'} - \mathrm{i}\sin\theta\, E_{y'} \overset{\text{polar represent.}}{=} \left(\cos^2\theta\, E_{x'}^2 + \sin^2\theta\, E_{y'}^2\right)^{\frac{1}{2}} \mathrm{e}^{\mathrm{i}\arctan\left(-\tan\theta \frac{E_{y'}}{E_{x'}}\right)}, \tag{2.279}$$

Where we have introduced the polar representation for the complex amplitude on the right side of the equation. Accordingly, for the y component, we obtain:

$$E_y \mathrm{e}^{\mathrm{i}\delta_y} = \sin\theta\, E_{x'} + \mathrm{i}\cos\theta\, E_{y'} \overset{\text{polar represent.}}{=} \left(\sin^2\theta\, E_{x'}^2 + \cos^2\theta\, E_{y'}^2\right)^{\frac{1}{2}} \mathrm{e}^{\mathrm{i}\arctan\left(\cot\theta \frac{E_{y'}}{E_{x'}}\right)}. \tag{2.280}$$

Again, we have used the polar representation of the complex amplitude to get an expression for the phase. From the polar representation of eqs. (2.279) and (2.280), we can now directly derive phase and amplitude of the most general elliptically polarized state of eq. (2.274). For the x component, we obtain from eq. (2.279) for phase δ_x and amplitude E_x:

$$\tan\delta_x = -\tan\theta\,\frac{E_{y'}}{E_{x'}} = -\tan\theta\,\tan\eta, \tag{2.281}$$

$$E_x^2 = \cos^2\theta\,E_{x'}^2 + \sin^2\theta\,E_{y'}^2. \tag{2.282}$$

For the y component, we obtain accordingly from eq. (2.280) for phase δ_y and amplitude E_y:

$$\tan\delta_y = \cot\theta\,\frac{E_{y'}}{E_{x'}} = \frac{\tan\eta}{\tan\theta}, \tag{2.283}$$

$$E_y^2 = \sin^2\theta\,E_{x'}^2 + \cos^2\theta\,E_{y'}^2. \tag{2.284}$$

The four eqs. (2.281)–(2.284) represent an unambiguous assignment in one direction.

The inverse relations are a little bit tricky as they only depend on the phase difference:

$$\delta = \delta_y - \delta_x \tag{2.285}$$

An explicit solution is given in eqs. (4.6) and (4.7), where $\theta = \psi$ as defined in Fig. (4.1).

3 Classical theory of dispersion

3.1 Lorentz theory of absorption and dispersion in insulators

In 1880, Hendrik Antoon Lorentz developed a simple atomic theory to explain the dispersion behavior in the vicinity of an absorption line. He assumed that bound electrons are held in equilibrium by Hooke's law. Introducing a damping γ proportional to the velocity and assuming a periodic electric field $\vec{E} = \vec{E}_0\, e^{i\omega t}$ acting on the medium along the x axis, we obtain the equation of motion for a bound electron:

$$m\ddot{x} + m\gamma\dot{x} + fx = qE_{0x}e^{i\omega t}, \tag{3.1}$$

where f is the spring constant of the restoring force, m the electron mass, and q is the charge, which is negative for an electron, that is, $q = -e$ where e is the electron charge. Of course, this equation of motion constitutes a brute simplification of reality. Besides being purely classical, it neglects the circular motion of the electron completely. In addition, it does not take into account the Coulomb binding force between the core and the electron. The reason why it nevertheless quite accurately describes what happens when light interacts with material is that the displacement of the electron due to the electric field is very small. Hence, the electric field just displaces the center of the electron motion slightly out of equilibrium. But the center of electron motion is also the center of the electron charge, that is, the position of the electron charge averaged over time. Because the centers of the negative electron charge and the positive nucleus charge coincide in a neutral atom, a slight displacement will separate the centers of charge inducing an electric dipole and hence a macroscopic polarization. Therefore, the variable x in eq. (3.1) should be looked upon as the displacement of the center of charge of the electron. Furthermore, the model can be treated simply one-dimensional. Solving the differential equation, by using the ansatz that the displacement $x(t)$ is proportional to the applied electric field, yields the displacement:

$$x(t) = \frac{qE_{0x}e^{i\omega t}}{m\left(\omega_0^2 - \omega^2 + i\omega\gamma\right)}, \tag{3.2}$$

where we introduced the resonance frequency $\omega_0 = \sqrt{f/m}$. We can now associate the displacement $x(t)$ with an induced dipole moment $qx(t)$ which generates an induced macroscopic polarization $P(t)$. Because $P(t)$ is induced by the electric field, it is proportional to the polarizability $\tilde{\alpha}_0$ according to eq. (2.8). Assuming a concentration of N electrons per unit volume, the total polarizability is

https://doi.org/10.1515/9783110635300-003

$$\tilde{\alpha} = N\tilde{\alpha}_0 = P/\left(E_{0x}\mathrm{e}^{\mathrm{i}\omega t}\right) = Nqx/\left(E_{0x}\mathrm{e}^{\mathrm{i}\omega t}\right). \tag{3.3}$$

Thus, we get the dielectric function:

$$\tilde{\varepsilon} = 1 + \frac{\tilde{\alpha}}{\varepsilon_0} = 1 + \frac{Nq^2}{\varepsilon_0 m}\frac{1}{\omega_0^2 - \omega^2 + \mathrm{i}\omega\gamma}. \tag{3.4}$$

By expanding with the complex conjugate of the denominator, we can split $\tilde{\varepsilon} = \varepsilon_1 - \mathrm{i}\,\varepsilon_2$ into real and imaginary part:

$$\varepsilon_1 = 1 + \frac{Nq^2}{\varepsilon_0 m}\frac{\omega_0^2 - \omega^2}{\left(\omega_0^2 - \omega^2\right)^2 + \omega^2\gamma^2}, \tag{3.5}$$

$$\varepsilon_2 = \frac{Nq^2}{\varepsilon_0 m}\frac{\omega\gamma}{\left(\omega_0^2 - \omega^2\right)^2 + \omega^2\gamma^2}. \tag{3.6}$$

The optical conductivity $\sigma_1 = \omega\varepsilon_2\varepsilon_0$ follows as

$$\sigma_1 = \frac{Nq^2}{m}\frac{\omega^2\gamma}{\left(\omega_0^2 - \omega^2\right)^2 + \omega^2\gamma^2}. \tag{3.7}$$

In Fig. 3.1(a)–(c) we plot the frequency dependence of $\tilde{\varepsilon}$, $\tilde{\sigma}$, n, and k. The absorptive properties are better described by the real part σ_1 of the optical conductivity than by the imaginary part ε_2 of the dielectric function. As can be proven by differentiation, σ_1 shows its maximum exactly at the resonance frequency ω_0 whereas the imaginary part ε_2 of the dielectric function does not. For a detailed treatment, see the problem in Section 3.7.1. In Fig. 3.1(a) we note that $n \to 1$ and $k \to 0$ for $\omega \to \infty$. On the other hand, it is well known that there exist strong excitations in the X-ray regime in solid-state materials: Core-level excitation take place at photon energies of several keV. In the problem in Section 3.7.2, we will demonstrate that the Lorentz theory allows at very high photon frequencies strong excitations despite the fact that at these frequencies n and k are close to 1 and 0, respectively. In Fig. 3.1(b) we plot the quantities on a logarithmic energy scale. Now, σ_1 is symmetrical with respect to the resonance frequency ω_0. The real and imaginary part of $\tilde{\sigma}$ is plotted in Fig. 3.1(c). Assuming small damping $\gamma \ll \omega_0$, the full width at half maximum (FWHM) of σ_1 is just γ. So, the width of a peak in the spectrum of σ_1 is a measure of the lifetime of the transition. Maximum and minimum of the real part ε_1 of the dielectric function occur at $\omega_0 - \gamma/2$ and $\omega_0 + \gamma/2$, respectively. At ω_0, the real part ε_1 of $\tilde{\varepsilon}$ equals 1. Going to infinity, ε_1 reaches 1 again. The limit of ε_1 at very low frequencies $\omega \ll \omega_0$, on the other hand, is $1 + Nq^2/\left(\varepsilon_0 m\omega_0^2\right)$. An estimate using parameters typical for the solid state such as $N \cong 10^{28}\ \mathrm{m}^{-3}$ and $\omega_0 = 1.5 \times 10^{15}\ \mathrm{s}^{-1}$ yields $\varepsilon_1(\omega \cong 0) = 15$ or $n(\omega \cong 0) = 3.9$ which is of the correct order of magnitude. In fact, it is surprising how well this simple classical theory – sometimes with minor changes – can explain a lot of optical properties of solids.

The reflectivity R and the absorption coefficient K are drawn in Fig. 3.1(d). We can distinguish four areas. Region 1: at low energies, R and K are small. Hence, the material is transparent or only weakly absorbing. It exhibits what we call a normal dispersion. It is defined by a positive slope $\partial\varepsilon/\partial\omega > 0$. Normal dispersion is the typical behavior of transparent materials, such as glass, at photon energies well below the fundamental energy gap. Region 2: strong absorption appears due to the anomalous dispersion of $\varepsilon_1(\omega)$. Anomalous dispersion is defined by a negative slope $\partial\varepsilon/\partial\omega < 0$. This is the region in transparent materials which is still below but close to the fundamental energy gap. Region 3: the absorption is still quite large, but the light does not pass through the sample since $\varepsilon_1 < 0$. This is the region of metallic reflection. As the name indicates, it is typical for metals: high reflectivity due to high absorption. Region 4: here, the sample is transparent once more but now ε_1 and n are smaller than 1. Consequently, the phase velocity of light is in this region greater than the speed of light in vacuum.

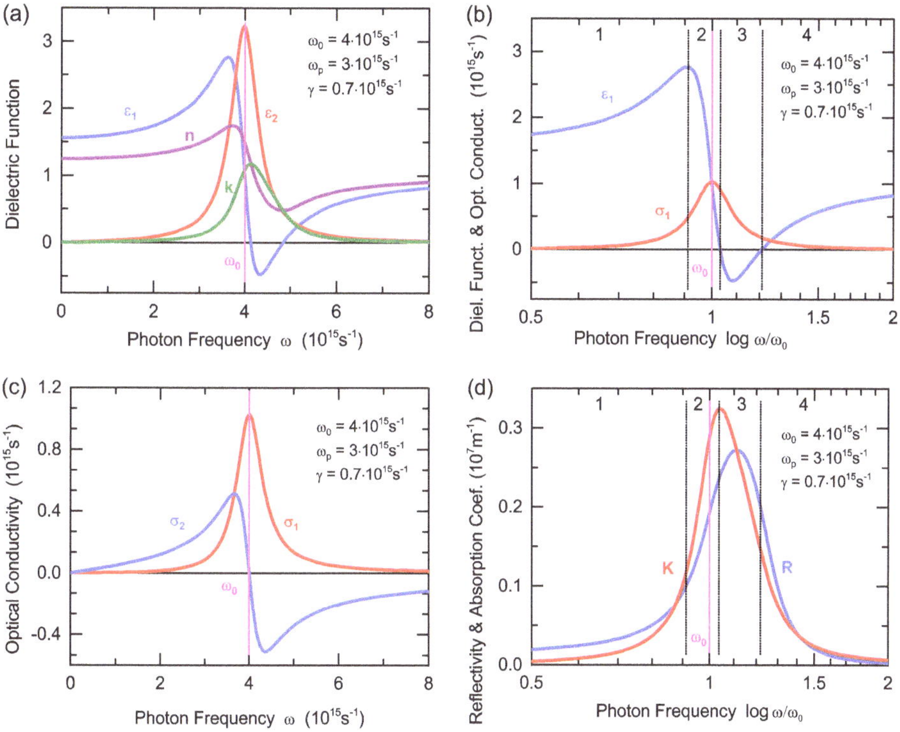

Fig. 3.1: The optical functions of a classical oscillator (resonance energy $\hbar\omega_0 = 2.6$ eV, damping $\hbar\gamma = 0.46$ eV) plotted on linear and logarithmic photon-frequency scale.(a) Real part ε_1 and imaginary part ε_2 of the complex dielectric function, real part n and imaginary part k of the complex refractive index. (b) Real part ε_1 and σ_1 of the complex dielectric function and the complex optical conductivity, respectively, on a logarithmic photon-frequency scale. (c) Real part σ_1 and imaginary part σ_2 of the complex optical conductivity. (d) Reflectivity R and absorption coefficient K on a logarithmic photon-frequency scale.

The Lorentz formula is correct only for electrons which do not interact, as in gases due to the low density of atoms or in metals, where the conduction electrons can be treated as a free gas. For strongly ionic compounds, however, the polarization generated by the external field must be considered in addition to the field itself. This leads to a correction of ε by the inner field as expressed by the relation of Clausius–Mosotti:

$$\frac{N\tilde{\alpha}_0}{3\varepsilon_0} = \frac{\tilde{\varepsilon}-1}{\tilde{\varepsilon}+2} \quad \Rightarrow \quad \tilde{\varepsilon} = 1 + 3\frac{N\tilde{\alpha}_0}{3\varepsilon_0}\left(1 - \frac{N\tilde{\alpha}_0}{3\varepsilon_0}\right)^{-1}. \tag{3.8}$$

A derivation of this formula can be found in textbooks on electrodynamics (85). On the left side, the total polarizability is expressed by $\tilde{\varepsilon}$. This is now to be used in eq. (3.4) instead of $\tilde{\varepsilon} = 1 + \tilde{\alpha}/\varepsilon_0$. One can see that the correction is irrelevant for $\tilde{\varepsilon} \cong 1$ or, which is equivalent, for small polarizabilities $\tilde{\alpha}$. For large $\tilde{\varepsilon}$, as in semiconductors, the correction can be significant, however. All the more remarkable is the experimental fact that, for example, in Si and Ge, the experimental data can be described much better without the local field correction, eq. (3.8).

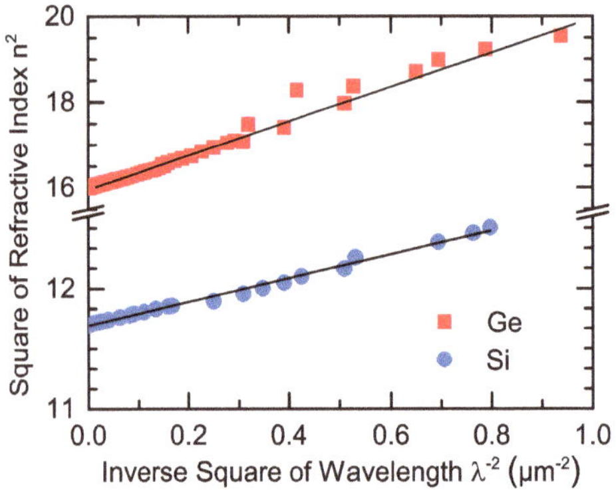

Fig. 3.2: Dispersion of germanium (Ge) and silicon (Si). Data is taken for germanium from Ref. (86) and for silicon from Ref. (87).

To prove this, we evaluate eq. (3.5) for energies small compared to the energy $\hbar\omega_0$ of the absorption edge. We get approximately $n^2 = 1 + \left(\omega_p^2/\omega_0^2\right)\left(1 + \lambda_0^2/\lambda^2\right)$, which is plotted for germanium and silicon in Fig. 3.2. As is evident, the experimental data follow a straight line as expected from Lorentz theory without incorporating a correction of the inner field according to Clausius–Mosotti.

From the slope of the straight line, the number N of electrons per unit volume can be calculated. In many cases, N calculated this way turns out to be wrong. This problem is solved with the introduction of a phenomenological oscillator strength f, motivated by quantum-mechanical treatment. Usually, not just one oscillator, that is, one electronic transition occurs but several with different resonance frequencies

ω_j, dampings γ_j, and oscillator strengths f_j. Thus, we can write the classical dispersion relation in a generalized form as follows:

$$\varepsilon_1 = 1 + \frac{Nq^2}{\varepsilon_0 m} \sum_j f_j \frac{\omega_j^2 - \omega^2}{\left(\omega_j^2 - \omega^2\right)^2 + \omega^2 \gamma_j^2}, \tag{3.9}$$

$$\varepsilon_2 = \frac{Nq^2}{\varepsilon_0 m} \sum_j f_j \frac{\omega \gamma_j}{\left(\omega_j^2 - \omega^2\right)^2 + \omega^2 \gamma_j^2}. \tag{3.10}$$

This representation is often used to fit an experimentally obtained optical spectrum using a reasonable number of Lorentz functions with individual oscillator strength, resonance frequency, and damping.

3.2 Lorentz theory including a homogeneous magnetic field – magnetic circular dichroism

In the next section, we will include in the Lorentz model a static magnetic field to describe the experimental setting used when observing a Faraday effect. To do this, we add to our approach of harmonically bound electrons in eq. (3.1) a homogeneous magnetic field $\vec{B}$ in addition to the periodic excitation by an electromagnetic wave. We obtain the following equation of motion:

$$m\ddot{\vec{r}} + m\gamma\dot{\vec{r}} + f\vec{r} = q\vec{E}_0 e^{i\omega t} + q\dot{\vec{r}} \times \vec{B}, \tag{3.11}$$

where again f is the spring constant, m is the electron mass, γ is the damping, and the charge q is negative for an electron: $q = -e$. The last term on the right side is the Lorentz force. Because of the additional force, the model must now be treated in three dimensions. Without loss of generality, the external field is taken along the z direction: $\vec{B} = (0, 0, B)$. Furthermore, we assume Faraday geometry. Hence, the electromagnetic wave is parallel to $\vec{B}$ and propagates in z direction. Therefore, $\vec{g} = (0, 0, g)$ and $\vec{E}_0 = \left(E_x, E_y, 0\right)$. For the individual components of the position vector $\vec{r} = (x, y, z)$, eq. (3.11) reads as follows:

$$\ddot{x} + \gamma\dot{x} + \omega_0^2 x = \frac{q}{m} E_x \, e^{i\omega t} + \omega_c \dot{y}, \tag{3.12}$$

$$\ddot{y} + \gamma\dot{y} + \omega_0^2 y = \frac{q}{m} E_y \, e^{i\omega t} - \omega_c \dot{x}, \tag{3.13}$$

$$\ddot{z} + \gamma\dot{z} + \omega_0^2 z = 0. \tag{3.14}$$

Here, we have introduced the resonance frequency of the harmonic oscillator

$$\omega_0^2 = \frac{f}{m} \tag{3.15}$$

and the cyclotron resonance frequency

$$\omega_c = \frac{qB}{m}. \tag{3.16}$$

Often, the Larmor frequency $\omega_L = \frac{1}{2}\omega_c$ is used instead. The z component eq. (3.14) is undisturbed, and the motion is that of a damped harmonic oscillator. The x and y components, eqs. (3.12) and (3.13), respectively, are now coupled due to the Lorentz force. We can solve the coupled differential equations with the ansatz of circularly polarized waves as the equation of motion eq. (3.11) is rotationally invariant with respect to the z axis. Thus,

$$\vec{r}_\pm = A_\pm \vec{E}_\pm, \tag{3.17}$$

where $\vec{E}_\pm$ is defined according to eqs. (2.121) and (2.122), neglecting space dependence, as $\vec{E}_\pm = \frac{1}{\sqrt{2}} E_{0\pm} e^{i\omega t} \left(\hat{e}_x \pm \mathrm{i}\hat{e}_y\right)$. The position vector is accordingly defined in a circular basis as $\vec{r}_\pm = \frac{1}{\sqrt{2}} r_{0\pm} \left(\hat{e}_x \pm \mathrm{i}\hat{e}_y\right)$. Inserting this into eq. (3.11) yields:

$$\frac{1}{\sqrt{2}}\left[\ddot{r}_{0\pm} + \gamma\dot{r}_{0\pm} + \omega_0^2 r_{0\pm}\right] \left(\hat{e}_x \pm \mathrm{i}\hat{e}_y\right) = \frac{1}{\sqrt{2}}\frac{q}{m} E_{0\pm} e^{\mathrm{i}\omega t} \left(\hat{e}_x \pm \mathrm{i}\hat{e}_y\right) + \frac{1}{\sqrt{2}} \omega_c \, \dot{r}_{0\pm} \left(\hat{e}_x \pm \mathrm{i}\hat{e}_y\right) \times \hat{e}_z. \tag{3.18}$$

Inserting the ansatz, eq. (3.17), and using the identity $\left(\hat{e}_x \pm \mathrm{i}\hat{e}_y\right) \times \hat{e}_z = \pm\ \mathrm{i}\left(\hat{e}_x \pm \mathrm{i}\hat{e}_y\right)$, we finally get for the amplitude $A_\pm$ of the displacement:

$$A_\pm E_{0\pm}\left(-\omega^2 + \mathrm{i}\gamma\omega + \omega_0^2\right) = \frac{q}{m} E_{0\pm} \pm \mathrm{i}^2 \omega\omega_c A_\pm E_{0\pm}. \tag{3.19}$$

This yields for the amplitude:

$$A_\pm = \frac{q}{m\left(\omega_0^2 - \omega^2 \pm \omega\omega_c + \mathrm{i}\omega\gamma\right)}. \tag{3.20}$$

The displacement generates again a macroscopic polarization in the material. Given a concentration of N electrons per unit volume, the total polarizability amounts to $\tilde{\alpha}_\pm = N\tilde{\alpha}_{0\pm} = \left|\vec{P}_\pm\right| / \left|\vec{E}_\pm\right| = Nq|\vec{r}_\pm| / \left|\vec{E}_\pm\right|$ and the dielectric function for rcp (+) and lcp (−) light is

$$\tilde{\varepsilon}_\pm = 1 + \frac{\tilde{\alpha}_\pm}{\varepsilon_0} = 1 + \frac{Nq^2}{\varepsilon_0 m} \frac{1}{\omega_0^2 - \omega^2 \pm \omega\omega_c + \mathrm{i}\omega\gamma}. \tag{3.21}$$

By expanding with the complex conjugate of the denominator, we can decompose $\tilde{\varepsilon}_\pm = \varepsilon_{1\pm} - \mathrm{i}\varepsilon_{2\pm}$ into real and imaginary part:

$$\varepsilon_{1\pm} = 1 + \frac{Nq^2}{\varepsilon_0 m} \frac{\omega_0^2 \pm \omega\omega_c - \omega^2}{\left(\omega_0^2 - \omega^2 \pm \omega\omega_c\right)^2 + \omega^2\gamma^2}, \tag{3.22}$$

$$\varepsilon_{2\pm} = \frac{Nq^2}{\varepsilon_0 m} \frac{\omega\gamma}{\left(\omega_0^2 - \omega^2 \pm \omega\omega_c\right)^2 + \omega^2\gamma^2}. \tag{3.23}$$

As can be seen from eqs. (3.21) to (3.23), the principal effect of applying a static magnetic field is that the resonance frequency is ***shifted*** contingent on the state of circular polarization. The shifted resonance frequency at $\omega = \omega_{0\pm}$ is defined by the condition

$$\omega_0^2 - \omega_{0\pm}^2 \pm \omega_{0\pm}\omega_c = 0. \tag{3.24}$$

It follows that

$$\pm\omega_{0\pm}\omega_c = \omega_{0\pm}^2 - \omega_0^2 = (\omega_{0\pm} - \omega_0)(\omega_{0\pm} + \omega_0) \cong 2\omega_{0\pm}(\omega_{0\pm} - \omega_0), \tag{3.25}$$

and finally:

$$\omega_{0\pm} - \omega_0 = \pm\frac{1}{2}\omega_c = \pm\omega_L, \tag{3.26}$$

where ω_L is the Larmor frequency. Note, that for an electron $\omega_c < 0$ because $q = -e < 0$. In conclusion, the resonance frequency shifts for rcp (lcp) light to smaller (larger) photon energies and the shift is proportional to the applied magnetic field.

We want to show next that the magnetic-field-induced change of $\tilde{\varepsilon}_\pm$ as a function of circular polarization can be interpreted as a ***macroscopic magnetic-field-induced anisotropy*** as assumed in our simple phenomenological model in Section 2.4.2. In an isotropic material, $\tilde{\varepsilon}$ reduces to a scalar complex number. In the presence of a magnetic field, the dielectric tensor is determined by eq. (3.21), that is, in a circular basis, it has a diagonal form

$$\overset{\leftrightarrow}{\tilde{\varepsilon}}^{\pm} = \begin{pmatrix} \tilde{\varepsilon}_+ & 0 & 0 \\ 0 & \tilde{\varepsilon}_- & 0 \\ 0 & 0 & \tilde{\varepsilon} \end{pmatrix}, \tag{3.27}$$

with eigenvalues $\tilde{\varepsilon}_\pm$ and the unaffected dielectric function $\tilde{\varepsilon}$, that is, the dielectric function in zero magnetic field. The electric displacement is given by $\vec{D}_\pm = \overset{\leftrightarrow}{\tilde{\varepsilon}}^{\pm}\vec{E}_\pm$. What is the interpretation of the splitting of the dielectric function in real space? To see that, we transform in the next step $\overset{\leftrightarrow}{\tilde{\varepsilon}}^{\pm}$ to the *xyz* basis. Following standard procedure of matrix transformation known from linear algebra, the dielectric tensor $\overset{\leftrightarrow}{\tilde{\varepsilon}}$ in *xyz* basis is obtained by

$$\overset{\leftrightarrow}{\tilde{\varepsilon}} = \mathbf{T}\,\overset{\leftrightarrow}{\tilde{\varepsilon}}^{\pm}\,\mathbf{T}^{-1}, \tag{3.28}$$

where the columns of the transformation matrix $\boldsymbol{T}$ are given by the coefficients of the expression of the old (circular) basis by the new (xyz) basis. So, we find

$$\mathbf{T}=\frac{1}{\sqrt{2}}\begin{pmatrix}1 & 1 & 0\\ \mathrm{i} & -\mathrm{i} & 0\\ 0 & 0 & \sqrt{2}\end{pmatrix},\quad \mathbf{T}^{-1}=\frac{1}{\sqrt{2}}\begin{pmatrix}1 & -\mathrm{i} & 0\\ 1 & \mathrm{i} & 0\\ 0 & 0 & \sqrt{2}\end{pmatrix}. \tag{3.29}$$

Inserting this into eq. (3.28), the dielectric tensor $\overset{\leftrightarrow}{\tilde{\varepsilon}}$ in the xyz system is obtained as

$$\overset{\leftrightarrow}{\tilde{\varepsilon}}=\begin{pmatrix}\frac{1}{2}(\tilde{\varepsilon}_{+}+\tilde{\varepsilon}_{-}) & -\frac{\mathrm{i}}{2}(\tilde{\varepsilon}_{+}-\tilde{\varepsilon}_{-}) & 0\\ \frac{\mathrm{i}}{2}(\tilde{\varepsilon}_{+}-\tilde{\varepsilon}_{-}) & \frac{1}{2}(\tilde{\varepsilon}_{+}+\tilde{\varepsilon}_{-}) & 0\\ 0 & 0 & \tilde{\varepsilon}\end{pmatrix}=\begin{pmatrix}\tilde{\varepsilon}_{xx} & \tilde{\varepsilon}_{xy} & 0\\ \tilde{\varepsilon}_{yx} & \tilde{\varepsilon}_{yy} & 0\\ 0 & 0 & \tilde{\varepsilon}_{zz}\end{pmatrix}. \tag{3.30}$$

It follows at once that $\overset{\leftrightarrow}{\tilde{\varepsilon}}$ has finite off-diagonal elements $\tilde{\varepsilon}_{yx}=-\tilde{\varepsilon}_{xy}$, where $\tilde{\varepsilon}_{xy}$ satisfies the relation

$$\tilde{\varepsilon}_{\pm}=\tilde{\varepsilon}_{xx}\pm\mathrm{i}\,\tilde{\varepsilon}_{xy} \tag{3.31}$$

and

$$\tilde{\varepsilon}_{xy}=-\tilde{\varepsilon}_{yx}=-\frac{\mathrm{i}}{2}(\tilde{\varepsilon}_{+}-\tilde{\varepsilon}_{-}). \tag{3.32}$$

The dielectric tensor takes the following form:

$$\overset{\leftrightarrow}{\tilde{\varepsilon}}=\begin{pmatrix}\tilde{\varepsilon}_{xx} & \tilde{\varepsilon}_{xy} & 0\\ -\tilde{\varepsilon}_{xy} & \tilde{\varepsilon}_{xx} & 0\\ 0 & 0 & \tilde{\varepsilon}_{zz}\end{pmatrix}. \tag{3.33}$$

As a major result, we learn from eq. (3.33) that the effect of an applied magnetic field is to ***induce a real anisotropy*** in the material, since the dielectric tensor has now non-vanishing off-diagonal elements. For the optical conductivity,

$$\overset{\leftrightarrow}{\tilde{\sigma}}=\begin{pmatrix}\tilde{\sigma}_{xx} & \tilde{\sigma}_{xy} & 0\\ -\tilde{\sigma}_{xy} & \tilde{\sigma}_{xx} & 0\\ 0 & 0 & \tilde{\sigma}_{zz}\end{pmatrix}, \tag{3.34}$$

an anisotropy is derived, as well. The scalar eq. (2.36) is also valid for the two tensors $\overset{\leftrightarrow}{\tilde{\varepsilon}}$ and $\overset{\leftrightarrow}{\tilde{\sigma}}$, except that the scalar 1 is replaced by the unit tensor $\overset{\leftrightarrow}{\mathbf{1}}$ (diagonal tensor with diagonal element 1):

$$\overset{\leftrightarrow}{\tilde{\varepsilon}}=\overset{\leftrightarrow}{\mathbf{1}}-\mathrm{i}\frac{1}{\varepsilon_0\omega}\overset{\leftrightarrow}{\tilde{\sigma}}. \tag{3.35}$$

Using this equation, we find the relations between the circular and the real components of the optical conductivity tensor:

$$\tilde{\sigma}_\pm = \tilde{\sigma}_{xx} \pm \mathrm{i}\tilde{\sigma}_{xy}, \tag{3.36}$$

$$\tilde{\sigma}_{xy} = -\frac{\mathrm{i}}{2}(\tilde{\sigma}_+ - \tilde{\sigma}_-). \tag{3.37}$$

From this equation, we can define in the circular basis again the same relation between the dielectric function and the optical conductivity:

$$\tilde{\varepsilon}_\pm = 1 - \mathrm{i}\frac{1}{\varepsilon_0 \omega}\tilde{\sigma}_\pm. \tag{3.38}$$

One can expand the optical conductivity tensor in a power series in the magnetic field. With respect to symmetry properties, the result is that the ***diagonal*** elements $\tilde{\sigma}_{jj}$, where $j = x,\ y,\ z$, only depend on ***even*** powers of the magnetic field and ***off-diagonal*** elements $\tilde{\sigma}_{jk}$, where $j,\ k = x,\ y,\ z$ and $j \neq k$, only on ***odd*** powers (88). In most cases, the power series is terminated after the linear term, so that $\tilde{\sigma}_{jj}$ is not dependent on and $\tilde{\sigma}_{jk}$ is proportional to an applied magnetic field. An alternative way to describe magneto-optic effects is given in the Appendix.

3.3 Microwave cyclotron resonance absorption

For classical, non-magnetic semiconductors, magneto-optics is the most widely used technique for determining the electronic structure. In particular, charge-carrier concentrations and band curvatures (corresponding to the effective mass) can be determined. We will discuss two effects, the microwave cyclotron resonance and the splitting of the reflectivity minimum, which can both be explained by classical Lorentz theory.

We will start with the microwave cyclotron resonance. It is observed at low temperatures near the absolute zero point as a resonant absorption of microwaves in doped semiconductors when a static magnetic field is applied. Usually, the measurement is performed in a microwave cavity to increase the sensitivity. We assume plane waves, as usual. Without loss of generality, the external field and the light-propagation direction are taken along the z direction: $\vec{B} = (0,\ 0,\ B)$ and $\vec{g} = (0,\ 0,\ g)$. The electric field is x polarized: $\vec{E}_0 = (E_0,\ 0,\ 0)$. We first consider the motion of a quasi-free electron with isotropic effective mass $m^\star$ in a homogeneous magnetic field lying along the z direction. We can describe the motion of a quasi-free electron using Lorentz theory as well, although the electron is not bound to an atom. We simply omit the restoring force responsible for the binding by setting $f = 0$ in eq. (3.11):

$$m^\star \ddot{\vec{r}} + m^\star \gamma \dot{\vec{r}} = q\vec{E}_0 e^{i\,\omega t} + q\dot{\vec{r}} \times \vec{B}. \tag{3.39}$$

If $f = 0$, then also the resonance frequency vanishes: $\omega_0 = 0$. Choosing again the approach that the displacement $\vec{r}$ is proportional to circularly polarized waves, which has been used in eq. (3.17), we obtain equivalent to eq. (3.20) for the amplitude of the displacement:

$$A_\pm = \frac{q}{m^*(-\omega^2 \pm \omega\omega_c + i\omega\gamma)}, \tag{3.40}$$

and for the dielectric function accordingly:

$$\tilde{\varepsilon}_\pm = 1 + \frac{\tilde{\alpha}_\pm}{\varepsilon_0} = 1 + \frac{Nq^2}{\varepsilon_0 m^*} \frac{1}{\omega(-\omega \pm \omega_c + i\gamma)}. \tag{3.41}$$

Using eqs. (3.36) and (3.38), it follows that

$$\tilde{\varepsilon}_\pm = 1 - i\frac{1}{\varepsilon_0\omega}\left(\tilde{\sigma}_{xx} \pm i\tilde{\sigma}_{xy}\right). \tag{3.42}$$

In the problem in Section 3.7.3, we discuss the frequency dependence of $\varepsilon_1(\omega)$ and $\sigma_1(\omega)$ for free electrons in more detail.

In experiment, the power loss of the microwaves due to absorption in the material is measured. Thus, we want to calculate the time-averaged power P absorbed per unit volume. According to eq. (2.48), section 2.1, and neglecting space dependence because it cancels during time averaging, the following applies:

$$P = \left\langle \mathrm{Re}\left\{\tilde{\vec{j}}\right\} \cdot \mathrm{Re}\left\{\vec{E}\right\}\right\rangle_t = \left\langle \mathrm{Re}\left\{\tilde{\overleftrightarrow{\sigma}}\vec{E}\right\} \cdot \mathrm{Re}\left\{\vec{E}\right\}\right\rangle_t = \left\langle \mathrm{Re}\{\tilde{\sigma}_{xx}E_x\} \cdot \mathrm{Re}\{E_x\} + \mathrm{Re}\left\{\tilde{\sigma}_{xy}E_y\right\} \cdot \mathrm{Re}\{E_x\}\right\rangle_t$$
$$= \sigma_{1xx}E_0^2 \underbrace{\left\langle \cos^2(\omega t)\right\rangle_t}_{=\frac{1}{2}} - \sigma_{2xx}E_0^2 \underbrace{\left\langle \sin(\omega t)\cos(\omega t)\right\rangle_t}_{=0} = \frac{1}{2}\sigma_{1xx}E_0^2. \tag{3.43}$$

As can be seen from eq. (3.43), the absorbed power depends merely on the real part σ_{1xx} of the diagonal element of the optical conductivity. In the next step, we will calculate σ_{1xx}. Combining eqs. (3.41) and (3.42), we get for the complex diagonal element of the optical conductivity $\tilde{\sigma}_{xx}$:

$$\begin{aligned} \tilde{\sigma}_{xx} &= \frac{i\varepsilon_0\omega}{2}(\tilde{\varepsilon}_+ + \tilde{\varepsilon}_- - 2) = \frac{iNq^2}{2m^*}\left(\frac{1}{-\omega + \omega_c + i\gamma} + \frac{1}{-\omega - \omega_c + i\gamma}\right) \\ &= \frac{iNq^2}{m^*} \frac{-\omega + i\gamma}{(-\omega + i\gamma)^2 - \omega_c^2} = \frac{Nq^2}{m^*} \frac{i\omega + \gamma}{(i\omega + \gamma)^2 + \omega_c^2}. \end{aligned} \tag{3.44}$$

By expanding eq. (3.44) with the complex conjugate of the denominator and inserting into eq. (3.43), we obtain for the time-averaged power absorbed per unit volume for light linearly polarized in x direction:

$$P = \frac{1}{2}\sigma_{1xx}E_0^2 = \frac{E_0^2 Nq^2}{2m^*}\,\frac{\gamma\left(\omega^2+\omega_c^2+\gamma^2\right)}{\left(\omega_c^2-\omega^2+\gamma^2\right)^2+4\omega^2\gamma^2}. \tag{3.45}$$

It is common to normalize the absorbed power to the DC power P_0, which is absorbed at $\omega = \omega_c = 0$, that is, if there is no magnetic field applied to the material:

$$P_0 = P(\omega = \omega_c = 0) = \frac{E_0^2 Nq^2}{2m^*\gamma}. \tag{3.46}$$

So, we finally get for the normalized absorbed power:

$$\frac{P}{P_0} = \frac{\gamma^2\left(\omega^2+\omega_c^2+\gamma^2\right)}{\left(\omega_c^2-\omega^2+\gamma^2\right)^2+4\omega^2\gamma^2}. \tag{3.47}$$

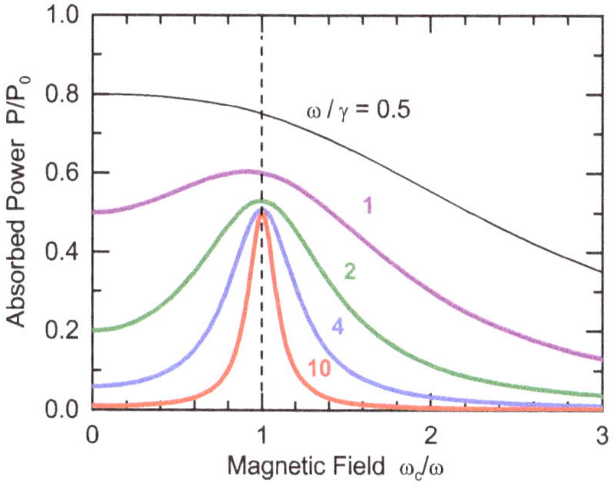

Fig. 3.3: Absorbed power P/P_0 of microwaves as a function of magnetic field in terms of ω_c/ω. A resonance is visible for $\omega/\gamma > 1$.

From eq. (3.47), it is easy to see that the absorbed power has for small dampings a maximum at the cyclotron resonance frequency $\omega = \omega_c$. What is not easy to evaluate is if the maximum is prominent or not. To discuss this, we plot in Fig. 3.3 the normalized absorbed power P/P_0 as a function of ω_c/ω for various values of ω/γ. This representation has been chosen because in experiment it is much simpler to vary the magnetic field as opposed to the microwave frequency. Especially, when a cavity is used for the microwaves, the frequency must be kept constant to meet the resonance conditions of the cavity. We notice from Fig. 3.3 that the requirement $\omega/\gamma > 1$ must be fulfilled for a resolvable resonance. This means that the charge carriers should cover between two collisions, on average, a distance of at least the sixth part of a full orbit. On the other hand, for quantization effects to be negligible, the magnetic field and, hence, $\hbar\omega_c$ must be small compared to the energy range where $\partial f/\partial E$ is finite, $f(E)$ being the Fermi function (E = energy). From this follows the condition $\hbar\omega_c < kT$ (k = Boltzmann constant, T = temperature). Because we have $\omega = \omega_c$ at the resonance, we can discuss the limits just for ω_c. To fulfill the two conditions $\omega_c/\gamma > 1$ and $\hbar\omega_c < kT$, we must choose ω_c to be small. This is satisfied by applying low magnetic fields. In addition, the damping γ must become as small as possible which implies a large relaxation time $\tau = \gamma^{-1}$. This can be achieved for Ge, Si, InSb, and other III–V semiconductors by going to low temperatures.

As an example, for Ge with $m^*/m = 0.1$, $q = e$, and $B = 0.1$ T, we have $\omega_c = eB/m^* \cong 1.8 \cdot 10^{11}\ \text{s}^{-1}$ or $\nu_c = \omega_c/2\pi = 28$ GHz. The condition $\omega_c/\gamma > 1$ implies for the relaxation time $\tau = \gamma^{-1} > 5 \times 10^{-12}$ s.

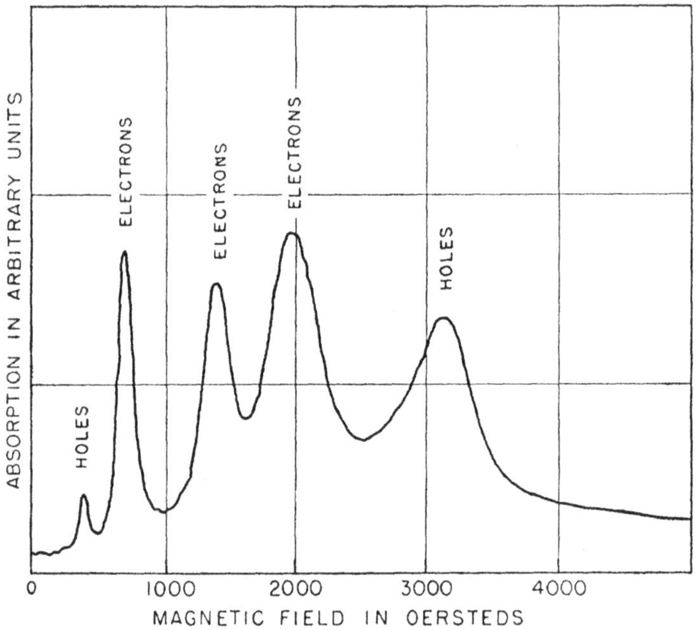

Fig. 3.4: Cyclotron resonance absorption in germanium at 24 GHz and 4 K. The static magnetic field is oriented in a (110) plane at 60° from the [100] direction. The electrons and holes have been excited by photons from an incandescent source. Figure from Ref. (52).

At room temperature, this cannot be achieved because γ is between 10^{13} and $10^{15}\ \text{s}^{-1}$. One must go to temperatures near the absolute zero point and use extremely pure crystals to reduce scattering on defects which also increases damping. The cyclotron resonance energy $\omega_c = 1.8 \times 10^{11}\ \text{s}^{-1}$ is equal to $E_c = \hbar\omega_c = 1.2 \times 10^{-4}$ eV. This corresponds to the thermal energy at a temperature of about 1 K. So, at 4 K a resonance should be visible. Figure 3.4 shows results for germanium at 4 K and at a measurement frequency of 24 GHz while the magnetic field is varied in the range of kG. Surprisingly, not just a single resonance can be distinguished but several. The reason is, first, that we have quasi-free charge carriers in various energy bands. So, not only electrons produce a resonance but holes as well. Second, the effective mass is measured as an average over the cross section between a surface of constant energy and a surface perpendicular to the magnetic field. Therefore, every closed orbit will lead to a resonance. To illustrate this, the band structure of germanium is reproduced in Fig. 3.5. The conduction-band minima lie along the <111> directions. The surfaces of constant energy at the conduction band minima have the shape of rotational ellipsoids aligned along the eight space diagonals, as shown in Fig. 3.6.

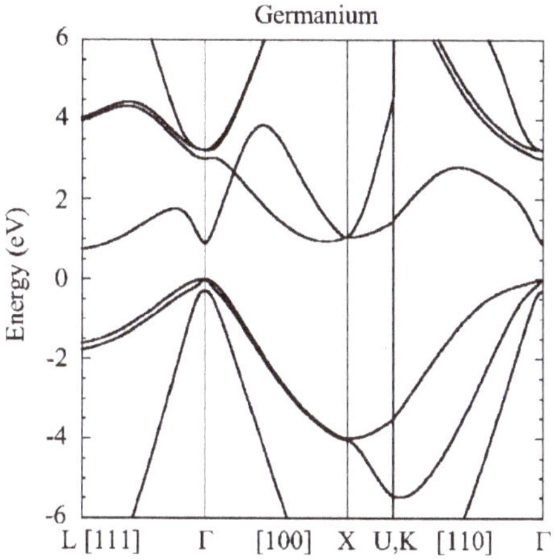

Fig. 3.5: Band structure of germanium near the band gap. The minimum of the conduction band is found at the *L* point in [111] direction. The small spin-orbit splitting of the valence band is visible at the *Γ* point, where the wave vector *k* is zero. Figure from Ref. (89).

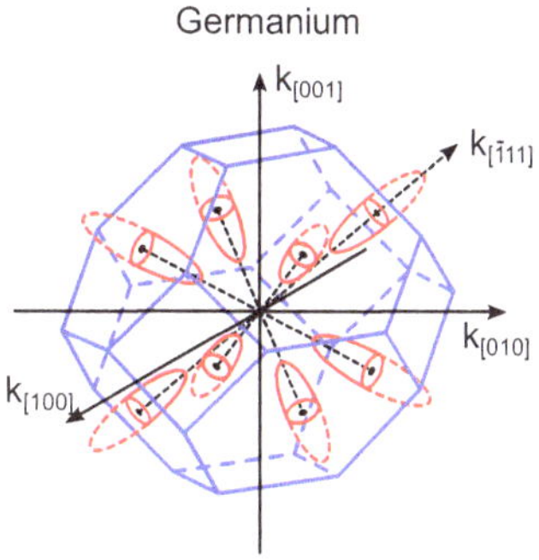

Fig. 3.6: Surfaces of constant energy (red ellipsoids) along the ⟨111⟩ directions for conduction-band electrons in germanium.

Depending on the angle between magnetic field and crystallographic orientation, there are different cross sections and thus different averaged effective masses (see Fig. 3.7).

By choosing a Cartesian coordinate system with z axis parallel to the symmetry axis of an ellipsoid, the energy can be described in reciprocal space near the band edge as:

$$E(k) = \hbar^2 \left(\frac{k_x^2 + k_y^2}{2m_t} + \frac{k_z^2}{2m_l} \right). \tag{3.48}$$

Here, m_l and m_t are the averaged effective masses parallel and perpendicular to the symmetry axis, respectively, and $\hbar$ is the reduced Planck constant. If the magnetic field encloses an angle φ with the symmetry axis, then the averaged effective mass is expressed as

$$\frac{1}{m^{*2}} = \frac{\cos^2\varphi}{m_t^2} + \frac{\sin^2\varphi}{m_t m_l}. \tag{3.49}$$

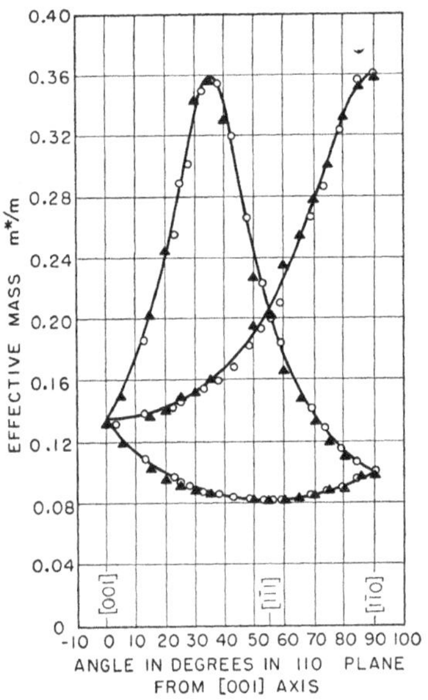

Fig. 3.7: Effective mass of the electrons in germanium for cyclotron resonance at 4 K and static magnetic field in a (110) plane. Figure from Ref. (52).

We can now determine the averaged effective masses by fitting the curves in Fig. 3.7. As a result, one finds experimentally in germanium:

$$\begin{aligned} m_t &= (0.0819 \pm 0.0003)\ m_e, \\ m_l &= (1.64 \pm 0.03)\ m_e, \end{aligned} \tag{3.50}$$

where m_e is the mass of the free electron. How the effective masses can be derived from Fig. 3.7 is discussed in the problem in Section 3.7.4. Using p doping, the effective masses of holes in the valence band can be determined as well.

3.4 Splitting of the reflectivity minimum in a magnetic field

Let us now look at the behavior of the reflectivity in a static magnetic field. We concentrate on a photon-energy range where free-charge-carrier interactions are prevailing. For doped semiconductors, this is in the infrared. The reflectivity at normal incidence is according to eq. (2.110) equal to:

$$R = \frac{(n-1)^2 + k^2}{(n+1)^2 + k^2}. \tag{3.51}$$

The frequency dependence of the optical functions for lcp and rcp light arises from eq. (3.41) but, for a realistic treatment, we must also consider the contribution of the bound electrons to the real part of the dielectric function $\tilde{\varepsilon}_\pm$ at higher energies. The real part of the dielectric function of every resonance that is described by a Lorentz oscillator is for photon energies below the resonance larger than for photon energies far higher than the resonance, where it asymptotically reaches a value of 1 (see Fig. 3.1). Therefore, starting from the optical transition with the highest resonance frequency and continuously decreasing the photon energy, the real part ε_1 of the dielectric function increases each time a resonance is passed. This is shown in Fig. 3.8, where we have plotted ε_1 (black solid line) for the sum of a free electron contribution and four resonances described by Lorentz oscillators according to eq. (3.9). After having passed the resonance at the lowest photon energy ω_{01}, only the contribution of the free charge carriers is left. However, ε_1 is now reaching at $\omega = 0$ the value ε_{opt} (black dash-dotted curve) which is much higher than the asymptotic limit of 1 that the free-charge-carrier contribution $\varepsilon_{\text{free}}$ would reach if it were the only transition in the material (grey dashed curve). Therefore, we can simply replace the constant 1 in eq. (3.41) by a constant $\varepsilon_{\text{opt}} > 1$ that takes care of the increase of the real part of the dielectric function due to all resonances at higher photon energies (m^* is the effective mass of electrons):

$$\tilde{\varepsilon}_\pm = \varepsilon_{\text{opt}} + \frac{Ne^2}{\varepsilon_0 m^*} \frac{1}{\omega(-\omega \pm \omega_c + i\gamma)}. \tag{3.52}$$

After separation into real and imaginary part, we get:

$$\varepsilon_{1\pm} = n_\pm^2 - k_\pm^2 = \varepsilon_{\text{opt}} - \frac{Ne^2}{\varepsilon_0 m^*} \frac{\omega \mp \omega_c}{\omega\left[(\omega \mp \omega_c)^2 + \gamma^2\right]}, \tag{3.53}$$

$$\varepsilon_{2\pm} = 2n_\pm k_\pm = \frac{Ne^2}{\varepsilon_0 m^*} \frac{\gamma}{\omega\left[(\omega \mp \omega_c)^2 + \gamma^2\right]}. \tag{3.54}$$

It is interesting to notice in eqs. (3.53) and (3.54) that only ε_{1-} and ε_{2-} show a cyclotron resonance due to the positive sign in front of ω_c (remember that $\omega_c < 0$ for electrons!).

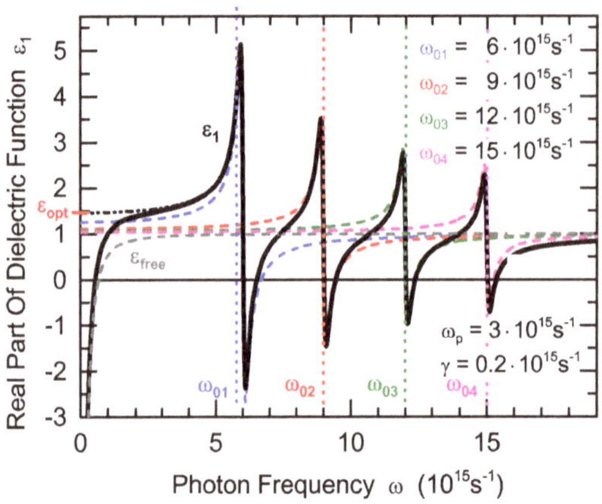

Fig. 3.8: Model to explain the contribution ε_{opt} of the bound electrons to the real part ε_1 of the dielectric function in the infrared. ε_1 (black solid curve) is plotted for a free electron contribution (grey dashed curve) plus four resonances (blue, red, green, and pink dashed curves) at higher energies. The sum of the four resonances (black dash-dotted curve) reaches in the infrared a value ε_{opt} which is larger than 1, the free-electron-limit.

So, only for lcp light we will observe a resonance behavior. In contrast, for rcp light, the absorption falls off monotonically with increasing energy.

We still consider the case of weak magnetic fields, that is, small ω_c. In addition, we assume a doping which is high enough that in the region of interest, where we want to measure, the light frequency ω is large with respect to ω_c and γ. Furthermore, we can neglect absorption because our region of interest is below the first resonance. But the first resonance is the fundamental energy gap of the semiconductor where it is completely transparent without doping. The contribution of the free charge carriers to the absorption is there very small because of the low number of carriers as opposed to the number of electrons in the valence band. So, we set in eq. (3.53) $k_{\pm} = 0$ and $\omega \gg \omega_c, \gamma$. Then it follows for $\varepsilon_{1\pm}$:

$$\varepsilon_{1\pm} \cong n_{\pm}^2 = \varepsilon_{\text{opt}} - \frac{Ne^2}{\varepsilon_0 m^*} \frac{1}{\omega(\omega \mp \omega_c)}. \tag{3.55}$$

We call the frequency where $\varepsilon_1 = 0$ the plasma frequency ω_p. It corresponds to a collective longitudinal oscillation of the charge carriers. The quantized oscillation is called plasmon. Plasmon-like collective excitations also exist for bound electrons as discussed in the problem in Section 3.7.5. From eq. (3.55), we get for the plasma frequency:

$$\omega_p = \sqrt{\frac{Ne^2}{\varepsilon_0 m^*}}. \tag{3.56}$$

Here, we are interested in the minimum of the reflectivity, which occurs at a slightly higher energy $\hbar\omega_{\text{min}}$ than the plasma energy. Still neglecting the absorption index k, we can derive a condition for the minimum from the reflectivity, eq. (3.51) by setting $R = 0$:

$$\sqrt{\varepsilon_1} = n = 1. \tag{3.57}$$

For circularly polarized light and still assuming $\omega \gg \gamma$ and zero magnetic field ($\omega_c = 0$), it follows from eq. (3.55) for the frequency ω_{min} of the reflectivity minimum:

$$\omega_{\text{min}}^2 = \frac{\omega_p^2}{\varepsilon_{\text{opt}} - 1}. \tag{3.58}$$

Inserting first ω_p using eq. (3.56) in eq. (3.55) and then replacing ω_p with ω_{min} using eq. (3.58) yields:

$$\varepsilon_{1\pm} = \varepsilon_{\text{opt}} - \left(\varepsilon_{\text{opt}} - 1\right) \frac{\omega_{\text{min}}^2}{\omega(\omega \mp \omega_c)}. \tag{3.59}$$

The minimum of reflectivity requires $\varepsilon_{1\pm} = n_{1\pm} = 1$. Therefore,

$$0 = \left(\varepsilon_{\text{opt}} - 1\right) \left[1 - \frac{\omega_{\text{min}}^2}{\omega(\omega \mp \omega_c)}\right]. \tag{3.60}$$

This yields for the frequency ω_{min} at the reflectivity minimum:

$$\omega_{\text{min}}^2 = \omega(\omega \mp \omega_c). \tag{3.61}$$

The four solutions of eq. (3.61) are:

$$\omega_{1,2\pm} = \frac{1}{2}\left(\pm\omega_c \pm \sqrt{\omega_c^2 + 4\omega_{\text{min}}^2}\right). \tag{3.62}$$

Since only positive frequencies are physically meaningful, the solutions with negative sign of the square root can be discarded. So, we are left with two solutions for the reflectivity minimum, one for lcp and one for rcp light which are shifted in frequency with respect to the position of the reflectivity minimum in zero magnetic field. In our approximation, we had assumed that the light frequency ω is large with respect to ω_c. So, at the reflectivity minimum, we have $\omega_{\text{min}}^2 \gg \omega_c^2$. Therefore, we get as a final result of the splitting of the reflectivity minimum in a static magnetic field:

$$\omega_{\pm} = \omega_{\text{min}} \pm \frac{\omega_c}{2} + \frac{\omega_c^2}{8\omega_{\text{min}}}. \tag{3.63}$$

Equation (3.63) states that the application of static magnetic fields yields a splitting of the reflectivity minimum (often called: plasma minimum). The size of the splitting between lcp and rcp light is exactly ω_c. Since only $m^\star$ enters the equation for ω_c and not N as in the case of ω_p, the effective mass $m^\star$ of the charge carriers can be determined from the splitting. In a second step, the charge-carrier concentration N can be determined from the averaged value $(\omega_+ + \omega_-)/2$ of the two shifted reflectivity minima.

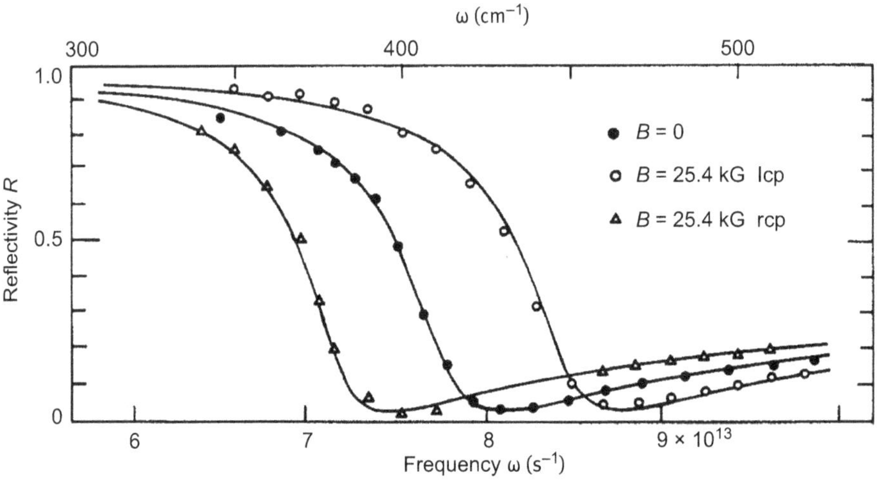

Fig. 3.9: Splitting of the minimum of reflectivity in a magnetic field for rcp and lcp light and determination of $m^\star/m$, N, and γ for InSb. Figure taken from Ref. (90).

Figure 3.9 shows, as an example, reflectivity measurements on InSb below the fundamental energy gap at 0.17 eV. The solid lines are fits to the data (symbols) by plotting eq. (3.51) with $k = 0$. The frequency dependence of the refractive index is obtained by combining eq. (3.59) with eq. (3.57). Using ω_c and ω_{min} as derived from the reflectivity minima, the damping γ can be determined from a fit to the experimental curves. In addition, we notice that the reflectivity approaches 1, that is 100% for photon energies $\hbar\omega \to 0$. This is surprising because InSb is a semiconductor with an energy gap. So, it should be transparent at low photon energies. However, due to the doping, the material is now conducting. This leads to a characteristic behavior at very low photon energies, the so-called Hagen-Rubens law (91). It states that for $\hbar\omega \to 0$, every conducting material will reach 100% reflectivity. Furthermore, the higher the DC conductivity is (and the lower the damping) the faster the increase of the reflectivity occurs for decreasing photon energy. In the problem in Section 3.7.6, we derive the Hagen–Rubens law.

3.5 Relation between experimental quantities and off-diagonal optical conductivity

In the last section of this chapter, we want to establish the explicit relation between the magnetic-field-induced off-diagonal element $\tilde{\sigma}_{xy}$ of the optical conductivity tensor and the quantities that one measures in an experiment. These are the Faraday rotation θ_F and Faraday ellipticity η_F in transmission and the Kerr rotation θ_K and Kerr ellipticity η_K in reflection. For this purpose, we define a ***complex Faraday rotation*** $\tilde{\theta}_\mathrm{F}$, whose real and imaginary part represent the rotation θ_F and the ellipticity η_F respectively:

$$\tilde{\theta}_\mathrm{F} = \theta_\mathrm{F} - \mathrm{i}\eta_\mathrm{F}. \tag{3.64}$$

Then, according to eq. (2.135) and (2.137), we get for small ellipticity η_F:

$$\tilde{\theta}_\mathrm{F} = \frac{\tilde{\delta}}{2} = \frac{\omega d}{2c}(\tilde{n}_+ - \tilde{n}_-). \tag{3.65}$$

On the other hand, according to eqs. (2.30), (3.37), and (3.38), we get:

$$\tilde{n}_+^2 - \tilde{n}_-^2 = \tilde{\varepsilon}_+ - \tilde{\varepsilon}_- = \frac{2}{\varepsilon_0 \omega}\tilde{\sigma}_{xy}. \tag{3.66}$$

We can decompose the left side into two factors:

$$\tilde{n}_+^2 - \tilde{n}_-^2 = (\tilde{n}_+ + \tilde{n}_-)(\tilde{n}_+ - \tilde{n}_-) \cong 2\tilde{n}(\tilde{n}_+ - \tilde{n}_-) \cong \frac{2}{\varepsilon_0 \omega}\tilde{\sigma}_{xy}. \tag{3.67}$$

Here, we have replaced $(\tilde{n}_+ + \tilde{n}_-)$ by $2\tilde{n}$, the refractive index in the absence of a magnetic field. Because the magnetic-field-induced splitting in the refractive index is very small, this is a good approximation. Substituting eq. (3.67) into eq. (3.65) finally yields:

$$\tilde{\theta}_\mathrm{F} = \frac{\omega d}{2c}(\tilde{n}_+ - \tilde{n}_-) = \frac{d}{2\varepsilon_0 c}\frac{\tilde{\sigma}_{xy}}{\tilde{n}}. \tag{3.68}$$

We notice that the complex Faraday rotation is directly proportional to the complex off-diagonal element of the conductivity tensor, which is the material property that appears as a result of the magnetic-field-induced anisotropy. We can decompose eq. (3.68) into real and imaginary part and get an expression for the Faraday rotation θ_F and the ellipticity η_F:

$$\theta_\mathrm{F} = \frac{d}{2\varepsilon_0 c}\frac{n\sigma_{1xy} - k\sigma_{2xy}}{n^2 + k^2}, \tag{3.69}$$

$$\eta_F = -\frac{d}{2\varepsilon_0 c}\frac{k\sigma_{1xy} + n\sigma_{2xy}}{n^2 + k^2}. \tag{3.70}$$

This way, we have derived an explicit relation between the experimentally accessible quantities of the Faraday effect and the magnetic-field-induced anisotropy in the optical conductivity. Note that the Faraday effect also depends on the refractive index and absorption index which are non-magnetic material properties.

For the polar Kerr effect similar formulas can be derived. Due to the magnetic-field-induced anisotropy in the complex refractive index, the complex reflection coefficients must also be different for rcp and lcp light as defined in eq. (2.109):

$$\tilde{\rho}_\pm = -\frac{\tilde{n}_\pm - 1}{\tilde{n}_\pm + 1}. \tag{3.71}$$

If we write $\tilde{\rho}_\pm$ in polar coordinates as $\tilde{\rho}_\pm = \rho_\pm e^{i\delta_\pm}$, the polar Kerr rotation θ_K and the polar Kerr ellipticity η_K can be expressed as:

$$\theta_K = -\frac{\delta_+ - \delta_-}{2}, \tag{3.72}$$

$$\tan\eta_K = -\frac{\rho_+ - \rho_-}{\rho_+ + \rho_-}. \tag{3.73}$$

These relations are exact and follow from a simple geometric consideration taking into account that linearly polarized light can be looked at as a combination of rcp and lcp light with equal amplitudes. In addition, the change of helicity after reflection must be considered. For detailed treatment, see part (a) of the problem in Section 3.7.7. We can define here, too, a complex polar Kerr rotation $\tilde{\theta}_K$ as:

$$\tilde{\theta}_K = \theta_K - i\eta_K, \tag{3.74}$$

Then, eqs. (3.72) and (3.73) can be combined as:

$$\tilde{\theta}_K = \theta_K - i\eta_K \cong i\frac{\tilde{\rho}_+ - \tilde{\rho}_-}{\tilde{\rho}_+ + \tilde{\rho}_-}. \tag{3.75}$$

This relation is proven in part (b) of the problem in Section 3.7.7. Substituting eq. (3.71) into (3.75) and using the approximations $(\tilde{n}_+ + \tilde{n}_-) \cong 2\tilde{n}$ and $\tilde{n}_+\tilde{n}_- \cong \tilde{n}^2$, it follows after some calculations:

$$\tilde{\theta}_K \cong -\frac{\tilde{\sigma}_{xy}}{\tilde{n}\tilde{\sigma}_{xx}} = i\frac{1}{\varepsilon_0\omega}\frac{\tilde{\sigma}_{xy}}{\tilde{n}\left(\tilde{n}^2 - 1\right)} = i\frac{1}{\varepsilon_0\omega}\frac{\tilde{\sigma}_{xy}}{A - iB}. \tag{3.76}$$

See part (c) of the problem in Section 3.7.7 for details of the calculation. Again, we notice that the complex polar Kerr rotation is directly proportional to the complex off-diagonal element of the conductivity tensor. Separating eq. (3.76) into real and imaginary part yields:

$$\theta_K = -\frac{1}{\varepsilon_0\omega}\frac{B\sigma_{1xy} + A\sigma_{2xy}}{A^2 + B^2}, \tag{3.77}$$

$$\eta_K = -\frac{1}{\varepsilon_0\omega}\frac{A\sigma_{1xy} - B\sigma_{2xy}}{A^2 + B^2}, \tag{3.78}$$

where the prefactors A and B are polynomial functions of n and k:

$$\tilde{n}\left(\tilde{n}^2 - 1\right) = A - \mathrm{i}B, \tag{3.79}$$

$$A = n^3 - 3nk^2 - n, \tag{3.80}$$

$$B = -k^3 + 3n^2k - k. \tag{3.81}$$

Again, we get an explicit relation between the experimentally accessible quantities of the polar Kerr effect and the magnetic-field-induced anisotropy in the optical conductivity. The polar Kerr effect also depends on the refractive index and absorption index but contrary to the Faraday effect in a nonlinear way because of the prefactors A and B. But there is another even more important difference to the Faraday effect: the complex refractive index used in the equations of the polar Kerr effect is not the one of the material. As we learnt in Section 2.3.2, it is the ***relative*** complex refractive index $\tilde{n} = \tilde{n}_2/\tilde{n}_1$, that is, the ratio between the refractive index $\tilde{n}_2$ of the material, where the light is being reflected from, and $\tilde{n}_1$ of the outside world, where the light is coming from. This has important consequences when measuring the polar Kerr effect and makes the interpretation of data from the polar Kerr effect much more difficult than the data from the Faraday effect as we will discuss in detail in Section 4.3.

3.6 Comprehension questions

3.6.1 Lorentz theory

(a) Discuss the basic assumptions of the Lorentz model.
(b) How is the displacement of an electron connected to the dielectric function of the material?
(c) Plot the typical frequency dependence of ε_1 and ε_2. What are the limiting values for $\omega \to 0$ and $\omega \to \infty$?
(d) Why is σ_{1xx} the most suitable quantity to discuss optical spectra?
(e) What can be concluded for n and k in the range where $\varepsilon_1 < 0$ and what is the consequence for the reflectivity R at normal incidence?
(f) How can the dielectric function ε be generalized to atoms with many electrons?

(g) How does the equation of motion change when a static magnetic field is added to the Lorentz model and what is the correct ansatz to solve it? How does the solution, that is, the dielectric function change when a static magnetic field is present?

(h) What is in real space the major consequence for the optical properties when a magnetic field is present?

(i) Write down the tensors for the dielectric function ε and the optical conductivity σ in a circular basis and in real space when a static magnetic field is present.

3.6.2 Cyclotron resonance, splitting of reflectivity minimum

(a) Which optical quantity is important when discussing the microwave cyclotron resonance? To which quantity measured in the experiment is it related?

(b) Which conditions must be fulfilled for the cyclotron resonance to be discernible in the spectra? Which electronic properties can be determined with this measurement?

(c) How does the reflectivity spectrum change when a magnetic field is switched on? What is the origin of that effect? Which electronic properties can be determined with this measurement?

3.6.3 Relation between experimental quantities and off-diagonal conductivity

(a) To which complex optical quantity (material property) is the complex Faraday effect as well as the complex polar Kerr effect directly proportional?

(b) Compare the Faraday and the Kerr effect for the same material: How do they differ at high photon energies? How do they depend on the index of refraction and absorption?

3.7 Problems

3.7.1 Classical theory of dispersion

(a) Within the context of the Lorentz theory, proof that the maximum of $\sigma_1(\omega)$ appears exactly at the resonance frequency ω_0.

(b) Calculate the full width at half maximum (FWHM) of the peak in $\sigma_1(\omega)$.

(c) Determine the maximum of $\varepsilon_2(\omega)$ as a function of the damping γ.

Hint: Assume $\omega_0 \gg \gamma$ and approximate the square roots to first order.

3.7.2 Excitations in the X-ray range

In the X-ray range ($\lambda \cong 1$ Å), there exist strong excitations in solid-state materials. The absorption coefficient reaches $K \cong 10^5\ \text{cm}^{-1}$ in core-level excitations. Explain why, nevertheless, the extrapolation $n \to 1$, $k \to 0$ is correct for photon energies $\hbar\omega \geq 30$ eV.

Hint: First, calculate the maximum of $\varepsilon_1(\omega)$ and $\varepsilon_2(\omega)$, which is a function of ω_0. Then, discuss the limiting values of the maxima $\varepsilon_{1max}(\omega_0)$ and $\varepsilon_{2max}(\omega_0)$ for $\hbar\omega_0 \to \infty$ and how they depend on powers of ω_0. Finally, calculate from ε_{1max} and ε_{2max} the optical properties $n(\omega_0)$, $k(\omega_0)$, and $K(\omega_0)$ using the relation $\tilde{\varepsilon} = \tilde{n}^2$ and assuming that ε_{1max} and ε_{2max} reach their maximum at the same frequency. Discuss then the behavior of $n(\omega_0)$, $k(\omega_0)$, and $K(\omega_0)$ for $\hbar\omega_0 \to \infty$ in terms of powers of ω_0.

3.7.3 Frequency dependence of free electrons at large frequencies

For free electrons, derive the frequency dependence of $\varepsilon_1(\omega)$ and $\sigma_1(\omega)$ for $\omega \to \infty$.

Hint: Consider ε_1 and σ_1 for $\omega \gg \gamma$ and discuss how the limiting values are reached for $\omega \to \infty$.

3.7.4 Cyclotron-resonance measurement

As shown in Fig. 3.6, the Fermi surface, which is a surface of constant energy, of the conduction band of germanium lies along the <111> directions in reciprocal space. It has an ellipsoidal shape. The rotational ellipsoid can be described in a local *xyz* coordinate system, with *z* axis parallel to the symmetry axis of the ellipsoid, as in eq. (3.48):

$$E(k) = \hbar^2 \left(\frac{k_x^2 + k_y^2}{2m_t} + \frac{k_z^2}{2m_l} \right). \tag{3.82}$$

(a) Consider the equation of motion of a free electron, eq. (3.39), with vanishing damping $\gamma = 0$ and vanishing electric field:

$$m^* \ddot{\vec{r}} = q \dot{\vec{r}} \times \vec{B}. \tag{3.83}$$

Show that for a static magnetic field $\vec{B}$ in the local *xy* plane the cyclotron resonance frequency is given by:

$$\omega_c = e\left|\vec{B}\right| \Big/ \sqrt{m_l m_t}. \tag{3.84}$$

Hint: Treat the electron as a free particle, that is, as a harmonic wave, and neglect the space dependence. Then, the velocity is described as

$$\vec{v}(t) = \vec{v}_0 e^{i\omega t}. \tag{3.85}$$

Now, solve the equation of motion for the velocity of the electron.

(b) Determine from the magnetic-field-dependent measurement of the effective masses of the electrons, Fig. 3.7, the value of the longitudinal m_l, and transversal m_t effective mass and compare with the results in eq. (3.50).

Hint: Evaluate the curves for distinguished magnetic field directions. For this, you must correlate the electron orbitals with the axes of the rotational ellipsoid of the Fermi surface. The electron orbital is the intersection of the plane perpendicular to the magnetic field direction with the Fermi surface that produces the extremal orbit.

3.7.5 Plasmons in germanium

The plasma frequency ω_p is defined as the frequency at which $\varepsilon_1(\omega = \omega_p) = 0$ and, simultaneously, $d\varepsilon_1/d\omega > 0$ in the limit of vanishing damping (i.e., $\gamma \to 0$). The plasma frequency ω_p, defined in such a way, is usually applied to free electrons, where $\omega_0 = 0$. It describes a collective oscillation of the free charge carriers against the positive charged background of the atomic cores. However, it can also be used for bound electrons. Then, it represents a collective oscillation within an energy band, which can be experimentally observed.

(a) Calculate the plasma frequency ω_p of the valence electrons in the semiconductor germanium.

Hint: Derive an equation for ω_p from the condition $\varepsilon_1(\omega_p) = 0$. Neglect ω_0 as it is much smaller than ω_p. Now, ω_p can be calculated by setting $m^* = m_e$ and by determining the number N of valence electrons per volume in germanium. Note that germanium crystallizes in diamond structure with lattice constant $a = 5.65$ Å.

(b) Determine the plasma frequency of free electrons in weakly doped germanium that contains 0.01% free electrons per germanium atom due to arsenic doping ($m^* = 0.15\, m_e$).

3.7.6 Law of Hagen-Rubens and DC conductivity

In classical dispersion theory, free electrons can be described by assuming a resonance frequency $\omega_0 = 0$.

(a) Calculate the DC conductivity $\sigma_0 = \sigma_1(\omega = 0)$.

(b) Derive the law of Hagen-Rubens:

$$R(\omega) = 1 - 2\sqrt{\frac{2\varepsilon_0 \omega}{\sigma_0}}. \tag{3.86}$$

It describes the reflectivity R as a function of frequency ω and of DC conductivity σ_0 in the limit $\omega \ll \gamma$ at normal incidence (i.e., angle of incidence $\alpha = 0$). Assume that the matrial is facing vacuum. Then we can set the relative refractive index n equal to the refractive index of the material n_2: $n = n_2$.

Hint: Expand $\varepsilon_1(\omega)$ and $\varepsilon_2(\omega)$ in a power series of ω/γ and consider only terms linear in ω/γ. Neglect the term 1 (respectively ε_{opt}) in $\varepsilon_1(\omega)$. Consequently, ε_1 and $\sigma_1 = \omega\varepsilon_0\varepsilon_2$ will become constant. Now, n and k can be expressed by ε_1 and ε_2 using the relations $n^2 = \frac{1}{2}\left(\varepsilon_1 + \sqrt{\varepsilon_1^2 + \varepsilon_2^2}\right)$ and $k^2 = \frac{1}{2}\left(-\varepsilon_1 + \sqrt{\varepsilon_1^2 + \varepsilon_2^2}\right)$. Expanding n and k in powers of ω/γ yields $n^2 \cong k^2$. Finally, the reflectivity R can be expressed as a function of frequency ω and DC conductivity σ_0. This is the law of Hagen–Rubens.

3.7.7 Complex polar Kerr effect

For rcp (+) and lcp (−) light, when reflected from a magneto-optically active sample at normal incidence, the reflection coefficients are defined analogous to linearly polarized light as:

$$\tilde{\rho}_\pm = -\frac{\tilde{n}_\pm - 1}{\tilde{n}_\pm + 1} = \rho_\pm e^{i\phi_\pm}, \tag{3.87}$$

where the reflected rcp wave becomes an lcp wave due to the same sense of rotation but opposite propagation direction and vice versa.

(a) First, verify geometrically that in x direction linearly polarized light, traveling in $-z$ direction, exhibits after reflection from a magneto-optically active sample (lying in the xy plane) a rotation θ_K and an ellipticity η_K which are given by

$$\theta_K = -\frac{\phi_+ - \phi_-}{2} \tag{3.88}$$

and

$$\eta_K = -\frac{\rho_+ - \rho_-}{\rho_+ + \rho_-}. \tag{3.89}$$

Hint: First, geometrically decompose the incident light into rcp and lcp light. After reflection according to eq. (3.87), build the geometrical superposition of the reflected rcp and lcp waves and compare to the expected state of polarization by looking at the extreme positions and comparing them to the semiaxes of the ellipse.

(b) Show further that the complex polar Kerr effect $\tilde{\theta}_K = \theta_K - \mathrm{i}\,\eta_K$ can be represented in accordance with eqs. (3.88) and (3.89) for small angles (θ_K, $\eta_K \ll 1$) approximately as follows:

$$\tilde{\theta}_K = \theta_K - \mathrm{i}\,\eta_K \cong \mathrm{i}\frac{\tilde{\rho}_+ - \tilde{\rho}_-}{\tilde{\rho}_+ + \tilde{\rho}_-}. \tag{3.90}$$

Hint: Start from the result, eq. (3.90), use the polar representation of $\tilde{\rho}_\pm$ from eq. (3.87) and apply the approximations $\rho_+\rho_- \cong \rho^2$, $\rho_+ + \rho_- \cong 2\rho$ and $|\phi_+ - \phi_-| \ll 1$.

(c) Derive eq. (3.76) to (3.81), the relationship between the off-diagonal optical conductivity $\tilde{\sigma}_{xy} = \sigma_{1xy} + i\sigma_{2xy}$ and the complex polar Kerr effect $\tilde{\theta}_K = \theta_K - i\eta_K$.

Hint: Start *from eq. (3.90), and express it with help of eq. (3.87) as a function of $\tilde{n}_\pm$. By using the approximations $(\tilde{n}_+ + \tilde{n}_-) \cong 2\tilde{n}$ and $\tilde{n}_+\tilde{n}_- \cong \tilde{n}^2$, the further course of the calculation is analogous to the derivation of the relation for the Faraday rotation in Section 3.5.*

3.8 Solutions

3.8.1 Classical theory of dispersion

(a) The maximum of a function is defined as the root of the first derivative, that is, we evaluate $d\sigma_1/d\omega = 0$. This works but is very tedious. A much shorter calculation results when we transform $\sigma_1(\omega)$ in such a way that the nominator is independent of ω. Then, the maximum of $\sigma_1(\omega)$ occurs where the denominator reaches its minimum. To make the nominator independent of ω, we extract a factor ω^2 from nominator and denominator:

$$\sigma_1 = \frac{Ne^2\gamma}{m}\frac{\omega^2}{\left(\omega_0^2-\omega^2\right)^2+\omega^2\gamma^2} = \frac{Ne^2\gamma}{m}\frac{\omega^2}{\omega^2\left[\omega_0^2\left(\frac{\omega_0}{\omega}-\frac{\omega}{\omega_0}\right)^2+\gamma^2\right]} = \frac{Ne^2\gamma}{m}\frac{1}{\omega_0^2\left(\frac{\omega_0}{\omega}-\frac{\omega}{\omega_0}\right)^2+\gamma^2}. \tag{3.91}$$

The denominator reaches its minimum when the term in the round brackets becomes zero:

$$\frac{\omega_0}{\omega}-\frac{\omega}{\omega_0} = 0. \tag{3.92}$$

From this follows the condition:

$$\frac{\omega}{\omega_0} = 1. \tag{3.93}$$

This proofs that the maximum of $\sigma_1(\omega)$ is exactly at the resonance frequency ω_0. Alternatively, we can take the first derivative of $1/\sigma_1(\omega)$ and search for the minimum:

$$\frac{d}{d\omega}\frac{\left(\omega_0^2-\omega^2\right)^2+\omega^2\gamma^2}{\omega^2} = \frac{d}{d\omega}\left[\frac{\left(\omega_0^2-\omega^2\right)^2}{\omega^2}+\gamma^2\right] = \left[\frac{-4\omega\left(\omega_0^2-\omega^2\right)}{\omega^2}-\frac{2\left(\omega_0^2-\omega^2\right)^2}{\omega^3}\right] = 0, \tag{3.94}$$

and then ($\omega > 0$):

$$\left[2 + \frac{(\omega_0^2 - \omega^2)}{\omega^2}\right](\omega_0^2 - \omega^2) = 0, \tag{3.95}$$

which yields again $\omega = \omega_0$ for the maximum.

(b) In order to calculate the FWHM of the peak in $\sigma_1(\omega)$, we determine the two positions ω_1 and ω_2 where $\sigma_1(\omega)$ reaches half the maximum and take the difference between them. The half maximum is:

$$\frac{1}{2}\sigma_{\max} = \frac{1}{2}\sigma(\omega_0) = \frac{Ne^2}{2m\gamma} = \sigma(\omega) = \frac{Ne^2\gamma}{m}\frac{\omega^2}{\left(\omega_0^2 - \omega^2\right)^2 + \omega^2\gamma^2}. \tag{3.96}$$

Simplifying the equation yields:

$$\frac{1}{2\gamma^2} = \frac{\omega^2}{\left(\omega_0^2 - \omega^2\right)^2 + \omega^2\gamma^2}, \tag{3.97}$$

and further:

$$2\gamma^2 = \frac{\left(\omega_0^2 - \omega^2\right)^2 + \omega^2\gamma^2}{\omega^2} = \frac{\left(\omega_0^2 - \omega^2\right)^2}{\omega^2} + \gamma^2. \tag{3.98}$$

Finally, we get an equation for the damping:

$$\gamma^2 = \frac{\left(\omega_0^2 - \omega^2\right)^2}{\omega^2}. \tag{3.99}$$

The two solutions for γ are:

$$\pm\gamma = \frac{\omega_0^2 - \omega^2}{\omega}. \tag{3.100}$$

From this, ω can be determined by solving a quadratic equation:

$$\omega^2 \pm \gamma\omega - \omega_0^2 = 0. \tag{3.101}$$

The four solutions, whereof two are not physical because they are negative, are:

$$\omega_{1,2} = \frac{1}{2}\left(\mp\gamma \underbrace{\pm}_{\text{"−" not physical}} \sqrt{\gamma^2 + 4\omega_0^2}\right) = \mp\frac{\gamma}{2} + \omega_0\sqrt{1 + \frac{\gamma^2}{4\omega_0^2}} \cong \mp\frac{\gamma}{2} + \omega_0\left(1 + \frac{\gamma^2}{8\omega_0^2}\right). \tag{3.102}$$

In eq. (3.102), we have approximated the square root to second order in γ using the well-known approximation $\sqrt{1+\varepsilon} \cong 1+\varepsilon/2$, when $\varepsilon << 1$. From the two solutions $\omega_{1,2}$, we can now calculate the FWHM by taking the difference between them:

$$\Delta\omega = \omega_2 - \omega_1 = \gamma. \tag{3.103}$$

In conclusion, up to second order in γ, the FWHM is indeed given by the damping γ.

(c) Now, we want to show that the maximum of $\varepsilon_2(\omega)$ is not at $\omega = \omega_0$ for finite damping $\gamma > 0$. The maximum is determined by the root of the first derivative of $\varepsilon_2(\omega)$:

$$\frac{d\varepsilon_2}{d\omega} = \frac{d}{d\omega}\frac{Ne^2}{\varepsilon_0 m}\frac{\gamma\omega}{\left(\omega_0^2-\omega^2\right)^2+\omega^2\gamma^2} = 0. \tag{3.104}$$

We can simplify eq. (3.104) by canceling all the constant factors before taking the derivative:

$$\frac{d}{d\omega}\frac{\omega}{\left(\omega_0^2-\omega^2\right)^2+\omega^2\gamma^2} = \frac{1}{\left(\omega_0^2-\omega^2\right)^2+\omega^2\gamma^2} - \frac{\omega\left(-4\omega\left(\omega_0^2-\omega^2\right)+2\omega\gamma^2\right)}{\left[\left(\omega_0^2-\omega^2\right)^2+\omega^2\gamma^2\right]^2} = 0. \tag{3.105}$$

Next, we make the expression of same denominator:

$$\frac{\left(\omega_0^2-\omega^2\right)^2+\omega^2\gamma^2+4\omega^2\left(\omega_0^2-\omega^2\right)-2\omega^2\gamma^2}{\left[\left(\omega_0^2-\omega^2\right)^2+\omega^2\gamma^2\right]^2} = 0. \tag{3.106}$$

For finding the solutions, we just must evaluate the numerator (and make sure that the denominator does not have a root at the solutions or a singularity at any frequency):

$$\left(\omega_0^2-\omega^2\right)^2-\omega^2\gamma^2+4\omega^2\left(\omega_0^2-\omega^2\right) = 0. \tag{3.107}$$

This yields a biquadratic equation:

$$\omega_0^4+\omega^4-2\omega_0^2\omega^2-\omega^2\gamma^2+4\omega^2\omega_0^2-4\omega^4 = 0, \tag{3.108}$$

which reduces to

$$3\omega^4+\omega^2\left(\gamma^2-2\omega_0^2\right)-\omega_0^4 = 0. \tag{3.109}$$

The four solutions are

$$\omega^2_{\max 1,2} = \frac{1}{6}\left[2\omega_0^2-\gamma^2 \underbrace{\pm}_{\text{"}-\text{" not physical}} \sqrt{\left(\gamma^2-2\omega_0^2\right)^2+12\omega_0^4}\right]. \tag{3.110}$$

Note that only the solutions with positive square root are physically meaningful. Assuming small damping $\gamma \ll \omega_0$, we obtain:

$$\omega_{max}^2 = \frac{1}{6}\left[2\omega_0^2 - \gamma^2 + \sqrt{\underbrace{\gamma^4}_{\cong 0} - 4\gamma^2\omega_0^2 + 16\omega_0^4}\right] \cong \frac{1}{6}\left[2\omega_0^2 - \gamma^2 + 4\omega_0^2\sqrt{1 - \frac{\gamma^2}{4\omega_0^2}}\right]$$
$$\cong \frac{1}{6}\left[2\omega_0^2 - \gamma^2 + 4\omega_0^2\left(1 - \frac{\gamma^2}{8\omega_0^2}\right)\right] = \omega_0^2 - \frac{\gamma^2}{4}. \qquad (3.111)$$

Taking the square root and approximating the right side to second order in γ results in the solution for ω_{max}:

$$\omega_{max} \cong \omega_0\sqrt{1 - \frac{\gamma^2}{4\omega_0^2}} \cong \omega_0\left(1 - \frac{\gamma^2}{8\omega_0^2}\right) = \omega_0 - \frac{\gamma^2}{8\omega_0}. \qquad (3.112)$$

The solution proofs that $\varepsilon_2(\omega)$ reaches its maximum not at $\omega = \omega_0$ but at $\omega = \omega_{max}$ for finite damping $\gamma > 0$. We can now also determine the maximum of $\varepsilon_2(\omega)$ explicitly:

$$\varepsilon_2(\omega = \omega_{max}) = \frac{Ne^2}{\varepsilon_0 m}\frac{\gamma\left(\omega_0 - \frac{\gamma^2}{8\omega_0}\right)}{\left(\frac{\gamma^2}{4} - \frac{\gamma^4}{64\omega_0^2}\right)^2 + \omega_0^2\gamma^2 - \frac{\gamma^4}{4} + \frac{\gamma^6}{64\omega_0^2}} \cong \frac{Ne^2}{\varepsilon_0 m}\frac{\gamma\left(\omega_0 - \frac{\gamma^2}{8\omega_0}\right)}{\omega_0^2\gamma^2 - \frac{3\gamma^4}{16}}$$
$$= \frac{Ne^2}{\varepsilon_0 m}\frac{\omega_0 - \frac{\gamma^2}{8\omega_0}}{\gamma\left(\omega_0^2 - \frac{3\gamma^2}{16}\right)}. \qquad (3.113)$$

3.8.2 Excitations in the X-ray range

In this problem, we must distinguish between the behavior of the optical functions $n(\omega)$ and $k(\omega)$ for $\omega \to \infty$ and the behavior of the resonances $n(\omega_0)$ and $k(\omega_0)$ for $\omega_0 \to \infty$. The behavior of the former is given by the frequency dependence of the dielectric function from Lorentz theory and is easy to determine by letting ω go to infinity. Then, we get $n(\omega \to \infty) \to 1$ and $k(\omega \to \infty) \to 0$. To discuss if strong transitions are still possible even at very high resonance frequencies, we must study the behavior of the maximum of $\varepsilon_1(\omega)$ and $\varepsilon_2(\omega)$. As we consider only very large resonance frequencies $\omega_0 \to \infty$, we can assume $\gamma \ll \omega_0$. The maxima of $\varepsilon_1(\omega)$ and $\varepsilon_2(\omega)$, which are also the maxima of $n(\omega)$ and $k(\omega)$, occur near ω_0. Therefore, it is sufficient to study the behavior of the maxima of $\varepsilon_1(\omega)$ and $\varepsilon_2(\omega)$ for $\omega_0 \to \infty$.

From eq. (3.113) of the solution to part (c) of the problem in Section 3.7.1, we can write the maximum of ε_2 for $\gamma \ll \omega_0$ as:

$$\varepsilon_2(\omega = \omega_{\max}) \cong \frac{Ne^2}{\varepsilon_0 m} \frac{\omega_0 - \frac{\gamma^2}{8\omega_0}}{\gamma\left(\omega_0^2 - \frac{3\gamma^2}{16}\right)} \cong \frac{Ne^2}{\varepsilon_0 m} \frac{\omega_0}{\gamma\, \omega_0^2}. \tag{3.114}$$

So, we get:

$$\varepsilon_{2\max}(\omega_0) = \varepsilon_2(\omega = \omega_{\max}) \cong \varepsilon_2(\omega_0) \propto \frac{1}{\omega_0}, \quad \omega_0 \to \infty. \tag{3.115}$$

From eq. (3.115), we see that $\varepsilon_{2\max}(\omega_0)$ goes to 0 as $1/\omega_0$ for $\omega_0 \to \infty$. We can proceed the same way for the real part $\varepsilon_1(\omega)$:

$$\varepsilon_1(\omega) = 1 + \frac{Ne^2}{\varepsilon_0 m} \frac{\omega_0^2 - \omega^2}{\left(\omega_0^2 - \omega^2\right)^2 + \omega^2\gamma^2}. \tag{3.116}$$

First, we determine the frequency of the maximum by setting the first derivative equal to zero:

$$\begin{aligned} \frac{d\varepsilon_1(\omega)}{d\omega} &= \frac{Ne^2}{\varepsilon_0 m}\left[\frac{-2\omega}{\left(\omega_0^2 - \omega^2\right)^2 + \omega^2\gamma^2} - \frac{\left(\omega_0^2 - \omega^2\right)\left[2\left(\omega_0^2 - \omega^2\right)(-2\omega) + 2\omega\gamma^2\right]}{\left[\left(\omega_0^2 - \omega^2\right)^2 + \omega^2\gamma^2\right]^2}\right] \\ &= \frac{Ne^2}{\varepsilon_0 m}\left[\frac{-2\omega\left[\left(\omega_0^2 - \omega^2\right)^2 + \omega^2\gamma^2\right] - \left(\omega_0^2 - \omega^2\right)\left[2\left(\omega_0^2 - \omega^2\right)(-2\omega) + 2\omega\gamma^2\right]}{\left[\left(\omega_0^2 - \omega^2\right)^2 + \omega^2\gamma^2\right]^2}\right] = 0. \end{aligned} \tag{3.117}$$

To find the roots of eq. (3.117), it is sufficient to find the roots of the numerator (and to make sure that the denominator does not have a root at the same frequency or a singularity at any frequency):

$$0 = -2\omega\left[\left(\omega_0^2 - \omega^2\right)^2 + \omega^2\gamma^2\right] - \left(\omega_0^2 - \omega^2\right)\left[-4\omega\left(\omega_0^2 - \omega^2\right) + 2\omega\gamma^2\right]. \tag{3.118}$$

In eq. (3.118), we have cancelled the constant factor in front of the bracket. In addition, we can cancel a factor 2ω

$$0 = -\left[\left(\omega_0^2 - \omega^2\right)^2 + \omega^2\gamma^2\right] + \left(\omega_0^2 - \omega^2\right)\left[2\left(\omega_0^2 - \omega^2\right) - \gamma^2\right]. \tag{3.119}$$

This simplifies to:

$$0 = \left(\omega_0^2 - \omega^2\right)^2 - \omega_0^2\gamma^2, \tag{3.120}$$

and further to:

$$\omega_0^2 - \omega^2 = \pm\, \omega_0\gamma. \tag{3.121}$$

From this we obtain the frequency of the extrema:

$$\omega^2 = \omega_0^2 \mp \omega_0 \gamma. \tag{3.122}$$

From Fig. 3.1(a), it is clear that the solution with the minus sign describes the maximum and the one with the plus sign the minimum. For $\gamma \ll \omega_0$, maximum and minimum frequency approach the resonance frequency ω_0. Now, we can obtain the maximum ε_1 ($\omega = \omega_{\max}$) as a function of the resonance frequency ω_0 by entering the value from eq. (3.122):

$$\begin{aligned}\varepsilon_1(\omega = \omega_{\max}) &= 1 + \frac{Ne^2}{\varepsilon_0 m} \frac{\omega_0^2 - \omega_{\max}^2}{\left(\omega_0^2 - \omega_{\max}^2\right)^2 + \omega_{\max}^2 \gamma^2} \\ &= 1 + \frac{Ne^2}{\varepsilon_0 m} \frac{\omega_0^2 - \omega_0^2 + \omega_0 \gamma}{\left(\omega_0^2 - \omega_0^2 + \omega_0 \gamma\right)^2 + \left(\omega_0^2 - \omega_0 \gamma\right)\gamma^2}.\end{aligned} \tag{3.123}$$

For $\gamma \ll \omega_0$, this simplifies to:

$$\begin{aligned}\varepsilon_{1\max}(\omega_0) = \varepsilon_1(\omega = \omega_{\max}) &= 1 + \frac{Ne^2}{\varepsilon_0 m} \frac{\omega_0 \gamma}{2\omega_0^2 \gamma^2 - \omega_0 \gamma^3} \\ &= 1 + \frac{Ne^2}{\varepsilon_0 m} \frac{1}{2\omega_0 \gamma - \gamma^2} \cong 1 + \frac{Ne^2}{\varepsilon_0 m} \frac{1}{2\omega_0 \gamma}.\end{aligned} \tag{3.124}$$

From eq. (3.124), we see that $\varepsilon_{1\max}(\omega_0)$ goes to 1 as $1/\omega_0$ for $\omega_0 \to \infty$.

In the next step, we calculate the optical functions n and k at the maxima of the dielectric functions. From eqs. (2.31) and (2.32), we can derive reversed formulas, expressing n and k with the help of ε_1 and ε_2. For n^2, we get:

$$n^2 = \frac{1}{2}\left(\varepsilon_1 + \varepsilon_1 \sqrt{1 + \left(\frac{\varepsilon_2}{\varepsilon_1}\right)^2}\right) \cong \frac{1}{2}\varepsilon_1 \left(1 + 1 + \frac{1}{2}\left(\frac{\varepsilon_2}{\varepsilon_1}\right)^2\right) = \varepsilon_1 + \frac{1}{4}\frac{\varepsilon_2^2}{\varepsilon_1}. \tag{3.125}$$

According to eqs. (3.115) and (3.124), the second term on the right goes to zero as $1/\omega_0^2$. Therefore, the first term on the right dominates and n^2 approaches 1 with a power of $1/\omega_0$ for $\omega_0 \to \infty$. For k^2, we get:

$$k^2 = \frac{1}{2}\left(-\varepsilon_1 + \varepsilon_1 \sqrt{1 + \left(\frac{\varepsilon_2}{\varepsilon_1}\right)^2}\right) \cong \frac{1}{2}\varepsilon_1 \left(-1 + 1 + \frac{1}{2}\left(\frac{\varepsilon_2}{\varepsilon_1}\right)^2\right) = \frac{1}{4}\frac{\varepsilon_2^2}{\varepsilon_1}. \tag{3.126}$$

According to eqs. (3.115) and (3.124), k^2 goes to zero as $1/\omega_0^2$ for $\omega_0 \to \infty$. Summarizing the results, n and k approach 1 and 0 as $1/\sqrt{\omega_0}$ and $1/\omega_0$, respectively, for $\omega_0 \to \infty$. On the other hand, the quantity that describes the absorption is the absorption coefficient K, which is defined in eq. (2.47). as $K = 2\frac{\omega}{c}k$. From this, in combination with eq. (3.126), it follows that K still shows all the excitations for $\omega_0 \to \infty$, because

$$K(\omega = \omega_0) = 2\frac{\omega_0}{c} k \propto 2\frac{\omega_0}{c} \frac{1}{\omega_0} \cong \text{const.} \tag{3.127}$$

The reason is that the additional factor ω in the definition of K cancels the $1/\omega_0$ decrease of k but only at the resonance frequency $\omega = \omega_0$.

3.8.3 Frequency dependence of free electrons at large frequencies

The real part $\varepsilon_1(\omega)$ of the dielectric function for free electrons is given by eq. (3.5), with $\omega_0 = 0$. We start with the high-frequency limit $\omega \gg \gamma$:

$$1 - \varepsilon_1(\omega) = \frac{Ne^2}{\varepsilon_0 m} \frac{1}{\omega^2 + \gamma^2} \cong \frac{Ne^2}{\varepsilon_0 m} \frac{1}{\omega^2} \propto \frac{1}{\omega^2} \quad (\omega \to \infty). \tag{3.128}$$

We notice that ε_1 approaches 1 with a power of $1/\omega^2$.

The real part of the optical conductivity σ_1 is given by eq. (3.7), where we set again $\omega_0 = 0$ and assume $\omega \gg \gamma$ for the high-frequency limit:

$$\sigma_1(\omega) = \frac{Ne^2\gamma}{m} \frac{1}{\omega^2 + \gamma^2} \cong \frac{Ne^2\gamma}{m} \frac{1}{\omega^2} \propto \frac{1}{\omega^2} \quad (\omega \to \infty), \tag{3.129}$$

Obviously, σ_1 goes to 0 with a power of $1/\omega^2$.

3.8.4 Cyclotron-resonance measurement

(a) The magnetic field in the xy plane is described as $\vec{B} = B_0(\cos\alpha,\ \sin\alpha,\ 0)$. We assume that the electron can be treated as a free particle. Then the velocity is described as a plane wave according to eq. (3.85):

$$\vec{v}(t) = \vec{v}_0 e^{i\omega t}, \tag{3.130}$$

with $\vec{v}_0 = (v_{0x},\ v_{0y},\ v_{0z})$. The equation of motion for the components of velocity of the electron is:

$$m_t i\omega v_{0x} = -q \sin\alpha\ v_{0z} B_0, \tag{3.131}$$

$$m_t i\omega v_{0y} = q \cos\alpha\ v_{0z} B_0, \tag{3.132}$$

$$m_l i\omega v_{0z} = q(\sin\alpha\ v_{0x} - \cos\alpha\ v_{0y}) B_0. \tag{3.133}$$

We can solve eq. (3.131) for v_{0x} and eq. (3.132) for v_{0y} and insert the result in eq. (3.133):

$$m_l i\omega v_{0z} = q(\sin\alpha\ v_{0x} - \cos\alpha\ v_{0y}) B_0 = \frac{q^2 B_0^2}{m_t i\omega}(-\sin^2\alpha - \cos^2\alpha) v_{0z} = -\frac{q^2 B_0^2}{m_t i\omega} v_{0z}. \tag{3.134}$$

This equation is only fulfilled if

$$\omega = \frac{qB_0}{\sqrt{m_l m_t}}. \tag{3.135}$$

(b) In order to derive the effective masses, we can use distinguished points such as extrema or degenerate values. They usually hint to highly symmetric orientations of the magnetic field with respect to the Fermi surface. Let us start with the [111] direction. Here we have just two values. The lower one is the overall minimum of the effective mass curve. Therefore, it must describe the orbit with the smallest radius which is the effective mass corresponding to the symmetry axis. This is corroborated by looking at Fig. 3.6. We see that a plane perpendicular to the [111] direction, that is, the space diagonal, will cut the Fermi energy perpendicular to the rotational axis creating a circular orbit with smallest radius. As a small radius implies a small effective mass, the value must then belong to the transversal effective mass m_t. From Fig. 3.7, we get $m_t = 0.0082\, m_e$.

In $[1\bar{1}0]$ direction, we have again only two values, where one is the maximum effective mass. Looking again at Fig. 3.6, we see that the orbit with the largest average radius is obtained by an intersection of the Fermi energy parallel to the symmetry axis. This happens for a plane containing two adjacent space diagonals and that is indeed perpendicular to the $[1\bar{1}0]$ direction. So, the value must describe the largest ellipse. Because this ellipse contains the semimajor and semiminor axis, the effective mass is the geometric average between the transversal and longitudinal mass:

$$m^\star = \sqrt{m_t m_l}. \tag{3.136}$$

From the measurement, we get $m^\star = 0.36\, m_e$. With the result for m_t, we finally obtain the value of the longitudinal mass: $m_l = 1.58\, m_e$.

3.8.5 Plasmons in germanium

(a) We write down the frequency dependence of the real part of the dielectric function according to eq. (3.5):

$$\varepsilon_1(\omega) = 1 - \frac{Ne^2}{\varepsilon_0 m} \frac{1}{\omega^2 + \gamma^2}. \tag{3.137}$$

For $\omega_0 \ll \omega_p$ and $\gamma \ll \omega_p$ and setting $\varepsilon_1(\omega = \omega_p) = 0$, we obtain:

$$0 = \varepsilon_1(\omega = \omega_p) = 1 - \frac{Ne^2}{\varepsilon_0 m} \frac{1}{\omega_p^2 + \gamma^2} \overset{\gamma \cong 0}{=} 1 - \frac{Ne^2}{\varepsilon_0 m} \frac{1}{\omega_p^2}. \tag{3.138}$$

This yields the plasma frequency:

$$\omega_p^2 = \frac{Ne^2}{\varepsilon_0 m} \tag{3.139}$$

Germanium has diamond structure, which comprises $n_{uc} = 8$ atoms per cubic unit cell. Furthermore, germanium has $n_v = 4$ valence electrons. The lattice constant is $a = 5.65$ Å. From this, we can calculate the number of valence electrons per volume:

$$N_e^V = \frac{n_v n_{uc}}{a^3} = \frac{8 \times 4}{(5.65 \times 10^{-8}\text{cm})^3} = 1.774 \times 10^{23}\text{cm}^{-3}. \tag{3.140}$$

Assuming an effective mass equal to the free electron mass $m^\star = m_e$, we get for the plasma energy according to eq. (3.139):

$$\hbar\omega_p = \hbar\sqrt{\frac{N_e^V e^2}{\varepsilon_0 m_e}} = e\hbar\sqrt{\frac{N_e^V}{\varepsilon_0 m_e}} = 15.65\,\text{eV} \mathrel{\hat{=}} 79.2\,\text{nm}, \tag{3.141}$$

that is, the plasmon energy corresponds to a wavelength of 792 Å, in good agreement with the measured reflectivity spectrum. The large value for ω_p justifies in retrospect our assumption that $\omega_0 \ll \omega_p$ because the lowest transition in germanium is $\hbar\omega_0 \cong 1.1$ eV $\ll \hbar\omega_p$.

(b) Doping germanium will increase the number of thermally excited charge carriers per volume by several orders of magnitude. A doping of 0.01% free electrons per germanium atom amounts to a relative electron concentration of $n_{conc} = 10^{-4}$. Assuming an effective mass $m^\star = 0.15\, m_e$, we can calculate from the density of germanium atoms, see eq. (3.140), the electron concentration:

$$N_e^C = n_{conc}\frac{n_{uc}}{a^3} = \frac{10^{-4} \times 8}{(5.65 \times 10^{-8}\text{cm})^3} = 4.44 \times 10^{18}\text{cm}^{-3} \tag{3.142}$$

From this, we can calculate the plasma frequency for doped germanium according to eq. (3.141):

$$\hbar\omega_p = \hbar\sqrt{\frac{N_e^C e^2}{\varepsilon_0 m^\star}} = e\hbar\sqrt{\frac{N_e^C}{\varepsilon_0\, 0.15\, m_e}} = 0.2\text{ eV} \mathrel{\hat{=}} 6.2\,\mu\text{m}. \tag{3.143}$$

The plasma frequency of the free carriers in doped Ge is roughly two orders of magnitude smaller than the plasma frequency of the valence band.

3.8.6 Law of Hagen-Rubens and DC conductivity

To treat free electrons in the context of the Lorentz theory, we simply set in the equation of motion, eq. (3.1), the spring constant $f = 0$, or the resonance frequency $\omega_0 = 0$. This is equivalent to setting the restoring force equal to zero.

(a) The DC conductivity σ_0 is defined as the real part of the (diagonal) optical conductivity σ_1 defined in eq. (3.7), at zero frequency:

$$\sigma_0 = \sigma_1(\omega_0 = 0,\ \omega = 0) = \frac{Ne^2\gamma}{m}\left.\frac{1}{\omega^2+\gamma^2}\right|_{\omega=0} = \frac{Ne^2}{m\gamma}. \tag{3.144}$$

(b) To derive the law of Hagen-Rubens, we must express the reflectivity R at normal incidence as a function of σ_1. The complex optical conductivity $\tilde{\sigma}$ connects to the dielectric function $\tilde{\varepsilon}$ via eq. (2.36). Therefore, we must express R as a function of real ε_1 and imaginary part ε_2 of $\tilde{\varepsilon}$. The frequency dependence of ε_1 and ε_2 is given in eqs. (3.5) and (3.6). For free electrons, we simply set $\omega_0 = 0$:

$$\varepsilon_1(\omega_0 = 0) = 1 - \frac{Ne^2}{\varepsilon_0 m}\frac{1}{\omega^2+\gamma^2}, \tag{3.145}$$

$$\varepsilon_2(\omega_0 = 0) = \frac{Ne^2}{\varepsilon_0 m}\frac{\gamma}{\omega(\omega^2+\gamma^2)}. \tag{3.146}$$

For very small frequencies $\omega \ll \gamma$, the real part ε_1 reduces to a constant proportional to the DC conductivity σ_0:

$$\varepsilon_1(\omega_0=0)=1-\frac{Ne^2}{\varepsilon_0 m\gamma^2}\frac{1}{1+\frac{\omega^2}{\gamma^2}}\cong 1-\frac{Ne^2}{\varepsilon_0 m\gamma^2}\left(1-\frac{\omega^2}{\gamma^2}+\cdots\right)\cong -\frac{Ne^2}{\varepsilon_0 m\gamma^2}=-\frac{1}{\varepsilon_0\gamma}\sigma_0\cong \text{const.} \tag{3.147}$$

The imaginary part ε_2 goes to zero. It can be expressed with the help of the low-frequency limit of ε_1:

$$\begin{aligned}\varepsilon_2(\omega_0 = 0) &= \frac{Ne^2}{\varepsilon_0 m}\frac{\gamma}{\omega(\omega^2+\gamma^2)}\\ &= \frac{Ne^2}{\varepsilon_0 m\gamma\omega}\frac{1}{1+\frac{\omega^2}{\gamma^2}} \cong \frac{Ne^2}{\varepsilon_0 m\gamma\omega}\left(1-\frac{\omega^2}{\gamma^2}+\cdots\right) \cong \frac{Ne^2}{\varepsilon_0 m\gamma\omega} \cong -\frac{\gamma}{\omega}\varepsilon_1 \gg |\varepsilon_1|.\end{aligned} \tag{3.148}$$

For simplicity, we now just write ε_1 and ε_2 for the low-frequency limit $\varepsilon_1(\omega_0 = 0, \omega \ll \gamma)$ and $\varepsilon_2(\omega_0 = 0, \omega \ll \gamma)$, respectively. Next, we need to express the refractive index n and absorption index k by ε_1 and ε_2. We start with the well-known expression from eqs. (2.31) and (2.32), $\varepsilon_1 = n^2 - k^2$ and $\varepsilon_2 = 2nk$, and construct the reverse expressions by solving the second equation for k and entering the result in the first equation. This yields a biquadratic equation for n:

$$\varepsilon_1 = n^2 - k^2 = n^2 - \left(\frac{\varepsilon_2}{2n}\right)^2 \quad \Rightarrow \quad n^4 - \varepsilon_1 n^2 - \frac{\varepsilon_2^2}{4} = 0. \tag{3.149}$$

We solve the biquadratic equation and enter the approximations for ε_1 and ε_2 for $\omega \ll \gamma$, defined in eqs. (3.147) and (3.148), respectively. As a result, we obtain for n^2:

$$n^2 = \frac{1}{2}\left(\varepsilon_1 + \sqrt{\underbrace{\varepsilon_1^2}_{\ll \varepsilon_2^2} + \varepsilon_2^2}\right) \cong \frac{1}{2}(\varepsilon_1 + \varepsilon_2) \cong \frac{1}{2}\varepsilon_2 \cong -\frac{\gamma}{2\omega}\varepsilon_1 \cong \frac{1}{2\varepsilon_0 \omega}\sigma_0 \gg 1. \tag{3.150}$$

We see that in the limit $\omega \ll \gamma$ the square of the refractive index is proportional to $1/\omega$ and becomes very large. For k, we solve the equation $\varepsilon_2 = 2nk$ for n and enter the result in $\varepsilon_1 = n^2 - k^2$. This yields a biquadratic equation for k:

$$\varepsilon_1 = n^2 - k^2 = \left(\frac{\varepsilon_2}{2k}\right)^2 - k^2 \quad \Rightarrow \quad k^4 + \varepsilon_1 k^2 - \frac{\varepsilon_2^2}{4} = 0. \tag{3.151}$$

We solve the biquadratic equation and enter again the approximations for ε_1 and ε_2 for $\omega \ll \gamma$. Thus, we obtain for k^2:

$$k^2 = \frac{1}{2}\left(-\varepsilon_1 + \sqrt{\underbrace{\varepsilon_1^2}_{\ll \varepsilon_2^2} + \varepsilon_2^2}\right) \cong \frac{1}{2}(\varepsilon_2 - \varepsilon_1) \cong \frac{1}{2}\varepsilon_2 \cong n^2 \gg 1. \tag{3.152}$$

We see that k^2 becomes equal to n^2 and very large, too. Assuming $n^2 \cong k^2 \gg 1$, the reflectivity can be simplified in the limit $\omega \ll \gamma$:

$$R = \frac{(n-1)^2 + k^2}{(n+1)^2 + k^2} = \frac{(n+1)^2 + k^2 - 4n}{(n+1)^2 + k^2} = 1 - \frac{4n}{(n+1)^2 + k^2} \cong 1 - \frac{4n}{n^2 + k^2} \cong 1 - \frac{4n}{2n^2}. \tag{3.153}$$

Entering the approximation for n^2 from eq. (3.137), the law of Hagen-Rubens is obtained:

$$R \cong 1 - \frac{2}{n} \cong 1 - 2\sqrt{\frac{2\varepsilon_0 \omega}{\sigma_0}}, \tag{3.154}$$

It is interesting to note that if $\sigma_0 > 0$ then R always approaches 100% for $\omega \to 0$. Therefore, it can be determined whether a material is conducting or not simply by a reflectivity measurement in the far infrared. Furthermore, we see that by keeping ω fixed, the reflectivity is the larger the larger the DC conductivity is. The law of Hagen-Rubens therefore describes the typical behavior of conducting materials in the infrared.

3.8.7 Complex polar Kerr effect

(a) The polar Kerr effect is measured using linearly polarized light. We know that this polarization can be looked at as being composed of equal contributions of rcp and lcp light. Therefore, the polar Kerr rotation can be treated as a linear combination of a reflection of rcp and lcp light with the same amplitude E_0 and different reflection coefficients. Let us make a drawing of the circular polarizations at the interface just after reflection. We assume that the amplitude of the reflection coefficient for rcp light ρ_+ is larger than the amplitude for lcp ρ_-: $\rho_+ > \rho_-$. Furthermore, we assume that the phase of the reflection coefficient for rcp light ϕ_+ is larger than the absolute value of the negative phase for lcp $|\phi_-|$: $\phi_+ > |\phi_-|$ and $\phi_- < 0$. This is depicted in Fig. 3.10 for time $t = 0$. The electric field vectors of rcp and lcp light run in opposite directions on circles with different radius $\rho_\pm$. The starting point at $t = 0$ is defined by an angle equal to the phase shift $\phi_\pm$. Be aware that the circular polarizations change their helicity through reflection. So, we must draw the outgoing lcp light $\vec{E}'_-$ as the reflected rcp light and vice versa:

$$E'_\mp = \rho_\pm E_0. \tag{3.155}$$

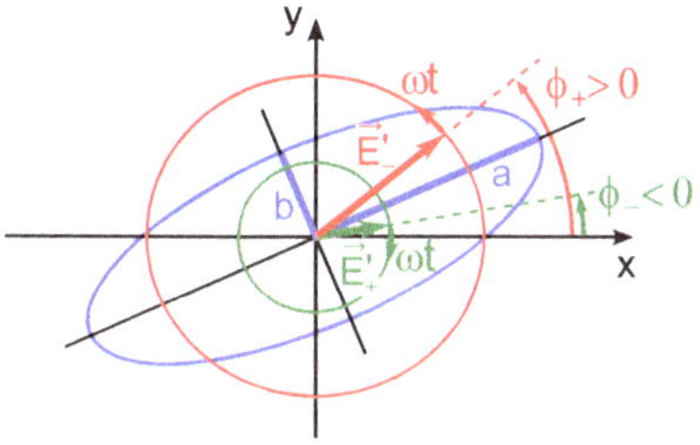

Fig. 3.10: Construction of the elliptical polarization of the reflected light in the polar Kerr effect. The incoming linearly polarized light is first decomposed into rcp and lcp light with equal amplitude. The different complex reflection coefficients lead then to an elliptical polarization. Note, that rcp light is reflected as lcp light and vice versa.

Now, let the time run and imagine how the electric field vectors for rcp and lcp light rotate in opposite directions. The outgoing light is the superposition of the two electric field vectors. The superposition reaches its maximum whenever the rcp field-vector $\vec{E}'_-$ and the lcp field-vector $\vec{E}'_+$ are parallel. This must be the direction of the semimajor axis of the elliptically polarized state after reflection because this is the direction of maximal electric field strength. The length of the semimajor axis a is then given by the sum of the rcp and lcp electric field vector.

$$a = (\rho_- + \rho_+)E_0. \tag{3.156}$$

The position, where the two electric field vectors become parallel, can be determined from the drawing. Traveling at the same angular velocity ω, they will meet at the angle bisector between the angles $\phi_\pm$ at $t=0$. Note, that for the reflected lcp light, the phase angle runs in the negative direction. So, a phase $\phi_- > 0$ represents mathematically a ***negative*** angle. Therefore, the angle bisector, which is indeed the polar Kerr rotation θ_K is obtained by averaging the mathematical angles, which is equivalent to taking the difference between ϕ_+ and ϕ_-:

$$\theta_K = -\frac{\phi_+ - \phi_-}{2}. \tag{3.157}$$

The minus sign assures the correct assignment of a positive polar Kerr rotation according to Faraday's sign convention: positive sign for clockwise rotation when looking into the reflected light and when the magnetic field is pointing in the direction of the incoming light, that is, into the material. Analogously, the direction and length of the semiminor axis can be found. It corresponds to the minimum electric field strength in elliptically polarized light and it is reached when the electric field vector $\vec{E}'_-$ of the rcp and $\vec{E}'_+$ of the lcp light are antiparallel to each other. This happens exactly a quarter turn (or 90°) after having met to align for the direction of the semimajor axis. The length b of the semiminor axis is then defined by the difference between the rcp and lcp electric field vector:

$$b = (\rho_+ - \rho_-)E_0. \tag{3.158}$$

The ration b/a defines, finally, the Kerr ellipticity of the reflected light:

$$\eta_K \cong \tan \eta_K = \frac{b}{a} = -\frac{\rho_+ - \rho_-}{\rho_+ + \rho_-}. \tag{3.159}$$

Note that the minus sign assures the correct assignment of a positive polar Kerr ellipticity according to Faraday's sign convention: positive sign for clockwise rotation when looking into the reflected light and when the magnetic field is pointing in the direction of the incoming light, that is, into the material.

(b) In this part of the problem, we want to proof eq. (3.90), the relation between the complex polar Kerr rotation $\tilde{\theta}_K = \theta_K - i\,\eta_K$ and the complex reflection coefficients $\tilde{\rho}_\pm$ in the circular basis:

$$\begin{aligned}\tilde{\theta}_K = \theta_K - i\,\eta_K &\overset{?}{\cong} i\frac{\tilde{\rho}_+ - \tilde{\rho}_-}{\tilde{\rho}_+ + \tilde{\rho}_-} = i\frac{\rho_+ e^{i\phi_+} - \rho_- e^{i\phi_-}}{\rho_+ e^{i\phi_+} + \rho_- e^{i\phi_-}}\\ &= i\frac{(\rho_+ e^{i\phi_+} - \rho_- e^{i\phi_-})(\rho_+ e^{-i\phi_+} + \rho_- e^{-i\phi_-})}{\rho_+^2 + \rho_-^2 + 2\rho_+\rho_- \cos(\phi_+ - \phi_-)}.\end{aligned} \tag{3.160}$$

Here, we have inserted the polar representation for $\tilde{\rho}_\pm$ and expanded with the complex conjugate of the denominator to obtain a real value for the denominator.

By expanding the brackets and applying the approximation $|\phi_+ - \phi_-| \ll 1$, we get rid of the sine and cosine function:

$$\tilde{\theta}_K = i\frac{\rho_+^2 - \rho_-^2 + i\,2\rho_+\rho_- \sin(\phi_+ - \phi_-)}{\rho_+^2 + \rho_-^2 + 2\rho_+\rho_- \cos(\phi_+ - \phi_-)} \simeq \frac{i\,(\rho_+^2 - \rho_-^2) - 2\rho_+\rho_-(\phi_+ - \phi_-)}{\rho_+^2 + \rho_-^2 + 2\rho_+\rho_-}. \tag{3.161}$$

The next simplification is achieved by taking the arithmetic and geometric mean of the amplitudes $\rho_\pm$, that is, we approximate $\rho_+\rho_- \cong \rho^2$ and $\rho_+ + \rho_- \cong 2\rho$, where ρ is the reflection coefficient in the absence of magnetism:

$$\tilde{\theta}_K \cong -\frac{2\overbrace{\rho_+\rho_-}^{\cong\rho^2}(\phi_+ - \phi_-)}{\underbrace{(\rho_+ + \rho_-)^2}_{\cong 4\rho^2}} + i\frac{\rho_+^2 - \rho_-^2}{(\rho_+ + \rho_-)^2} \cong -\frac{\phi_+ - \phi_-}{2} + i\frac{(\rho_+ - \rho_-)(\rho_+ + \rho_-)}{(\rho_+ + \rho_-)^2}. \tag{3.162}$$

Reducing the equation finally yields the complex polar Kerr effect as derived from the geometric arguments in part (a) of this problem:

$$\tilde{\theta}_K \cong \underbrace{-\frac{\phi_+ - \phi_-}{2}}_{\theta_K} + \underbrace{i\frac{\rho_+ - \rho_-}{\rho_+ + \rho_-}}_{-\eta_K}. \tag{3.163}$$

(c) In the last part of the problem, we will derive the relation between $\tilde{\theta}_K = \theta_K - i\,\eta_K$ and the complex off-diagonal optical conductivity $\tilde{\sigma}_{xy} = \sigma_{1xy} + i\,\sigma_{2xy}$. We start again with eq. (3.90) and replace $\tilde{\rho}_\pm$ with the help of the Fresnel formula of eq. (3.87):

$$\tilde{\theta}_K = \theta_K - i\,\eta_K \cong i\frac{\tilde{\rho}_+ - \tilde{\rho}_-}{\tilde{\rho}_+ + \tilde{\rho}_-} = i\frac{\frac{\tilde{n}_+ - 1}{\tilde{n}_+ + 1} - \frac{\tilde{n}_- - 1}{\tilde{n}_- + 1}}{\frac{\tilde{n}_+ - 1}{\tilde{n}_+ + 1} + \frac{\tilde{n}_- - 1}{\tilde{n}_- + 1}} = i\frac{(\tilde{n}_+ - 1)(\tilde{n}_- + 1) - (\tilde{n}_+ + 1)(\tilde{n}_- - 1)}{(\tilde{n}_+ - 1)(\tilde{n}_- + 1) + (\tilde{n}_+ + 1)(\tilde{n}_- - 1)}$$
$$= i\frac{\tilde{n}_+\tilde{n}_- + (\tilde{n}_+ - \tilde{n}_-) - 1 - [\tilde{n}_+\tilde{n}_- - (\tilde{n}_+ - \tilde{n}_-) - 1]}{\tilde{n}_+\tilde{n}_- + (\tilde{n}_+ - \tilde{n}_-) - 1 + [\tilde{n}_+\tilde{n}_- - (\tilde{n}_+ - \tilde{n}_-) - 1]} = i\frac{(\tilde{n}_+ - \tilde{n}_-)}{(\tilde{n}_+\tilde{n}_- - 1)}. \tag{3.164}$$

By expanding the fraction in order to get a difference of the square of $\tilde{n}_\pm$, and with the help of the arithmetic and geometric mean of $\tilde{n}_\pm$, that is, we approximate $(\tilde{n}_+ + \tilde{n}_-) \cong 2\tilde{n}$ and $\tilde{n}_+\tilde{n}_- \cong \tilde{n}^2$, where $\tilde{n}$ is the refractive index in the absence of magnetism, we get:

$$\tilde{\theta}_K = i\frac{(\tilde{n}_+ - \tilde{n}_-)(\tilde{n}_+ + \tilde{n}_-)}{(\underbrace{\tilde{n}_+\tilde{n}_-}_{\cong\tilde{n}^2} - 1)\underbrace{(\tilde{n}_+ + \tilde{n}_-)}_{\cong 2\tilde{n}}} = i\frac{\tilde{n}_+^2 - \tilde{n}_-^2}{2\tilde{n}\left(\tilde{n}^2 - 1\right)}. \tag{3.165}$$

Combining the relations between $\tilde{n}_\pm$, $\tilde{\varepsilon}_\pm$, $\tilde{\sigma}_\pm$, and $\tilde{\sigma}_{xy}$ according to eqs. (2.30), (3.36), (3.38), and (3.66), we obtain:

$$\tilde{\varepsilon}_{\pm} = \tilde{n}_{\pm}^2 = 1 - \mathrm{i}\frac{1}{\varepsilon_0 \omega}\tilde{\sigma}_{\pm} = 1 - \mathrm{i}\frac{1}{\varepsilon_0 \omega}\left(\tilde{\sigma}_{xx} \pm \mathrm{i}\,\tilde{\sigma}_{xy}\right). \tag{3.166}$$

This can now be inserted into eq. (3.165) for $\tilde{\theta}_K = \theta_K - \mathrm{i}\,\eta_K$:

$$\tilde{\theta}_K = \theta_K - \mathrm{i}\,\eta_K \cong \mathrm{i}\frac{\tilde{n}_+^2 - \tilde{n}_-^2}{2\tilde{n}\left(\tilde{n}^2 - 1\right)} = \mathrm{i}\frac{\overbrace{\tilde{\varepsilon}_+ - \tilde{\varepsilon}_-}^{= 2\frac{1}{\varepsilon_0 \omega}\tilde{\sigma}_{xy}}}{2\tilde{n}\underbrace{\left(\tilde{n}^2 - 1\right)}_{= \tilde{\varepsilon}_{xx} - 1}} = \mathrm{i}\frac{1}{\varepsilon_0 \omega}\frac{\tilde{\sigma}_{xy}}{\tilde{n}\underbrace{\left(\tilde{\varepsilon}_{xx} - 1\right)}_{= -\mathrm{i}\frac{1}{\varepsilon_0 \omega}\tilde{\sigma}_{xx}}} = -\frac{\tilde{\sigma}_{xy}}{\tilde{n}\,\tilde{\sigma}_{xx}}. \tag{3.167}$$

By expressing the denominator by $\tilde{n}$, as in the second term of eq. (3.167), and replacing it by a complex function $\tilde{n}\left(\tilde{n}^2 - 1\right) = A - \mathrm{i}\,B$ according to eq. (3.79), we get:

$$\tilde{\theta}_K = \theta_K - \mathrm{i}\,\eta_K \cong \mathrm{i}\frac{1}{\varepsilon_0 \omega}\frac{\tilde{\sigma}_{xy}}{\tilde{n}\left(\tilde{n}^2 - 1\right)} = \mathrm{i}\frac{1}{\varepsilon_0 \omega}\frac{\sigma_{1xy} + \mathrm{i}\,\sigma_{2xy}}{A - \mathrm{i}\,B}, \tag{3.168}$$

Finally, we can separate the equation into real and imaginary part to obtain expressions for θ_K and η_K:

$$\tilde{\theta}_K = \theta_K - \mathrm{i}\,\eta_K \cong -\frac{1}{\varepsilon_0 \omega}\left[\frac{B\,\sigma_{1xy} + A\sigma_{2xy}}{A^2 + B^2} - \mathrm{i}\frac{A\,\sigma_{1xy} - B\sigma_{2xy}}{A^2 + B^2}\right]. \tag{3.169}$$

From $\tilde{n}\left(\tilde{n}^2 - 1\right) = A - \mathrm{i}\,B$ follows by expanding and separating into real and imaginary part:

$$A = n^3 - 3nk^2 - n, \tag{3.170}$$

$$B = -k^3 + 3n^2k - k. \tag{3.171}$$

4 Experimental aspects of magneto-optics

4.1 Jones matrix theory

In magneto-optics, it is often absolutely not intuitive how a polarization state changes when the light ray passes an optical element. It would be advantageous if one could describe the polarization state of light after passing through an optical element or a complete experimental setup by some mathematical formalism. For that purpose, matrix methods have been developed in which each optical element that is placed in the beam path is described by a matrix, including the sample itself. We want to introduce in the following the simplest formalism, a 2 × 2 matrix description invented by Robert Clark Jones in 1941 (92) which is valid for fully polarized light. For unpolarized or partially polarized light, there exists a similar more elaborate formalism based on four-dimensional Stokes vectors and 4 × 4 Mueller matrices.

In eq. (2.123), we represented elliptically polarized light as follows:

$$\vec{E} = \frac{1}{\sqrt{2}} E_0 \left(\hat{e}_x + e^{i\delta}\,\hat{e}_y\right) e^{i\,(\omega t - \tilde{g} z)}, \tag{4.1}$$

assuming a propagation direction $\vec{\tilde{g}}||\hat{e}_z$. Let us now write eq. (4.1) in vector form by separating the phase δ in δ_x and δ_y, where

$$\delta = \delta_y - \delta_x, \tag{4.2}$$

and setting

$$v = \omega t - \tilde{g} z. \tag{4.3}$$

Here, we have neglected an overall phase $-\delta_x$ in phase v because it will cancel out when we compute the light intensity. This way, we get the ***two-dimensional complex Jones vector*** describing the elliptical polarization state from eq. (4.1):

$$\vec{E} = \frac{1}{\sqrt{2}} E_0\, e^{iv} \begin{pmatrix} e^{i\,\delta_x} \\ e^{i\,\delta_y} \end{pmatrix}. \tag{4.4}$$

As we know, this equation describes elliptically polarized light with a semimajor axis rotated by 45° or 135° with respect to the x-axis and with an ellipticity $\eta = \delta/2$. As we have discussed in the problem in Section 2.6.14, the ***most common Jones vector*** is formed by allowing in eq. (4.4) different amplitudes for the projections in x and y directions:

$$\vec{E} = e^{iv} \begin{pmatrix} E_x\, \mathrm{e}^{\mathrm{i}\,\delta_x} \\ E_y\, \mathrm{e}^{\mathrm{i}\,\delta_y} \end{pmatrix}. \tag{4.5}$$

https://doi.org/10.1515/9783110635300-004

In Fig. 4.1, we plot the quantities of elliptically polarized light for the most general case. As is obvious from eq. (4.5), E_x and E_y are the maximum amplitudes that the electric field vector reaches in x and y directions, respectively, when traveling around the ellipse. This can be visualized by plotting a rectangle with width $2E_x$ and height $2E_y$ that is tangent to the ellipse. What cannot be visualized is the phase shift δ between the x and y components. On the other hand, the ellipse can be unequivocally described by two angles as we know from the magneto-optic effects. The first angle, ψ, describes the direction of the semimajor axis with respect to the x-axis. The second angle, η, describes the ellipticity. This can be done in two ways. First, η is the arc tangent of the ratio of semiminor b and semimajor axis a. The second description is more geometrical. For this, we draw an outer rectangle around the ellipse that is aligned with the semiaxes and, consequently, has length $2a$ and width $2b$. Then, η is the angle that the diagonal of the rectangle forms with the semimajor axis. Both sets can describe the elliptical polarization state completely and independently. Nevertheless, there is a difference. The set involving the two angles can only define the form and orientation but not the size of the ellipse whereas the other set can describe the size as well. Between the two sets the following relations hold:

$$\tan(2\psi) = 2\cos\delta \frac{E_x E_y}{E_x^2 - E_y^2}, \tag{4.6}$$

$$\tan(2\eta) = 2\sin\delta \frac{E_x E_y}{E_x^2 + E_y^2}, \tag{4.7}$$

with $\delta = \delta_y - \delta_x$. The derivation of the relations is not straightforward and takes some time. There exist, of course, inverse relations, too. They are derived in the solutions to the problem in Section 2.6.14.

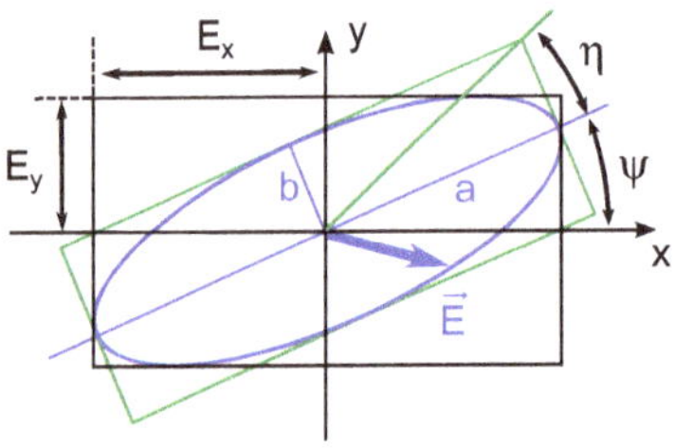

Fig. 4.1: Definition of the different quantities of elliptically polarized light.

From eq. (4.5), all states of polarization arise as special cases. If only vectors normalized to 1 are considered and all phase and intensity factors left unconsidered, we obtain simplified Jones vector representations for the various polarization states:

$$\begin{pmatrix} \cos\psi \\ \sin\psi \end{pmatrix} \quad \text{linearly polarized with azimuth } \psi, \tag{4.8}$$

$$\frac{1}{\sqrt{2}}\begin{pmatrix}1\\ i\end{pmatrix} \quad \text{rcp,} \tag{4.9}$$

$$\frac{1}{\sqrt{2}}\begin{pmatrix}1\\ -i\end{pmatrix} \quad \text{lcp.} \tag{4.10}$$

This is the Jones vector commonly used to describe polarization states. So, from now on we call them just "Jones vector" and leave out the specification "simplified." The disadvantage of this simplified representation is the partial loss of phase information. The reason is that the complex components of the Jones vector affect the phase term v in the exponential phase factor of the wave. To make this clear, let us make an example: If one had chosen a negative time dependence $e^{-i\omega t}$ in place of $e^{+i\omega t}$, the simplified Jones vector for rcp light would look like the one for lcp light with positive time dependence. So, just from looking at the Jones vector, we are not able to tell for sure which circular polarization is meant unless we know the presupposed time dependence of the wave. This yields unresolvable misunderstandings between the ***optical*** representation of polarization states, using a ***positive*** time dependence for the wave, and the ***theoretical*** one, utilized in ***quantum mechanics*** as well, that uses a ***negative*** time dependence. This is also the reason why in quantum mechanics rcp light, which is still defined according to eq. (4.9), has a ***negative*** helicity because it represents in reality lcp light looking like a left-handed helix in space! We resume this problem in Section 5.4, where we make the connection between the experimental measurement and quantum mechanics.

The assignment of a simplified Jones vector to a polarization state is thus unique except for the problem regarding the initial choice of the phase

$$v = \omega t - \vec{g} \cdot \vec{r} \tag{4.11}$$

in the exponential term discussed before. Referring now to the passage of polarized light through optical elements, the wave is described before and after the transit by a simplified Jones vector. Thus, the effect of an optical element can be represented by a **complex 2 × 2 Jones matrix *M***. So, we get:

$$\vec{E}' = \mathbf{M}\,\vec{E}, \tag{4.12}$$

where $\boldsymbol{M}$ is the complex 2×2 Jones matrix representing the action of the optical element on the electric field vector and $\vec{E}$ and $\vec{E}'$ are the two-dimensional electric field vectors before and after the optical element, respectively. The electric field vectors are always defined in a plane perpendicular to the propagation direction. Note that the Jones formalism does not give any information about the propagation direction. It is always assumed that it lies along the z-axis. We will now specify the matrix description of the most common optical elements.

4.1.1 Polarizer

A polarizer decomposes the incident light into two orthogonal components whereof one is transmitted while the other is (ideally) completely blocked. Corresponding to orthogonal pairs of linearly, circularly, and elliptically polarized light, there are linear, circular, and elliptical polarizers. Some examples of orthogonal pairs of polarized light are given in Fig. 4.2.

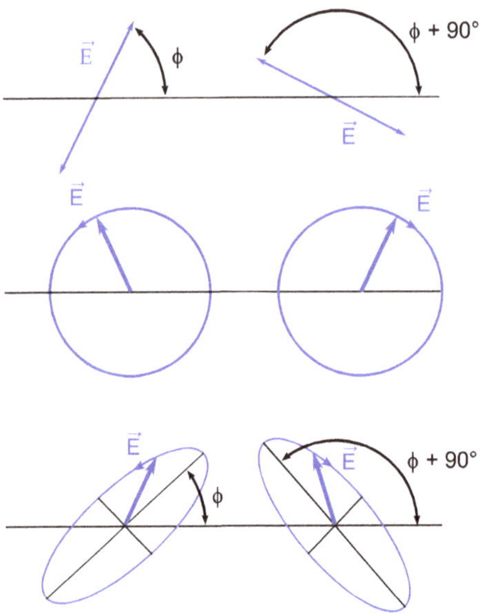

Fig. 4.2: Orthogonal pairs of (a) linearly, (b) circularly, and (c) elliptically polarized light.

The most common is the linear polarizer. We consider an incoming linearly polarized wave represented by eq. (4.5) with $\delta_x = \delta_y = 0$. A linear polarizer (assumed to be ideal) with the polarization axis at an angle ϕ with respect to the x-axis will transmit the projection of the electric field vector $\vec{E}$ in the direction ϕ of the polarizer whereas the perpendicular component is blocked. The projection of $\vec{E}$ is easily performed geometrically via the two components E_x and E_y, as explained in Fig. 4.3.

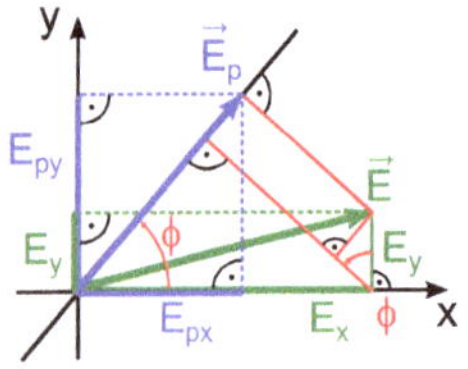

Fig. 4.3: Construction of the projection of the electric field vector onto the polarization axis, positioned at an angle ϕ with respect to the x axis.

The projection, E_p, of the electric field vector onto the polarization axis is the sum of the two projections of the components E_x and E_y of the electric field vector:

$$E_p = E_x \cos\phi + E_y \sin\phi. \tag{4.13}$$

The projection back to the coordinate axes is calculated in a similar way by projecting E_p back to the x and y axes yielding the transmitted components E_{px} and E_{py}, respectively. This finally yields the transmitted component in x direction,

$$E_{px} = \cos\phi\, E_p = \cos\phi\,[E_x \cos\phi + E_y \sin\phi], \tag{4.14}$$

and in y direction,

$$E_{py} = \sin\phi\, E_p = \sin\phi\,[E_x \cos\phi + E_y \sin\phi]. \tag{4.15}$$

From the prefactors to the vector components in eqs. (4.14) and (4.15), we obtain the Jones matrix $\boldsymbol{P}$ for a linear polarizer with azimuth ϕ with respect to the x-axis:

$$\mathbf{P} = \begin{pmatrix} \cos^2\phi & \sin\phi\cos\phi \\ \sin\phi\cos\phi & \sin^2\phi \end{pmatrix}. \tag{4.16}$$

As an example, let the incoming electric field vector be $\vec{E} = E_0(1, 0)$ and $\phi = 0$. Then we get $\mathbf{P}\vec{E} = \vec{E}$. If we choose $\phi = 90°$, we will get $\mathbf{P}\vec{E} = 0$, as expected. Cheap linear polarizers are built from highly anisotropic polymer foils containing long polymer chains that are aligned in one direction. The absorption of a polymer chain is highly anisotropic due to its linear shape. Much better linear polarizers are obtained by means of total internal reflection. Here, a birefringent transparent crystal is cut in such a way that for two orthogonal polarization directions, one direction is total internally reflected at the interface, whereas the other is fully transmitted. The disadvantage of these polarizers is that complete polarization is obtained only with light rays having a very small opening angle (around 10°, depending on the wavelength). As another example, a circular polarizer is treated in the problem in Section 4.5.1.

4.1.2 Phase shifter

A phase shifter, also called birefringent plate or compensator, is defined as an optical element that splits a polarized light beam into two orthogonal components and generates between them a retardation Δ due to its birefringent nature, that is, different index of refraction for two orthogonal directions. The retardation leads to a phase shift between the two components. The two polarization directions are called the eigenvectors of the element. The direction belonging to the smaller index of refraction n_f is called fast axis, the one belonging to the larger index n_s slow axis referring to the different speed of light. Depending on the orthogonal polarization states, there are linear, circular, and elliptical phase shifters. A circular phase shifter, also called rotator, we have already met because the Faraday rotation results from a phase shift between rcp and lcp light. We will specify the respective matrix in the next section.

The linear phase shifter, which we will discuss here, is necessary to measure ellipticity, as we shall see later. A linear phase shifter with an 180° or π phase shift is called half-wave plate or $\lambda/2$ plate. With 90° ($\pi/2$), it is called quarter-wave plate or $\lambda/4$ plate. A phase shifter with continuously adjustable retardation is called ***Soleil–Babinet compensator***. Its layout is shown in Fig. 4.4.

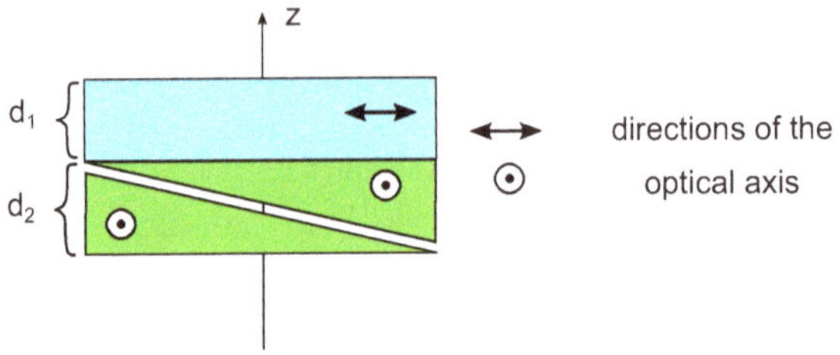

Fig. 4.4: Longitudinal section through a Soleil–Babinet compensator.

The effect of the compensator is based on the natural linear birefringence of the material used. Depending on the wavelength of light, this is crystalline quartz (SiO_2), calcite ($CaCO_3$), or magnesium fluoride (MgF_2). The natural birefringence is based on the anisotropy of the crystal structure inducing different index of refraction along different crystal axes. Hexagonal and tetragonal structures have a distinguished direction, called the optical axis, and, consequently, two refractive indices, namely parallel and perpendicular to the optical axis (ordinary and extraordinary refractive index, respectively). Besides the above-mentioned materials, other optically uniaxial crystals are corundum (Al_2O_3) and rutile (TiO_2). Crystals with rhombic, monoclinic, or triclinic symmetry have two optical axes and, therefore, three different refractive indices. Among this group are mica and sugar.

In the arrangement shown in Fig. 4.4, the optical axis is in the plate and in the wedges perpendicular to the z-axis but rotated by 90° with respect to each other. Consider now light incident along the z-axis that is linearly polarized at an angle of 45° with respect to the optical axes. The fast component in the plate will then be the slow one in the wedges and vice versa. This generates a phase difference Δ between the two components of the incident light by passing through the device:

$$\Delta = \frac{2\pi}{\lambda}(n_s - n_f)\,(d_2 - d_1) = \frac{2\pi}{\lambda}\delta n\,(d_2 - d_1), \tag{4.17}$$

where $\delta n = (n_s - n_f)$ is the difference between the refractive index n_s and n_f along the slow and the fast axes, respectively, d_1 is the fixed thickness of the plate, and d_2 is the variable thickness obtained by sliding the two wedges along each other.

To determine the associated Jones matrix, we consider a compensator with an angle ϕ between the fast axis and the x-axis and a retardation Δ. We assume again that the incoming light is linearly polarized and is represented by eq. (4.5) with $\delta_x = \delta_y = 0$. The situation is sketched in Fig. 4.5.

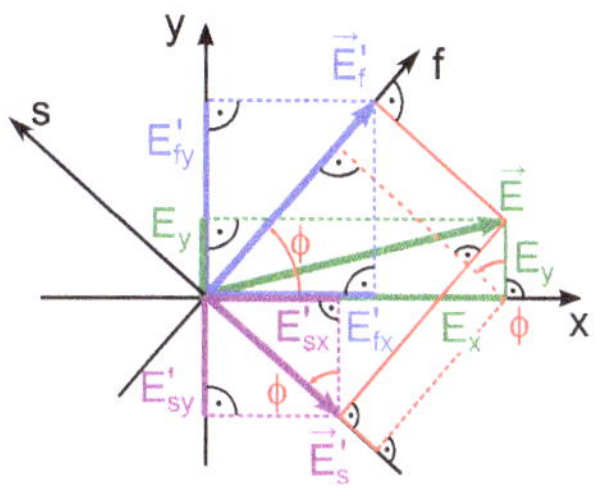

Fig. 4.5: Construction of the projection of the electric field vector onto the fast (f) and the slow (s) axis of the compensator, positioned at an angle ϕ with respect to the x-axis.

The components E_f and E_s of the linearly polarized incident light along, respectively, the fast and slow axes are:

$$E_f = \cos\phi\, E_x + \sin\phi\, E_y, \tag{4.18}$$

$$E_s = -\sin\phi\, E_x + \cos\phi\, E_y. \tag{4.19}$$

After passing the compensator, the two components are out of phase:

$$E_f' = e^{i\Delta} E_f = e^{i\Delta} \cos\phi\, E_x + e^{i\Delta} \sin\phi\, E_y, \tag{4.20}$$

$$E_s' = E_s = -\sin\phi\, E_x + \cos\phi\, E_y. \tag{4.21}$$

Note that E_s and E_s' point in the negative direction of the s-axis. By projecting back E_f' and E_s' to the x- and y-axes, we obtain the components E_{fx}' and E_{fy}',

$$E_{fx}' = \cos\phi\, E_f' = e^{i\Delta}\cos^2\phi\, E_x + e^{i\Delta} \cos\phi \sin\phi\, E_y, \tag{4.22}$$

$$E_{fy}' = \sin\phi\, E_f' = e^{i\Delta} \sin\phi \cos\phi\, E_x + e^{i\Delta}\sin^2\phi\, E_y, \tag{4.23}$$

and E_{sx}' and E_{sy}',

$$E_{sx}' = -\sin\phi\, E_s' = \sin^2\phi\, E_x - \sin\phi \cos\phi\, E_y, \tag{4.24}$$

$$E_{sy}' = \cos\phi\, E_s' = -\cos\phi \sin\phi\, E_x + \cos^2\phi\, E_y. \tag{4.25}$$

Note the minus sign in the projection of E_s' onto the x-axis. It must be added because $E_s' < 0$, as it is a component in the negative s-axis direction while we have $E_{sx}' > 0$, as it is a component in the positive x-axis direction. Finally, the phase-shifted vector $\vec{E}'$ is obtained by adding the two vectors on the local axis of the compensator:

$$\vec{E}' = \vec{E}_f' + \vec{E}_s'. \tag{4.26}$$

The projections E_x' and E_y' of the phase-shifted vector $\vec{E}'$ we get by simply adding the components of the transformed fast and slow vector $\vec{E}_f'$ and $\vec{E}_s'$, respectively:

$$E_x' = E_{fx}' + E_{sx}' = \left[e^{i\Delta}\cos^2\phi + \sin^2\phi\right] E_x + \left[\left(e^{i\Delta} - 1\right)\sin\phi\cos\phi\right] E_y. \tag{4.27}$$

$$E_y' = E_{fy}' + E_{sy}' = \left[\left(e^{i\Delta} - 1\right)\sin\phi\cos\phi\right] E_x + \left[e^{i\Delta}\sin^2\phi + \cos^2\phi\right] E_y. \tag{4.28}$$

From them we can derive the Jones matrix $\boldsymbol{C}$ for a ***compensator*** with a retardation Δ and an angle ϕ between fast axis and x-axis:

$$\mathbf{C} = \begin{pmatrix} e^{i\Delta}\cos^2\phi + \sin^2\phi & \left(e^{i\Delta} - 1\right)\sin\phi\cos\phi \\ \left(e^{i\Delta} - 1\right)\sin\phi\cos\phi & e^{i\Delta}\sin^2\phi + \cos^2\phi \end{pmatrix}. \tag{4.29}$$

Let us again check the result by making a simple example. Assuming the incident light as E_0 (1, 1), a retardation $\Delta = \pi$ (half-wave plate), and $\phi = 45°$, we get $E_x' = -E_0$ and $E_y' = E_0$. This characterizes a rotation of the linear polarization by 90° as one would expect for a half-wave plate. If we set $\Delta = \pi/2$ (quarter-wave plate) and $\phi = 90°$, we obtain $E_x' = E_0$ and $E_y' = iE_0$. This pictures rcp light as we expect from a quarter-wave plate.

4.1.3 Rotator

As mentioned above and calculated in the simple phenomenological model in Section 2.4.2, a circular phase shifter is just a rotator. According to eq. (2.131), the Jones matrix, $\boldsymbol{R}$, of a ***rotator*** with retardation Δ between rcp and lcp states can be described as a rotational matrix with rotation angle $\delta = \Delta/2$:

$$\mathbf{R} = \begin{pmatrix} \cos\frac{\Delta}{2} & -\sin\frac{\Delta}{2} \\ \sin\frac{\Delta}{2} & \cos\frac{\Delta}{2} \end{pmatrix}. \tag{4.30}$$

4.1.4 Jones matrix of a magneto-optically active sample

In the previous section, we have got to know the Jones matrix of a pure rotator. However, a sample showing Faraday effect does not just rotate the plane of polarization but, additionally, changes the polarization of the light to elliptical. As we have seen in eqs. (2.135) and (2.136), this effect can be described by a complex rotation angle, $\frac{1}{2}\tilde{\delta} = \frac{\omega d}{2c}(\tilde{n}_+ - \tilde{n}_-) = \theta_F - i\eta_F$, where d is the distance traversed by the light in the sample. Thus, the Jones matrix, $\boldsymbol{F}$, describing the ***complex Faraday effect*** of light with energy $\hbar\omega$ after traversing a distance d in an absorbing material is defined as follows (in accordance with the sign convention for a magnetic field pointing in the positive z direction):

$$\mathbf{F} = \begin{pmatrix} \cos\tilde{\theta}_F & -\sin\tilde{\theta}_F \\ \sin\tilde{\theta}_F & \cos\tilde{\theta}_F \end{pmatrix} = \begin{pmatrix} \cos(\theta_F - i\eta_F) & -\sin(\theta_F - i\eta_F) \\ \sin(\theta_F - i\eta_F) & \cos(\theta_F - i\eta_F) \end{pmatrix}. \tag{4.31}$$

It is also possible to define a Jones matrix for a sample showing complex polar Kerr effect. To characterize the complex polar Kerr effect, we must remember that it is essentially nothing else than a reflection from the surface of a sample showing magnetic circular birefringence that manifests itself as a splitting of the complex refractive index $\tilde{n}_\pm$. Therefore, the starting point is the Fresnel formula at normal incidence, eq. (3.71). In a circular basis, the Jones matrix $\boldsymbol{K}^\pm$ for the reflection is diagonal with elements $\tilde{\rho}_\pm$. In addition, we must consider that rcp light becomes lcp after reflection and vice versa. Finally, we must respect the correct sign convention for a magnetic field pointing in positive z direction, which means here that it points out of the sample surface. This leads to an inverting of the magnetic circular birefringence that we assumed in our phenomenological model to explain the Faraday rotation in chapter 2.4.2. Consequently, $\tilde{n}_+$ and $\tilde{n}_-$ swap and so do $\tilde{\rho}_+$ and $\tilde{\rho}_-$. This simply compensates the swap induced by the reflection and we get the following diagonal matrix:

$$\mathbf{K}^\pm = \begin{pmatrix} \tilde{\rho}_+ & 0 \\ 0 & \tilde{\rho}_- \end{pmatrix}. \tag{4.32}$$

This matrix must then be transformed back into the xy coordinate system according to eq. (3.28) using the transformation matrix $\boldsymbol{T}$ of eq. (3.29), neglecting the z dimension:

$$\mathbf{K} = \mathbf{T}\,\mathbf{K}^\pm\mathbf{T}^{-1}. \tag{4.33}$$

As a result, a reflection from a sample with a magnetic circular birefringence can be represented in xy base by the following matrix:

$$\mathbf{K} = \frac{1}{2}(\tilde{\rho}_+ + \tilde{\rho}_-)\begin{pmatrix} 1 & -i\frac{\tilde{\rho}_+ - \tilde{\rho}_-}{\tilde{\rho}_+ + \tilde{\rho}_-} \\ i\frac{\tilde{\rho}_+ - \tilde{\rho}_-}{\tilde{\rho}_+ + \tilde{\rho}_-} & 1 \end{pmatrix}. \tag{4.34}$$

Substituting first eq. (3.75) and then eq. (3.74) and replacing $\frac{1}{2}(\tilde{\rho}_+ + \tilde{\rho}_-)$ by the reflection coefficient $\tilde{\rho}$ in zero magnetic field, finally yields the Jones matrix $\boldsymbol{K}$ describing the ***complex polar Kerr effect***:

$$\mathbf{K} = \tilde{\rho}\begin{pmatrix} 1 & -\tilde{\theta}_K \\ \tilde{\theta}_K & 1 \end{pmatrix} = \tilde{\rho}\begin{pmatrix} 1 & -(\theta_K - i\eta_K) \\ \theta_K - i\eta_K & 1 \end{pmatrix}, \tag{4.35}$$

where we have assumed small angles θ_K and η_K. We note that this is for small angles again a rotational matrix (except for the prefactor), rotating in the same direction as the Faraday effect, eq. (4.31). This is not surprising at all, as the polar Kerr rotation also rotates the polarization of the incoming linearly polarized light.

4.1.5 Principle of Senarmont

To determine the complete polarization state of light, which is, in general, elliptically polarized, the location of the semiaxes with respect to the coordinate system as well as the ellipticity must be determined. Using a linear phase shifter, elliptically polarized light with ellipticity $\tan\eta \cong \eta$ can be converted into linearly polarized light, whose polarization plane is rotated by an angle η with respect to the semimajor axis of the incident light. This method is called the ***principle of Senarmont***, which will now be briefly discussed. As circular polarization is a special case of elliptical, this works for circular polarization as well.

Let the elliptical polarization be described by the angle Ψ between semimajor axis and x-axis and by the ellipticity η, as shown in Fig. 4.6. What distinguishes circular or elliptical polarization from linear polarization is a 90° or $\pi/2$ phase shift between two orthogonal directions. In case of elliptically polarized light, the orthogonal directions are the semiaxes. To convert elliptically or circularly polarized light to linearly polarized, the phase shift must be compensated. Therefore, the retardation of a linear phase shifter is set to $\Delta = -\pi/2$. Then, it is oriented in such a way that the slow axis coincides with the semimajor axis of the polarization ellipse, that is, at an angle Ψ with respect to the x-axis, as shown on the left of Fig. 4.6. When an elliptically polarized light wave passes the phase shifter, the semiminor axis is parallel to the fast axis and, therefore, will experience a phase shift of $-\pi/2$ relative to the semimajor axis thus canceling the phase shift of $\pi/2$ of the elliptical polarization. This produces linearly polarized light with the direction of polarization along the diagonal of the rectangle spanned by semimajor and semiminor axis, that is, along a straight line which is rotated by an angle η with respect to the semimajor axis. Hence, the linear polarization is oriented at a total angle $\eta + \Psi$ with respect to the x-axis as illustrated on the right of Fig. 4.6. Alternatively, the linear phase shifter can be aligned with its fast axis coinciding with the semimajor axis of the polarization ellipse and the retardation set to $\Delta = \pi/2$. This will yield the same result.

4.2 Experimental methods

We have now learnt enough about magneto-optics to be able to discuss how to determine in experiment the rotation θ and ellipticity η. Since the measurement of η can be reduced to an angular measurement using the principle of Senarmont, a distinction between η and θ is made only if η is determined by other means.

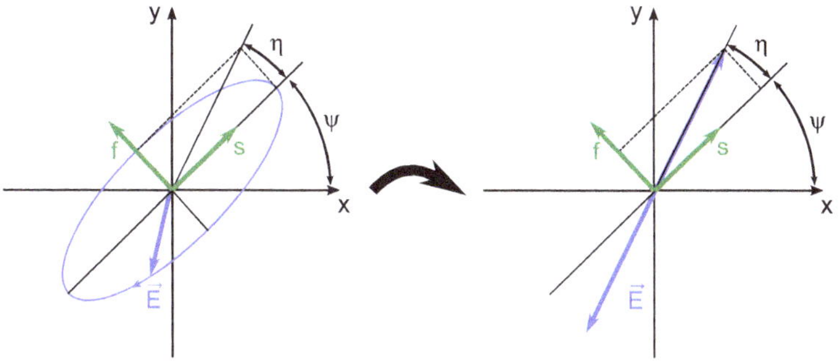

Fig. 4.6: Senarmont's principle. A phase shifter with retardation $\Delta = -\pi/2$ is aligned with its fast axis along the semimajor axis of the ellipse. The retardation compensates the phase shift between the semiminor and the semimajor axis, and the elliptical polarization transforms to linear.

The methods for measuring an angle θ can be divided into three groups: ***Angle-sensitive methods*** are defined as methods where θ is measured by rotating a polarizer for zeroing a signal. The angular position of the polarizer directly yields the angle θ. In ***phase-sensitive methods***, a phase difference, which is proportional to θ, is measured between two signals with a phase meter. Finally, there are ***intensity methods*** where the measured intensity of a signal is proportional to θ.

4.2.1 Angle-sensitive methods

The simplest method to measure the direction of polarization is to let light fall onto a linear polarizer and to look by eye at which position of the polarizer the transmitted intensity is minimal. Thus, a resolution of approximately 3° can be achieved. This method can be refined by using a photomultiplier to measure the intensity as a function of the polarizer position ϕ. According to Malus's law, we can fit the intensity in the vicinity of the minimum by a function $f = c \cdot \phi^2$. This way, we obtain a precision of roughly 0.5°.

Considerably more sensitive are methods in which the polarization of the light is modulated. The signal is then measured using a lock-in amplifier. This is a device that performs a Fourier expansion of the measured signal and considers only the first or second harmonic of the modulation frequency. Modern lock-in amplifiers can resolve even much higher harmonics or simultaneously measure two harmonics. With such a modulation method, an angular resolution of 0.001° can be achieved.

The most common polarization modulators are Faraday modulator and photoelastic modulator. In a Faraday modulator, as imaged in Fig. 4.7, the light passes through a transparent, strainless material (usually a glass) which is placed within an air-core coil. If an alternating current is applied to the coil, a sinusoidal magnetic

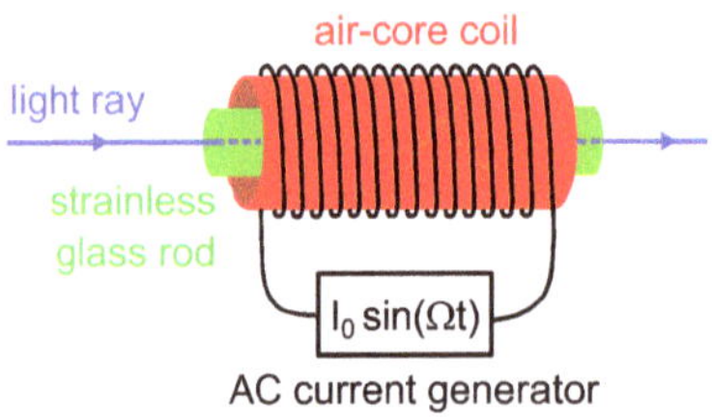

Fig. 4.7: Principle of a Faraday modulator. A transparent, strainless glass rod is positioned inside of an air-core coil. An AC current produces an oscillating magnetic field inside the coil. The magnetic field modulates the polarization of the traversing light by means of the Faraday effect in the glass rod.

field $B = B_0 \sin(\Omega t)$ is generated. Hence, the polarization gets a sinusoidal modulation as well, according to eq. (2.138):

$$\theta_F = VdB_0 \sin(\Omega t) = \theta_0 \sin(\Omega t). \tag{4.36}$$

The modulation just adds to the polarization state independent of its nature. So, for elliptically polarized light, the ellipse would oscillate around the zero-field orientation.

The other type of modulator is based on a piece of strainless quartz glass where a birefringence is induced by uniaxial mechanical stress. The stress is generated via the piezoelectric effect in a crystalline quartz that is attached to the quartz glass. Hence, the name photoelastic modulator (PEM). The operating principle is plotted in Fig. 4.8.

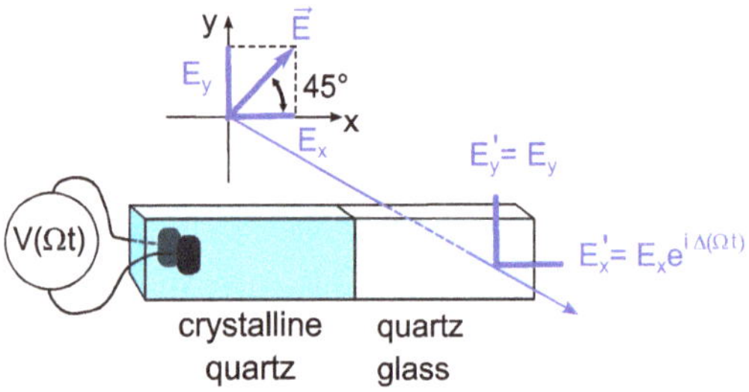

Fig. 4.8: Principle of a photoelastic modulator. A bar of crystalline quartz is fused to a bar of strainless quartz glass. The light passes through the quartz glass. By applying an oscillating high voltage to the crystalline quartz, an oscillating uniaxial mechanical stress is induced in the quartz glass leading to a small oscillating change of the refractive index in one direction. This generates an oscillating uniaxial birefringence in the quartz glass.

In the amorphous part of the quartz, the refractive index changes along the x direction proportional to the induced mechanical stress that results from the piezoelectric effect in crystalline quartz when applying an AC voltage $V = V_0 \sin(\Omega t)$. This generates a phase difference $\Delta = \Delta_0 \sin(\Omega t)$ between light that is linearly polarized in x and y directions. Hence, the modulator acts as a linear phase shifter with a time-dependent retardation. In contrast to the Faraday modulator, where only the polarization ***direction*** is changed, here the polarization ***state*** (ellipticity) of the light is modulated. In the following, we want to calculate an experimental setup involving a photoelastic modulator using the Jones matrix formalism.

Figure 4.9 shows the typical assembly for an angle-sensitive measurement using a Faraday modulator. Each wavelength must be measured point by point, which is time-consuming, but the angular resolution of 0.001° is excellent. As a broadband light source, a halogen lamp is normally used because it offers a very stable intensity and it is rather cheap. For measurements in the ultraviolet photon-energy region, a high-pressure xenon lamp can be used. It offers a bright intensity in the ultraviolet range but it is far less stable than a halogen lamp due to thermal fluctuations of the arc generating the light. Another disadvantage is the sharp atomic lines from xenon that are superimposed to the pressure-broadened background. By accidentally hitting such a line, the detector might go into overload.

For polarizing the beam, grating monochromators are used nowadays, where the available photon-energy range is selected by different gratings. Two decades back, prism monochromators were prevalent but they became too expensive. The disadvantage of a grating as opposed to a prism monochromator is that it transmits higher harmonics. Therefore, one must use low-pass filter glasses to suppress higher harmonics. As an alternative to the combination of lamp plus monochromator, tunable solid-state lasers can be used. However, they are expensive and the stability of the intensity is worse.

The monochromatic light passes a focusing lens and gets linearly polarized in a polarizer, which is usually a Glan–Thompson or a Glan prism polarizer. The latter is shorter and has no glue between the two prisms constituting the polarizer, which is advantageous for ultraviolet light.

In the next step, the ray is modulated by a Faraday modulator or a PEM. For the Faraday modulator, the usage of strainless glasses is very important. For a glass to be strainless, it must be cooled over a long period (several months). If the glass contains mechanical strain, it behaves not as a pure rotator but will produce elliptical polarization, thus lowering the resolution of the measurement. For modulation, heavy lead glasses, such as Schott SF57, are used in the infrared because of the high Verdet constant. Highly transparent quartz rods are used in the ultraviolet because of the low absorption.

The modulated light is then focused on the sample, which is placed in a strong magnet with central bore. With laboratory electromagnets, DC magnetic fields up to 2 Tesla can be reached, with superconducting magnets and Bitter magnets more than 10 Tesla. For temperature-dependent measurements, the sample is additionally put inside a cryostat.

After having passed the sample, the light is focused on a second polarizer, usually called analyzer. Finally, a detector will measure the transmitted intensity. Various detectors must be used for different photon-energy ranges. In the visible and ultraviolet, a photomultiplier tube is commonly used as it is the most sensitive detector. In the near-infrared and visible, a photodiode is best and for the mid-infrared, an InGaAs diode is most suitable.

The measured signal of the detector and the reference signal of the modulator are fed into a lock-in amplifier. Depending on the modulator used, either the first or the second harmonic of the modulation frequency is measured (see later).

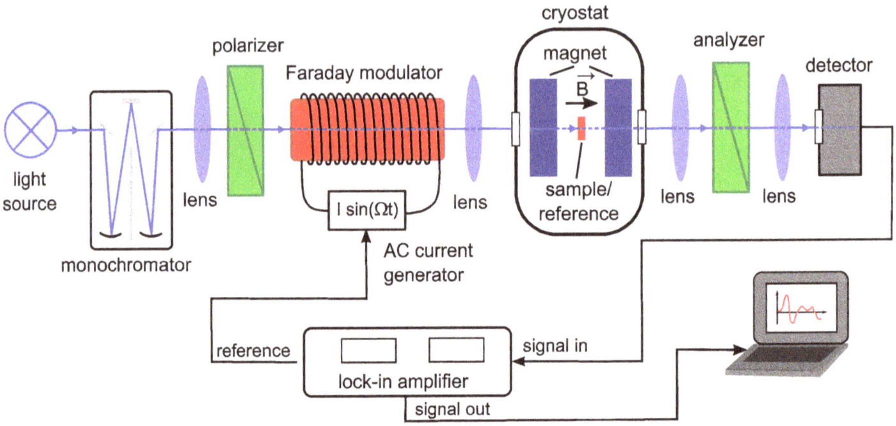

Fig. 4.9: Typical setup for an angle-sensitive method with modulator in Faraday geometry.

For our calculation, we replace the Faraday modulator by a photoelastic modulator. We choose the fast and slow axis of the PEM as the x and y coordinate axes, respectively. The light is assumed to be polarized at an angle of 45° with respect to these axes. After the polarizer, the light is, thus, described by the normalized Jones vector as:

$$\vec{E} = \frac{1}{\sqrt{2}} E_0 \begin{pmatrix} 1 \\ 1 \end{pmatrix}. \tag{4.37}$$

According to eq. (4.29), the PEM is described in our coordinate system by

$$\mathbf{M} = \begin{pmatrix} e^{i\Delta} & 0 \\ 0 & 1 \end{pmatrix} \tag{4.38}$$

with a time-dependent retardation $\Delta = \Delta_0 \sin(\Omega t)$. The electric-field vector of the light ray is expressed after the modulator as:

$$\vec{E}' = \mathbf{M}\vec{E} = \frac{1}{\sqrt{2}} E_0 \begin{pmatrix} e^{i\Delta} \\ 1 \end{pmatrix}. \tag{4.39}$$

According to eqs. (4.5) and (4.8)–(4.10), this vector describes for $\Delta = 0°$ and $\Delta = 180°$ linearly polarized light, for $\Delta = \pm 90°$ circularly polarized light, and for all other values of Δ elliptically polarized light.

We assume the magneto-optically active sample to be a pure rotator with no absorption in our model calculation. Hence, it will be described by

$$\mathbf{F} = \begin{pmatrix} \cos\theta_\mathrm{F} & -\sin\theta_\mathrm{F} \\ \sin\theta_\mathrm{F} & \cos\theta_\mathrm{F} \end{pmatrix} \tag{4.40}$$

according to eq. (4.31). After traversing the sample, the electric-field vector is:

$$\vec{E}'' = \mathbf{F}\vec{E}' = \frac{1}{\sqrt{2}} E_0 \begin{pmatrix} e^{i\Delta}\cos\theta_\mathrm{F} - \sin\theta_\mathrm{F} \\ e^{i\Delta}\sin\theta_\mathrm{F} + \cos\theta_\mathrm{F} \end{pmatrix}. \tag{4.41}$$

Finally, after passing the analyzer, which is a polarizer and thus expressed according to eq. (4.16) by

$$\mathbf{A} = \begin{pmatrix} \cos^2\phi & \sin\phi\cos\phi \\ \sin\phi\cos\phi & \sin^2\phi \end{pmatrix}, \tag{4.42}$$

the electric-field vector reaching the detector is written as:

$$\vec{E}''' = \mathbf{A}\vec{E}'' = \frac{1}{\sqrt{2}} E_0 \left[\cos\phi\left(e^{i\Delta}\cos\theta_\mathrm{F} - \sin\theta_\mathrm{F}\right) + \sin\phi\left(e^{i\Delta}\sin\theta_\mathrm{F} + \cos\theta_\mathrm{F}\right)\right] \begin{pmatrix} \cos\phi \\ \sin\phi \end{pmatrix}. \tag{4.43}$$

Here, we have excluded a constant factor from the vector. Before we proceed, it is important to simplify eq. (4.43). This will help to interpret the result and to reduce errors in the calculation. Bundling terms with and without phase factor and applying trigonometric formulas, the prefactor can be simplified by introducing combinations of angles:

$$\vec{E}''' = \frac{1}{\sqrt{2}} E_0 \left[e^{i\Delta}\cos(\phi - \theta_\mathrm{F}) + \sin(\phi - \theta_\mathrm{F})\right] \begin{pmatrix} \cos\phi \\ \sin\phi \end{pmatrix}, \tag{4.44}$$

The detector measures the intensity and not the electric field vector. According to eq. (2.46), the intensity is proportional to the time-averaged square of the ***real part*** of the electric-field vector. For simplicity, we consider in the following a reduced intensity by dividing the intensity by all the constant prefactors so that the reduced intensity is equal to the time-averaged square of the ***real part*** of the electric-field vector. We must then not just take the simplified Jones vector but the complete electric-field vector of the plane wave including the exponential term with the phase $v = \omega t - \tilde{g}z$ as in eq. (4.3); that is, the reduced intensity is $I = \left\langle \mathrm{Re}\left\{\vec{E}''' e^{i(\omega t - \tilde{g}z)}\right\} \cdot \mathrm{Re}\left\{\vec{E}''' e^{i(\omega t - \tilde{g}z)}\right\}\right\rangle_t$. Regarding time-averaging, we must distinguish two timescales: the fast timescale of the angular frequency of the light and the slow timescale of the polarization modulation. The detector is not able to resolve the fast timescale but it can resolve the slow one. Therefore, we must only average the time dependence of the phase factor of the plane wave. Because the prefactor in eq. (4.44) is complex, it must be multiplied with the complex phase factor before taking the real part. Neglecting space dependence, because it does not contribute to the time-averaging, we get for the real part of the electric field vector two terms, one proportional to the real part of the prefactor times $\cos(\omega t)$ and one proportional to the imaginary part of the prefactor times $\sin(\omega t)$. The reduced intensity then contains three terms, one proportional to $\cos^2(\omega t)$ and to $\sin^2(\omega t)$ and a cross term proportional to $\cos(\omega t)\cdot\sin(\omega t)$. When time averaging, the first two terms yield a factor ½ and the last term vanishes. The result is then one-half times the square of the absolute value of the prefactor. But this result is, up to a factor ½, equivalent to just calculating the reduced intensity as $I=\vec{E}'''\cdot\vec{E}'''^{*}$, where "*" indicates the complex conjugated value, and neglecting the phase $v=\omega t-\tilde{g}z$ and its time averaging completely. This calculation is, of course, much simpler and saves time-averaging. The missing factor ½ is taken care of by normalizing the complex Jones vector.

Now, we are ready to calculate the reduced intensity from eq. (4.44):

$$\begin{aligned} I=\vec{E}'''\cdot\vec{E}'''^{*} &= \frac{1}{2}E_0^2\left[e^{i\Delta}\cos(\phi-\theta_F)+\sin(\phi-\theta_F)\right] \\ &\quad \left[e^{-i\Delta}\cos(\phi-\theta_F)+\sin(\phi-\theta_F)\right]\left(\cos^2\phi+\sin^2\phi\right) \\ &= \frac{1}{2}E_0^2\left[\cos^2(\phi-\theta_F)+\sin(\phi-\theta_F)\cos(\phi-\theta_F)\left(e^{i\Delta}+e^{-i\Delta}\right)+\sin^2(\phi-\theta_F)\right] \\ &= \frac{1}{2}E_0^2\left[1+\sin[2(\phi-\theta_F)]\frac{1}{2}\left(e^{i\Delta}+e^{-i\Delta}\right)\right], \end{aligned}$$

$$I=\vec{E}'''\cdot\vec{E}'''^{*}=\frac{1}{2}E_0^2\{1+\cos\Delta\sin[2(\phi-\theta_F)]\}. \tag{4.45}$$

In eq. (4.45), there is a time-dependent phase $\Delta=\Delta_0\sin(\Omega t)$. This leads to an awkward nesting of trigonometric functions: $\cos\Delta=\cos[\Delta_0\sin(\Omega t)]$. This expression can no longer be described by a finite sum of trigonometric functions. But there exists a series expansion using Bessel functions:

$$\cos(\Delta_0\sin(\Omega t))=J_0(\Delta_0)+2\sum_{n=1}^{\infty}J_{2n}(\Delta_0)\cos(2n\Omega t), \tag{4.46}$$

where J_n is the Bessel function of order n. With this expansion, the reduced intensity is finally expressed up to second order as:

$$I=\frac{1}{2}E_0^2\{1+J_0(\Delta_0)\sin[2(\phi-\theta_F)]+2J_2(\Delta_0)\sin[2(\phi-\theta_F)]\cos(2\Omega t)+\cdots\}. \tag{4.47}$$

For maximum modulation, an amplitude $\Delta_0 = 180°$ is used because the Bessel function $J_2(\Delta_0)$ reaches a maximum for this value. If the lock-in amplifier is set to determine the second harmonic, it will only output the second term on the right of eq. (4.47) which is the 2Ω signal $E_0^2 J_2(\Delta_0)\sin[2(\phi - \theta_F)]\cos(\Omega t)$. Furthermore, we see that for $\phi = \theta_F$ the 2Ω signal equals zero and the DC component minimizes to $\frac{1}{2}I_0$. For zeroing procedure, the analyzer is rotated in the experiment until the 2Ω signal is zero. The position ϕ of the analyzer is then equal to θ_F. This way, we can determine the Faraday rotation θ_F with high accuracy. Of course, the lock-in amplifier still sees on its input the DC component $\frac{1}{2}E_0^2\{1 + J_0(\Delta_0)\sin[2(\phi - \theta_F)]\}$. We notice that the DC component on the detector is 50% of the total intensity. This load will ultimately limit the resolution because the instrument must determine the zero crossing relative to the total input signal. This is the major disadvantage of a modulation by a linear phase shifter. When using a Faraday modulator, the fraction of the DC light on the detector can be made very small. The advantage of a PEM is, however, the large modulation amplitude, extending from rcp over linearly polarized up to lcp. With a Faraday modulator, on the other hand, it is very difficult to achieve a high modulation amplitude in the near-infrared. A detailed discussion of the Faraday modulator is given in the problem in Section 4.5.2.

One question remains to be answered: We said in the beginning that the ellipticity can be transformed into an angle using the principle of Senarmont. Therefore, in our calculation, we neglected the ellipticity generated by the sample completely. But what happens if the sample has absorption and produces ellipticity? Do we still measure the correct Faraday rotation with our method if the polarization is no longer just rotated and linear but elliptical with the semimajor axis rotated with respect to the input polarization? This is indeed hard to judge without calculation. So, let us take in the reduced intensity, eq. (4.45), into account a complex Faraday angle, $\tilde{\theta}_F = \theta_F - i\eta_F$:

$$
\begin{aligned}
I = \vec{E}''' \cdot \vec{E}'''^{*} &= \frac{1}{2}E_0^2\left[e^{i\Delta}\cos(\phi - \theta_F + i\eta_F) + \sin(\phi - \theta_F + i\eta_F)\right] \\
&\qquad \left[e^{-i\Delta}\cos(\phi - \theta_F - i\eta_F) + \sin(\phi - \theta_F - i\eta_F)\right] \\
&= \frac{1}{2}E_0^2[\cos(\phi - \theta_F + i\eta_F)\cos(\phi - \theta_F - i\eta_F) + e^{i\Delta}\cos(\phi - \theta_F + i\eta_F)\sin(\phi - \theta_F - i\eta_F) \\
&\quad + e^{-i\Delta}\sin(\phi - \theta_F + i\eta_F)\cos(\phi - \theta_F - i\eta_F) + \sin(\phi - \theta_F + i\eta_F)\sin(\phi - \theta_F - i\eta_F)] \\
&= \frac{1}{2}E_0^2\left[\cos^2(\phi - \theta_F)\cosh^2\eta_F + \sin^2(\phi - \theta_F)\sinh^2\eta_F\right. \\
&\quad + e^{i\Delta}[\sin(\phi - \theta_F)\cos(\phi - \theta_F) - i\sinh\eta_F\cosh\eta_F] \\
&\quad + e^{-i\Delta}[\sin(\phi - \theta_F)\cos(\phi - \theta_F) + i\sinh\eta_F\cosh\eta_F] \\
&\quad \left.+ \sin^2(\phi - \theta_F)\cosh^2\eta_F + \cos^2(\phi - \theta_F)\sinh^2\eta_F\right]
\end{aligned}
$$

$$= \frac{1}{2} E_0^2 \left[\cosh^2\eta_F + \sinh^2\eta_F + \sin(\phi - \theta_F)\cos(\phi - \theta_F)\left[e^{i\Delta} + e^{-i\Delta}\right]\right.$$
$$\left. - i \sinh\eta_F \cosh\eta_F \left[e^{i\Delta} - e^{-i\Delta}\right]\right]$$
$$= \frac{1}{2} E_0^2 \left[\cosh^2\eta_F + \sinh^2\eta_F + \sin(2[\phi - \theta_F])\cos\Delta + \sinh(2\eta_F)\sin\Delta\right]$$

$$I = \vec{E}''' \cdot \vec{E}'''^{*} = \frac{1}{2} E_0^2 \left[\cosh^2\eta_F + \sinh^2\eta_F + \sin(2[\phi - \theta_F])\cos\Delta + \sinh(2\eta_F)\sin\Delta\right]. \tag{4.48}$$

We easily see from eq. (4.48) that it reduces to the non-absorbing case, eq. (4.45), if we set the ellipticity $\eta_F = 0$. We also see that the calculation becomes very wearisome by just changing one angle to a complex one. So, what is the effect of switching on absorption? We get a rather small change in the DC component but an additional modulation term proportional to $\sin\Delta = \sin[\Delta_0 \sin(\Omega t)]$. Similar to eq. (4.46), there exists another series expansion using Bessel functions:

$$\sin(\Delta_0 \sin(\Omega t)) = 2\sum_{n=0}^{\infty} J_{2n+1}(\Delta_0)\sin([2n+1]\,\Omega t), \tag{4.49}$$

where J_n is again the Bessel function of order n. The reduced intensity is then expressed up to second order as:

$$I = \tfrac{1}{2} E_0^2 \{\cosh^2\eta_F + \sinh^2\eta_F + J_0(\Delta_0)\sin(2[\phi - \theta_F]) + 2J_1(\Delta_0)\sinh(2\eta_F)\sin(\Omega t)$$
$$+ 2J_2(\Delta_0)\sin(2[\phi - \theta_F])\cos(2\Omega t) + \cdots\}. \tag{4.50}$$

Now, we can finally discuss the effect of the ellipticity on the measurement of the Faraday rotation θ_F which is now the angle of the semimajor axis with respect to the x-axis. The 2Ω component has not changed at all. Therefore, if the lock-in amplifier measures the second harmonic component, it will output the same signal as for the non-absorbing case with linearly polarized light. The only effect of the ellipticity is to increase the DC component because $\cosh^2\eta_F + \sinh^2\eta_F > 1$. In addition, the non-vanishing first harmonic component will increase the load on the detector as well. Hence, it will further reduce the resolution.

The setup for measuring the polar Kerr rotation is substantially the same except that the light is now reflected from the sample. To ensure polar geometry, the angle of incidence should be kept as small as possible (≤ 5°). If not, contributions will be measured that depend quadratically on the magnetic field and the light will be slightly depolarized lowering the angular resolution. A test of the correct sign of the polar Kerr effect can be easily done using a reference mirror, which does not exhibit a Kerr effect such as Au or Al, on a glass substrate. If one reflects from the mirror through the substrate, the reflected beam has encountered twice a Faraday rotation by the substrate. A positive polar Kerr rotation is then a rotation in the same direction as the rotation measured from the glass substrate.

In practice, the experiment is performed in the following way: (1) For Faraday geometry, a measurement without sample (for polar Kerr geometry: with a reference mirror, which does not exhibit a Kerr effect such as Au or Al) is made to determine the rotation of the apparatus. For this, the sample must be moved out of the magnetic field in Faraday geometry. In Kerr geometry, it will be replaced by the reference mirror, which is usually mounted on the same holder as the sample. The rotation of the apparatus consists primarily of contributions to the Faraday or polar Kerr rotation from lenses, cryostat windows, from the substrate, and other optical elements which are exposed to magnetic stray fields. If the cryostat windows are positioned within the magnet, the rotation can amount to several degrees at large magnetic fields. In this case, the input polarization should be adjusted by rotating it in the opposite direction by half the amount of the rotation. This ensures that the polarization is kept on the sample always along the same direction avoiding spurious effects due to sample anisotropies. (2) The Faraday or polar Kerr rotation is determined with the sample positioned in the beam path. (3) The previously determined rotation of the apparatus is subtracted to obtain the effective rotation of the sample. (4) In order to cancel out birefringence of non-magnetic origin (stress-induced birefringence, intrinsic anisotropy), which is virtually always present, a complete measurement, that is, steps (1)–(3), is successively performed in positive and negative magnetic field and the two values are subtracted and divided by two. Stress-induced birefringence can be present in thin film samples due to the deposition process or in any sample due to the fixation to the sample holder. Furthermore, any strain in optical elements such as lenses, polarizers, or modulator glasses will produce a rotation or ellipticity Thus, in total, for each wavelength or magnetic field, ***four*** zeroing measurements are needed to determine ***one single*** data point (i.e., the Faraday or polar Kerr rotation). Assuming two minutes per zeroing measurement to achieve a resolution of 0.001°, the measurement of one data point will last 8 min. Thus, this is a very slow measurement!

To determine the ellipticity, the principle of Senarmont is applied. For this purpose, a linear phase shifter with retardation $\Delta = -\pi/2$ must be introduced into the beam path. Since complex matrices are not necessarily commutative, it plays a crucial role where the phase shifter is placed in relation to the modulator. In case of a Faraday modulator, the phase shifter must be placed in the optical path ***before*** the modulator as can be shown by calculating the Jones matrices with appropriate computational effort. See the problem in Section 4.5.3 for a detailed calculation and discussion. Again, a reference measurement (without sample or with reference mirror) is made first to be able to align the principal axes of the phase shifter. Then introducing the sample in the beam path, the ellipticity of the sample is determined by means of the principle of Senarmont.

Therefore, to determine rotation and ellipticity of the sample, eight zeroing measurements must be carried out which takes, assuming again two minutes per zeroing measurement, a total of 16 min per point. To obtain a full spectrum (0.8–5.5 eV), around 100 data points are needed. Thus, the measurement of a spectrum takes longer than one full day. This shows that the highest standards must be met by the apparatus to perform high-resolution magneto-optic measurements. The measurement time can be shortened considerably (by a factor of ~2–3), when one measures without cryostat since then the influence of the apparatus is much smaller.

4.2.2 Phase-sensitive methods

For a phase-sensitive measurement, two phase-shifted beams must be measured simultaneously, a reference beam and a sample beam. The typical setup is outlined in Fig. 4.10. First, monochromatic light is produced as described in the angle-sensitive method above. To obtain a reference and a sample beam, the light is then split with the help of a beam splitter into two parallel beams of equal intensity. Both beams pass the same polarizer and become linearly polarized. It is important to polarize after the beam splitter because the beam splitter will considerably depolarize the light. However, the disadvantage of this order is, that for divergent light, as it is generated by a monochromator, rather large dimensions of the polarizer are needed, which makes them very expensive. Using two single polarizers is not a good idea either because no polarizer is perfect and completely strainless. Therefore, a wavelength-dependent phase shift would be generated leading to a systematic error. Using a tunable laser as light source would be favorable for this method as the shared polarizer can be kept small. A lens focuses the beam onto the sample holder which must contain the sample as well as the reference. Both are placed simultaneously within a magnetic field and eventually a cryostat as described before. However, the requirements for the homogeneity of the magnetic field are much higher here due to the larger space occupied by sample and reference. In the next step, the two rays are focused on a rotating polarizer which serves at the same time as analyzer and modulator. Two matched detectors simultaneously measure reference and sample signal. The detectors must be as similar as possible which is very difficult to obtain. The phase difference between the reference and the sample signal is usually directly measured by a lock-in amplifier. Let the angle between the rotating analyzer axis and the fixed polarization direction be equal to $\phi = \Omega t$, where Ω is the angular frequency of the rotating polarizer. Then, the intensity of the reference beam after the analyzer will be proportional to $\cos^2 \phi$. The sample beam, however, has an extra angle θ_F due to the Faraday effect. Thus, the intensities of the two beams are given by:

$$I_{sample} \propto cos^2[\phi(t) - \theta_F] \propto cos[2[\phi(t) - \theta_F]] = cos\,(2\Omega t - 2\theta_F), \tag{4.51}$$

$$I_{ref} \propto cos^2\phi \propto cos\,(2\Omega t), \tag{4.52}$$

where we have used the fact that a square of a cosine can be expressed as a cosine of the double angle. Furthermore, we have neglected the DC components in eqs. (4.51) and (4.52). We see that the phase shift of the 2Ω component of the sample signal corresponds

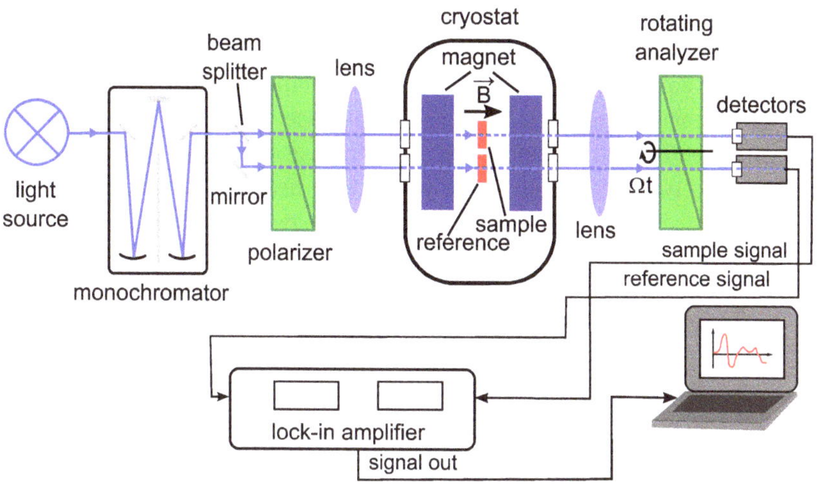

Fig. 4.10: Typical setup of a phase-sensitive method using double-beam detection.

to twice the Faraday rotation. With this method, a continuous and fast measurement of θ_F as a function of photon energy is relatively easily possible because it must neither be measured point-by-point, nor do changes in intensity have any influence on the result. However, the problem is that the two beams do not travel at the same place through the optical elements and the magnetic field. Therefore, the rotation of the reference does not cancel completely the rotation caused by windows and by the substrate due to inhomogeneities in the magnetic field and due to mechanical stress in the optical elements. Further problems arise from a beam displacement in the rotating analyzer. Since a Glan or Glan–Thompson polarizer is composed of two prims with a small gap, there is always a small beam displacement. In addition, a small beam divergence cannot be avoided because the two prisms are not aligned perfectly parallel. Both effects cause a precession of the beam on the detector leading to variations in the zero position. This limits the precision. Finally, the sensitivity is also limited by the finite phase sensitivity of the lock-in amplifier. Nevertheless, an angular resolution of 0.1–0.01° can be reached.

4.2.3 Intensity methods

The angle-sensitive method using a Faraday or photoelastic modulator, illustrated in Fig. 4.9, may also be used as an intensity method. In this case, the analyzer is locked into position with angle $\phi = 0$. The reduced intensity of the 2Ω signal is then given according to eq. (4.47) by

$$I(2\Omega) = E_0^2 J_2(\Delta_0) \sin(2\theta_F) \cos(2\Omega t), \tag{4.53}$$

where we set $\phi = 0$ so that $I(2\Omega) = 0$ for $\theta_F = 0$. With this, a continuous measurement is possible if the strongly wavelength-dependent lamp intensity $I_0(\lambda) = E_0^2(\lambda)$ is kept constant with the help of a feedback loop. Of course, sample and reference measurement must be subtracted from each other also in this case.

Another frequently used method is based on the use of a prism that divides an incident beam into two orthogonal linearly polarized beams. One example is the Wollaston prism, as shown in Fig. 4.11, which produces two rays diverging symmetrically with respect to the direction of the incoming beam.

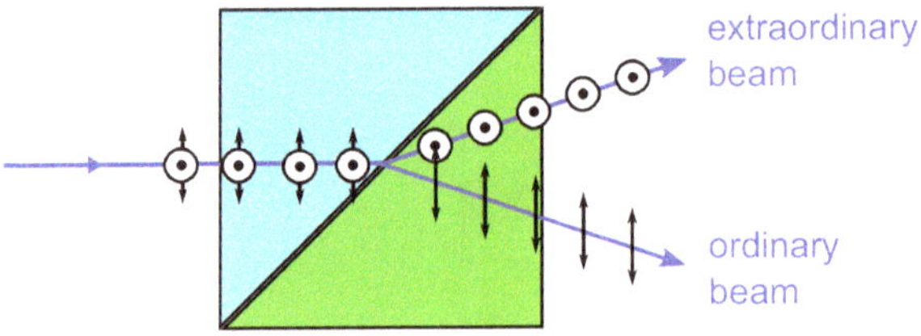

Fig. 4.11: Sketch of a Wollaston prism.

If the incident beam is linearly polarized at an angle of 45° with respect to the two orthogonal polarizing axes of the Wollaston prism, the two emergent rays will have equal intensities. To establish an intensity-sensitive method, the setup for angular-sensitive measurements (Fig. 4.9) must be slightly changed as shown in Fig. 4.12 for measuring the Faraday rotation. The Faraday modulator or PEM is removed as no modulation is necessary. It may, however, be replaced by a chopper that modulates the light intensity. This will slightly increase the sensitivity. In addition, the analyzer is replaced by the Wollaston prism and the two emerging orthogonally polarized rays are measured with two matched detectors similar as in the phase-sensitive method. First, a measurement with no sample (or with a reference mirror for the polar Kerr effect) must be made. If subsequently the sample is measured, it will rotate the polarization be an angle θ_F. With respect to the polarizing axes of the Wollaston prism, the angle is then $45° + \theta_F$ and the intensities on the two detectors 1 and 2, defined by the projection on the polarizing axes, are:

$$I_1 = I_0 \cos^2(45^\circ + \theta_F) = \frac{I_0}{2}(1 + \cos(2[45^\circ + \theta_F])) = \frac{I_0}{2}(1 - \sin(2\theta_F)) \cong \frac{I_0}{2}(1 - 2\theta_F), \tag{4.54}$$

$$I_2 = I_0 \cos^2(45^\circ - \theta_F) = \frac{I_0}{2}(1 + \cos(2[45^\circ - \theta_F])) = \frac{I_0}{2}(1 + \sin(2\theta_F)) \cong \frac{I_0}{2}(1 + 2\theta_F). \tag{4.55}$$

According to the sketch depicted in Fig. 4.12, the normalized difference of the two intensities is measured, which can be calculated from eqs. (4.54) and (4.55) as:

$$\frac{I_2 - I_1}{I_2 + I_1} \cong 2\theta_F. \tag{4.56}$$

As shown in eq. (4.56), by normalizing with the sum of the partial beams, the measurement is, to first approximation, independent of the intensity I_0 of the beam. This is true only if the two detectors have exactly the same sensitivity, which is difficult to achieve. Introducing after the polarizer a chopper, that modulates the intensity, will yield a slight improvement of the sensitivity. Using lasers at a fixed wavelength as a light source, a very high precision can be achieved with this method (approximately 10^{-6} deg).

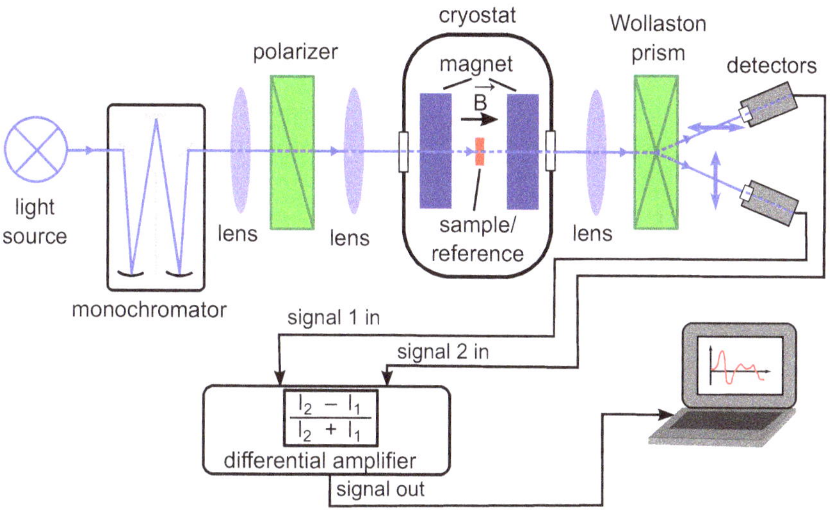

Fig. 4.12: Typical setup of an intensity method using a Wollaston prism.

Another very fast intensity method uses essentially the arrangement of Fig. 4.9 with a photoelastic modulator with modulation frequency Ω. The incident light is polarized at an angle of 45° with respect to the axes of the PEM while the analyzer is fixed parallel to the fast axis of the PEM. To determine the intensity and transmission of the apparatus, a chopper is placed before the polarizer, modulating the light intensity at a frequency $\Omega_0 < \Omega$. Using now up to three lock-in amplifiers and measuring simultaneously the Ω, the 2Ω, and the Ω_0 component, I_Ω, $I_{2\Omega}$, and I_0, respectively, one can mathematically determine for a polar Kerr geometry the rotation θ_K and ellipticity η_K of the sample in a ***single*** measurement (93):

$$\frac{I_\Omega}{I_0} = \frac{2\left(\tan^2\psi - 1\right) J_1(\pi/2)}{1 + \tan^2\psi - 2\tan\psi \sin\delta\, J_0(\pi/2)}, \tag{4.57}$$

$$\frac{I_{2\Omega}}{I_0} = \frac{-4\tan\psi \sin\delta\, J_2(\pi/2)}{1 + \tan^2\psi - 2\tan\psi \sin\delta\, J_0(\pi/2)}, \tag{4.58}$$

where J_n is the Bessel function of order n, $\tan\psi = \rho_+/\rho_-$, $\delta = \delta_+ - \delta_-$, and $\rho_\pm$ and $\delta_\pm$ are defined by the complex reflection coefficient in the circular basis $\tilde{\rho}_\pm = \rho_\pm e^{i\delta_\pm}$. For small angles δ and small differences between ρ_+ and ρ_-, we can set $\tan\psi \cong 1$ and $\sin\delta \cong 0$. Then, it follows:

$$\frac{I_\Omega}{I_0} = 2J_1(\pi/2)\frac{\tan^2\psi - 1}{\tan^2\psi + 1} \quad \Rightarrow \quad \tan\psi = \sqrt{\frac{2J_1(\pi/2) + \frac{I_\Omega}{I_0}}{2J_1(\pi/2) - \frac{I_\Omega}{I_0}}}, \tag{4.59}$$

$$\frac{I_{2\Omega}}{I_0} = -4J_2(\pi/2)\frac{\tan\psi\sin\delta}{\tan^2\psi + 1} \quad \Rightarrow \quad \sin\delta = -\frac{I_{2\Omega}}{I_0}\frac{\tan^2\psi + 1}{4J_2(\pi/2)\tan\psi}. \tag{4.60}$$

Finally, we get the polar Kerr rotation, θ_K, and ellipticity, η_K, according to eq. (3.72) and (3.73), section 3.5, as follows:

$$\theta_K = -\frac{\delta}{2} \text{ and } \tan\eta_K = -\frac{\rho_+ - \rho_-}{\rho_+ + \rho_-} = -\frac{\rho_+/\rho_- - 1}{\rho_+/\rho_- + 1} = -\frac{\tan\psi - 1}{\tan\psi + 1}. \tag{4.61}$$

The resolution of this method is limited by the dynamic range of the lock-in amplifiers which cannot be fully utilized as there is always a large DC signal on the detector which changes only slightly as a function of the magneto-optical activity of the sample. Furthermore, the phase shift of the PEM must be kept constant at $\Delta = \pi/2$ as a function of photon energy, which further limits the resolution. A resolution better than 0.02° can be obtained.

A last method to be mentioned here is measuring the so-called dynamic Faraday rotation (104). In this method, the magnetic field is modulated sinusoidally, that is, $B_{mod} = B_{mod0}\sin(\Omega t)$, by using either a small coil in Helmholtz configuration or modulating the magnet directly. This will modulate the Faraday rotation of the sample, $\theta_F = \theta_{F0}\sin(\Omega t)$, as long as magnetic saturation is not reached. To ensure this, very small modulation fields B_{mod0} are used, typically 0.001 T. If in addition a static magnetic field $B_0 \gg B_{mod0}$, much larger than the modulation field, is applied, the first derivative of the rotation with respect to the magnetic field can be measured for each magnetic field. This is particularly useful when measuring a magnetic phase diagram. With this method, a precision of 0.0001° is achieved due to the small size of the modulation field having only a small impact on the optics.

4.3 Enhancement effects (optical constants effects)

Enhancement effects play a major role in technical applications since materials can be optimized relatively easily therewith. As can be seen from eq. (3.77) and (3.78), the size of the Kerr effect depends significantly on the optical functions $n(\omega)$ and $k(\omega)$ via the prefactors A and B, which are third-degree polynomials of n and k. In addition, the complex refractive index, $\tilde{n} = n - ik$, is the ***relative*** refractive index as defined in eq. (2.105). This implicates that the polar Kerr effect is influenced not only by the material properties, where the light is reflected from, but also by the optical properties of the outside world, where the light is coming from. We will consider in the next sections three basic enhancement effects which are attributable to these facts.

4.3.1 Interface effect

If the magneto-optically active surface is not directed against vacuum (with $n_{vac} = 1$) or air but against a sufficiently transparent material, the complex ***relative*** refractive index, $\tilde{n}$, at the interface differs from the complex refractive index $\tilde{n}_{MO}$ of the magneto-optically active material. This is depicted in Fig. 4.13.

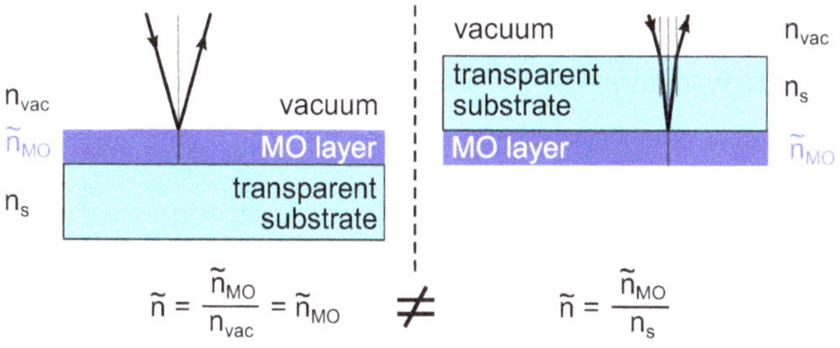

Fig. 4.13: Model of the interface effect. The relative refractive index $\tilde{n}$ is different if the light is reflected directly from the surface of the MO layer with refractive index $\tilde{n}_{MO}$ (vacuum/MO layer interface), or through a transparent substrate with refractive index n_s (substrate/MO layer interface). The refractive index of vacuum is $n_{vac} \cong 1$.

For simplicity, we assume that the covering material is transparent ($k = 0$) with a real refractive index, n_s, as e.g., a glass substrate. In this case, the relative refractive index at the interface is given by:

$$\tilde{n} = \frac{\tilde{n}_{MO}}{n_s} = \frac{n_{MO}}{n_s} - i\frac{k_{MO}}{n_s}. \tag{4.62}$$

Of course, the fact that we must consider the relative refractive index is independent of the basis in which we represent the refractive index. Therefore, also in a circular basis, we still have a relative refractive index, $\tilde{n}_\pm$. If the relationship between $\tilde{\theta}_K$ and $\tilde{\sigma}_{xy}$ is calculated as in Section chapter 3.5, modified prefactors A_s and B_s are obtained. We start with eq. (3.167):

$$\tilde{\theta}_{Ks} \cong i\frac{\tilde{\rho}_+ - \tilde{\rho}_-}{\tilde{\rho}_+ + \tilde{\rho}_-} = i\frac{\tilde{n}_+^2 - \tilde{n}_-^2}{2\tilde{n}\left(\tilde{n}^2 - 1\right)} = \frac{i}{n_s^2}\frac{\tilde{n}_{MO+}^2 - \tilde{n}_{MO-}^2}{2\tilde{n}\left(\tilde{n}^2 - 1\right)} = \frac{i}{\varepsilon_0 \omega n_s^2}\frac{\tilde{\sigma}_{xy}}{\frac{\tilde{n}_{MO}}{n_s}\left(\frac{\tilde{n}_{MO}^2}{n_s^2} - 1\right)}$$

$$= \frac{\mathrm{i}}{\varepsilon_0 \omega n_\mathrm{s}^2} \frac{\tilde{\sigma}_{xy}}{A_\mathrm{s} - \mathrm{i}B_\mathrm{s}} \cong n_\mathrm{s} \mathrm{i} \frac{1}{\varepsilon_0 \omega} \frac{\tilde{\sigma}_{xy}}{A - \mathrm{i}B} \cong n_\mathrm{s} \tilde{\theta}_\mathrm{K}, \tag{4.63}$$

where we have defined the modified complex function $A_\mathrm{s} - \mathrm{i}B_\mathrm{s}$ of n_MO and k_MO:

$$A_\mathrm{s} = \left(\frac{n_\mathrm{MO}}{n_\mathrm{s}}\right)^3 - 3\frac{n_\mathrm{MO}}{n_\mathrm{s}}\left(\frac{k_\mathrm{MO}}{n_\mathrm{s}}\right)^2 - \frac{n_\mathrm{MO}}{n_\mathrm{s}} \cong \frac{1}{n_\mathrm{s}^3}\left(n_\mathrm{MO}^3 - 3n_\mathrm{MO}k_\mathrm{MO}^2 - n_\mathrm{MO}\right) = \frac{1}{n_\mathrm{s}^3} A(n_\mathrm{MO}, k_\mathrm{MO}), \tag{4.64}$$

$$B_\mathrm{s} = -\left(\frac{k_\mathrm{MO}}{n_\mathrm{s}}\right)^3 + 3\left(\frac{n_\mathrm{MO}}{n_\mathrm{s}}\right)^2 \frac{k_\mathrm{MO}}{n_\mathrm{s}} - \frac{k_\mathrm{MO}}{n_\mathrm{s}} \cong \frac{1}{n_\mathrm{s}^3}\left(-k_\mathrm{MO}^3 + 3n_\mathrm{MO}^2 k_\mathrm{MO} - k_\mathrm{MO}\right) = \frac{1}{n_\mathrm{s}^3} B(n_\mathrm{MO}, k_\mathrm{MO}), \tag{4.65}$$

and A and B are the polynomial functions of the complex refractive index of the material alone as defined in eqs. (3.79) – (3.81). From eqs. (3.77) and (3.78), it follows that the size of the polar Kerr rotation and polar Kerr ellipticity measured through a transparent substrate is approximately amplified by the refractive index of the substrate n_s:

$$\theta_\mathrm{Ks} = -\frac{1}{\varepsilon_0 \omega n_\mathrm{s}^2} \frac{B_\mathrm{s}\sigma_{1xy} + A_\mathrm{s}\sigma_{2xy}}{A_\mathrm{s}^2 + B_\mathrm{s}^2} \cong n_\mathrm{s}\theta_\mathrm{K}, \tag{4.66}$$

$$\eta_\mathrm{Ks} = -\frac{1}{\varepsilon_0 \omega n_\mathrm{s}^2} \frac{A_\mathrm{s}\sigma_{1xy} - B_\mathrm{s}\sigma_{2xy}}{A_\mathrm{s}^2 + B_\mathrm{s}^2} \cong n_\mathrm{s}\eta_\mathrm{K}. \tag{4.67}$$

Thus, by measuring through a transparent substrate, one can enhance the Kerr effect uniformly across the whole photon-energy range by a factor equal to the refractive index of the substrate.

We call this the interface effect. It can easily be verified in experiment as demonstrated in Fig. 4.14 (94). On the left side, the polar Kerr rotation θ_K of crystalline UAs (solid line), measured directly at the surface, is plotted together with the polar Kerr rotation calculated for the case of a measurement through a glass substrate with $n_\mathrm{s} = 1.4$ (dashed line) according to eq. (4.66). On the right side of Fig. 4.14, the polar Kerr spectra of an amorphous U-As-thin film is shown for comparison, which was measured both, directly at the surface of the layer (▲) and through a glass substrate (●). As can be seen, the agreement with the calculation is remarkable. The deviations at higher energies (≥ 2.5 eV) are attributed to the increasing absorption of the glass substrate above 2.5 eV.

4.3.2 Interference effect

Another enhancement effect, which is called interference effect, occurs if a dielectric layer adjacent to the magneto-optic (MO) material or the MO material itself is just a few nanometers thick. This can happen unintentionally if the surface of a thin

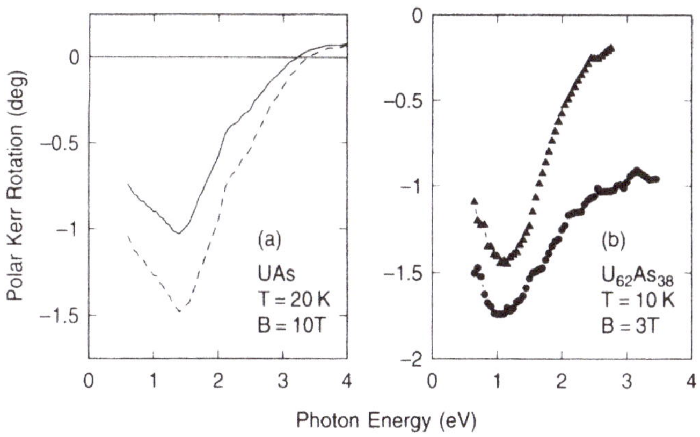

Fig. 4.14: (a) The polar Kerr rotation of crystalline UAs as measured on the surface [solid line, data taken from Ref. (95)] and as calculated through a glass substrate with $n_s = 1.4$ (dashed line). (b) The polar Kerr rotation of an amorphous $U_{62}As_{38}$ film as measured from the film side (▲) and through a glass substrate (●). Figure from Ref. (94).

metal film oxidizes or if a thin protecting layer is deposited on top of the MO material to prevent oxidation. In this case, an enhancement can evolve caused by superposition of multiple reflections. In a simple model, the necessary condition for this to happen is a thickness d fulfilling the requirement of ***destructive interference*** between the rays reflected from the front side of the capping layer and from the MO material or, in the case of a very thin MO layer, from the front and backside of the MO layer as shown in Fig. 4.15. This is associated with a significant reduction in reflectivity. According to eq. (2.109) and (2.110), this implies for the relative complex refractive index the condition $\tilde{n} \cong 1$. The Kerr rotation can this way be increased to a large extent because the denominator in eq. (3.76), will go to zero: $\tilde{\theta}_K \propto \tilde{\sigma}_{xx}^{-1} \propto \left(\tilde{n}^2 - 1\right)^{-1}$.

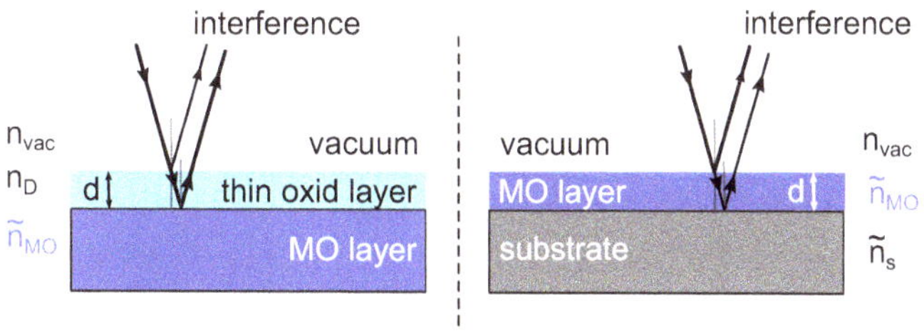

Fig. 4.15: Model of the interference effect. (Left) The ray reflected from the surface of a thin transparent oxide layer (vacuum/oxide layer interface) interferes with the ray reflected from the surface of the MO layer (oxide layer/MO layer interface). (Right) The ray reflected from the front surface (vacuum/MO layer interface) of a thin MO layer interferes with the ray reflected from the back surface (MO layer/substrate interface) of the MO layer. In both cases, an enhancement of the Kerr rotation is observed for destructive interference.

The condition for destructive interference in a thin dielectric layer is an optical phase difference Δ which equals half the wavelength λ:

$$\Delta = 2n_D d = \lambda/2 + m\lambda,\ m \in \mathbb{N}_0, \tag{4.68}$$

where d is the thickness and n_D the refractive index of the thin dielectric layer.

In Fig. 4.16, the polar Kerr rotation of two double layers is shown as an example (96). A TbFeCo layer (a magneto-optic storage material) of 100 nm thickness was covered with an AlN film of thickness 20 nm (■) and 50 nm (□). The thicker layer exhibits a significant enhancement of the Kerr signal by a factor of two at 2 eV. The thinner layer does not show an enhancement. Let us see, if our simple model gives a reasonable prediction of the energetic position of the enhancement due to the interference effect. At 2 eV, the refractive index of AlN is $n_D = 2.15$. According to eq. (4.68), this yields for the thicker AlN layer an enhancement at a wavelength of 430 nm, which corresponds to a photon energy of 2.88 eV. That is quite far away from the measured resonance at 2 eV. For the thinner layer, we get $\lambda = 172$ nm, corresponding to 7.2 eV, which is outside the measurement range. This explains, why it does not show an enhancement. Also shown in Fig. 4.16 is a calculated curve (●) which gives a very good prediction of the findings. The main reason why our simple model does not work is that it neglects the optical properties of the AlN/TbFeCo interface completely, which was not the case in the calculation. Nevertheless, the simple model provides a rough estimate where an enhancement could be found.

Instead of creating the interference by a transparent cover layer, the magneto-optically active material can be itself deposited as a thin layer onto a substrate. Again, the condition of eq. (4.68) holds but now in the magnetic material. For iron layers deposited with different thickness on a thick copper film, the result is shown in Fig. 4.17 (97). We see again a significant increase in polar Kerr rotation for an iron-layer thickness of 131 Å, which exceeds the polar Kerr rotation of bulk iron metal by far. In this case, the origin of the enhancement is more complicated since the Faraday rotation through the magneto-optically active material and the Kerr reflection on the backside must be considered.

The last example is a system where an interference effect occurs within the MO layer itself. The sample is the magnetic semiconductors EuO which exhibits a high polar Kerr rotation ($\theta_K \sim 10°$) but only at very low temperatures (see chapter 6). A thin EuO film was deposited on a silver mirror in such a way that the anti-reflection condition was satisfied (98). As seen in Fig. 4.18, the result is a polar Kerr rotation of incredible 70°. The drawback is that the reflectivity is now less than 1%. With this method, Kerr rotations of almost any size can be achieved but only at the expense of a very low reflectivity, limiting their benefit for technical applications.

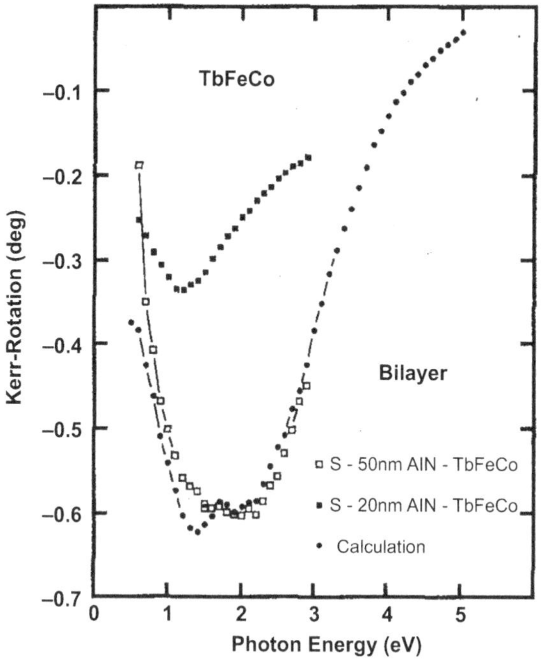

Fig. 4.16: Interference effect in an AlN/TbFeCo bilayer. The polar Kerr rotation has been measured through the AlN layer. Figure from Ref. (96).

4.3.3 Plasma-edge effect

The last enhancement effect, we want to treat here, is the plasma-edge effect. It is again connected to a minimum in the reflectivity, the plasma minimum. This time, however, the minimum is not caused by an interference effect but occurs intrinsically in the magneto-optically active material due to the behavior of the free charge carriers. A typical reflectivity spectrum of a metal is shown on the left side of Fig. 4.19 using the example of NdS (99). One can clearly distinguish the law of Hagen–Rubens in the far infrared, that is, the region where the reflectivity approaches 100%, and the pronounced plasma minimum at about 3 eV. The law of Hagen-Rubens is derived in the problem in Section 3.7.6. At slightly lower photon energy, the spectrum reaches maximum negative slope coinciding roughly with the so-called plasma edge. We consider first the normalized prefactors C and D, which relate to the prefactors A and B from eqs. (3.77) and (3.78) as follows:

$$C = \frac{1}{\hbar\omega}\,\frac{A}{A^2 + B^2}\,. \tag{4.69}$$

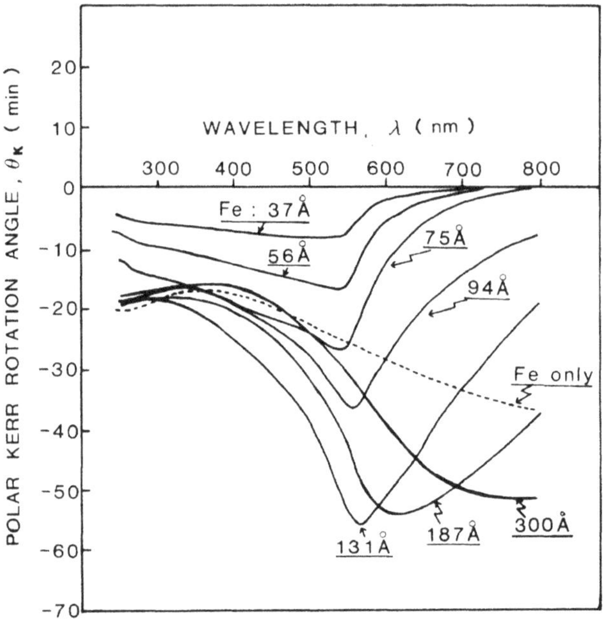

Fig. 4.17: Wavelength dependence of θ_K in various kinds of Fe/Cu bilayered films in which Fe is deposited on a Cu film of 2400 Å thickness. Figure from Ref. (97).

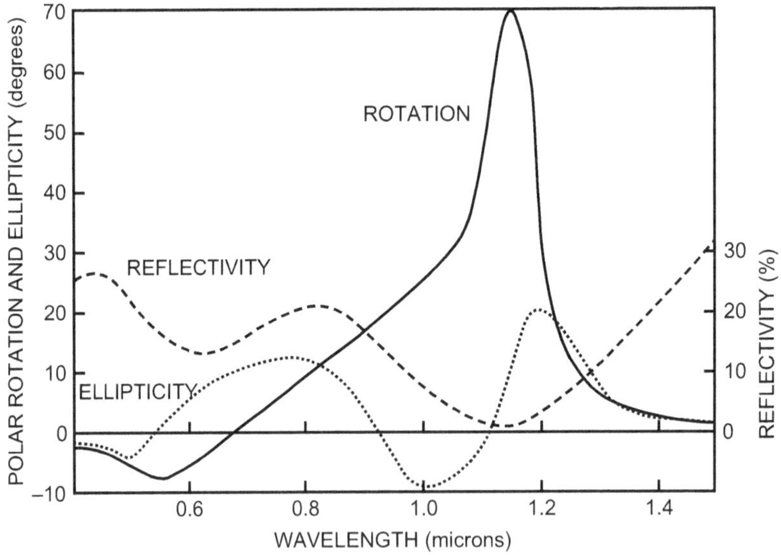

Fig. 4.18: Polar Kerr rotation, polar Kerr ellipticity, and (intensity) reflectivity for an EuO film on top of a silver film on glass. The measurement temperature was 5 K and the applied magnetic field normal to the film was 20 kOe. Figure from Ref. (98).

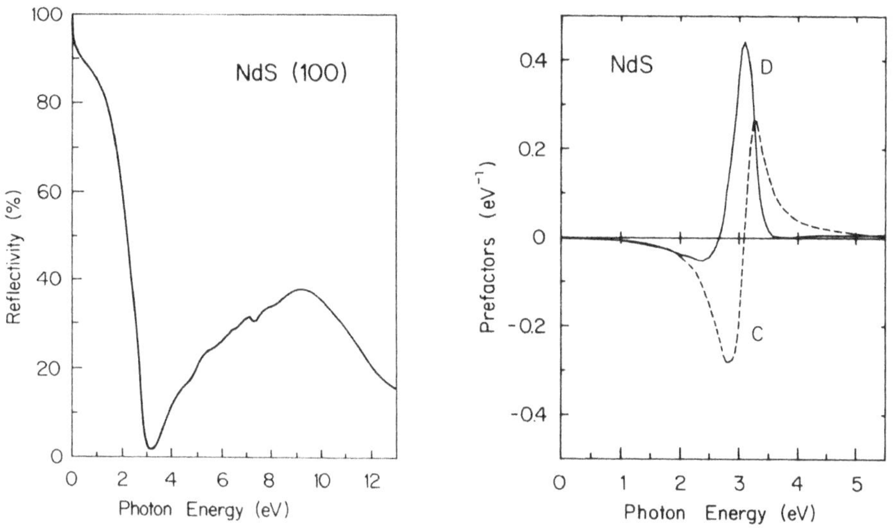

Fig. 4.19: *Left*: Reflectivity of NdS. Note the plasma minimum at approximately 3 eV. *Right*: Normalized prefactors *C* and *D* of NdS. Both quantities show a resonance at the energy of the plasma minimum. Figure from Ref. (99).

$$D = \frac{1}{\hbar\omega}\frac{B}{A^2 + B^2}, \tag{4.70}$$

They exhibit a resonance at the plasma minimum. This is shown for NdS on the right side of Fig. 4.19. Obviously, the resonance of the normalized prefactors C and D leads to a resonance in the polar Kerr rotation θ_K and ellipticity η_K which are plotted on the left side of Fig. 4.20 (99). However, the off-diagonal optical conductivity, which is the material property responsible for the magneto-optic activity, has no distinct resonance at this photon energy. This is pointed out on the right side of Fig. 4.20. The small features around 3 eV in σ_{1xy} and σ_{2xy} are artifacts due to the imperfect determination of the refractive index n and absorption index k in this photon-energy range. The true spin-polarized electronic transition is around 4.5 eV. Due to the plasma-edge effect, it is, however, hardly visible in the polar Kerr spectra and shows up as a small feature around 4.5 eV in Fig. 4.20, on the left.

In summary, we note that in the vicinity of a reflectivity minimum an enhancement effect in the magneto-optic spectrum is very likely to occur due to the above-described plasma-edge effect and, therefore, needs to be considered in the interpretation of the Kerr effect.

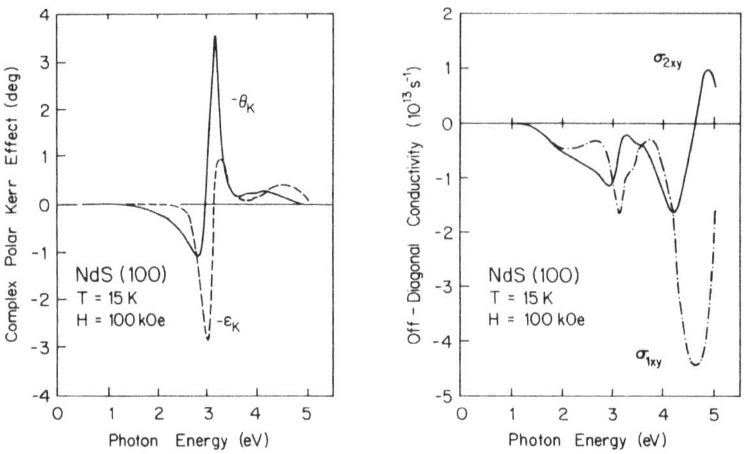

Fig. 4.20: *Left*: Complex polar Kerr effect of NdS. Note the resonance at the energy of the plasma minimum at approximately 3 eV. *Right*: The off-diagonal optical conductivity of NdS does not show a resonance at the plasma minimum, that is, the large polar Kerr effect is just an optical enhancement effect. Figure from Ref. (99).

4.4 Comprehension questions

4.4.1 Jones matrix theory

(a) What is a Jones vector? Write a Jones vector for light linearly polarized at 30° with respect to the x-axis, for lcp light, and for elliptically polarized light with an ellipticity of 0.5 and a semimajor axis parallel to the x-axis.
(b) What is a Jones matrix? Derive the Jones matrix for a linear polarizer.
(c) How can a phase shifter with variable phase shift be realized (Soleil–Babinet compensator)?
(d) Write the Jones matrix for an absorbing sample exhibiting Faraday effect.

4.4.2 Experimental methods

(a) Name three different measurement techniques to measure polarization states in magneto-optics.
(b) Using Jones matrix theory, show how to derive the resulting polarization state at the detector for one of the three methods (just formally, no calculation).
(c) What is the advantage of using a lock-in method?
(d) Describe the two types of modulators used for polarization modulation. How do they differ (advantages, disadvantages)?

(e) Why is it better to use a polarization modulator than using a chopper (for intensity modulation)?
(f) How can ellipticity be measured?
(g) Describe the typical setup for a measurement of the Faraday effect.

4.4.3 Enhancement effects

(a) Why do enhancement effects predominantly show up in the magneto-optic Kerr effect (MOKE)?
(b) Name three different enhancement effects. Discuss for each of them the effect on the MOKE spectra. Which of them produces strong peaks in the MOKE spectra?
(c) For which of the three enhancement effects is the effect in the MOKE spectra correlated to spectral features of the reflectivity spectra of the magneto-optic material, and how do they manifest themselves in the reflectivity spectra?

4.5 Problems

4.5.1 Circular polarizer

(a) A circular polarizer fully transmits rcp light and fully blocks lcp light. Construct the Jones matrix of a circular polarizer in an *xy* coordinate system.

Hint: Assume *a* 2×2 *matrix* $\mathbf{P_c} = \begin{pmatrix} a & b \\ c & d \end{pmatrix}$ *with the four unknowns a, b, c, d. Write down how the circular polarizer acts on rcp and lcp light represented in an xy coordinate system. This yields four equations for the four unknowns.*

(b) How can you realize such a polarizer using the optical elements that were described in Section 4.1?

Hint: You know just one polarizing element: a linear polarizer according to eq. (4.16). *Therefore, you must use optical elements to transform the polarization first to a linear state, where you can polarize the beam, and then back again to a circular state.*

4.5.2 Faraday modulator

Light, which is assumed to be linearly polarized along the *x* direction, is first passing through a Faraday modulator, then through an analyzer and is finally detected. An analyzer is a linear polarizer at an angle ϕ with respect to the *x*-axis. The Faraday modulator generates a modulation $\theta_F = \theta_0 \sin(\Omega t)$ of the polarization direction.

(a) Calculate the time dependence of the intensity $I(t)$ measured after passing the analyzer as a function of modulation amplitude θ_0 and position ϕ of the analyzer. To do this, expand $I(t)$ in a Fourier series up to second order in Ωt.

(b) Discuss the Fourier components proportional to Ω and 2Ω as well as the DC contribution as a function of ϕ and θ_0. For which positions ϕ of the analyzer, the Fourier component proportional to Ω will vanish? How do these positions differ with respect to the value of θ_0? What is the effect of this on the quality of the measurement (signal-to-noise ratio)?

Hint: $I(t)$ describes the intensity on the detector, that is, time averaged over the fast light frequency ω but time-dependent on the slow modulation frequency Ω. Remember that $I(t) \propto \vec{E} \cdot \vec{E}^*$. Use the Jones matrices and the following expansions:

$$\cos[a \sin(\Omega t)] = J_0(a) + 2\sum_{n=1}^{\infty} J_{2n}(a)\cos(2n\Omega t), \tag{4.71}$$

$$\sin[a \sin(\Omega t)] = 2\sum_{n=0}^{\infty} J_{2n+1}(a)\sin[(2n+1)\,\Omega t]. \tag{4.72}$$

4.5.3 Ellipticity measurement

With the help of Senarmont's principle, the ellipticity η of an elliptically polarized plane wave shall be measured. The elliptically polarized state of the light with the semimajor axis parallel to the x-axis, can be described as:

$$\vec{E} = E_0 \begin{pmatrix} 1 \\ \mathrm{i}\tan\eta \end{pmatrix} e^{\mathrm{i}v}, \tag{4.73}$$

where the phase v is defined according to eq. (4.3). To measure η, the plane wave is first transmitted through a Soleil–Babinet compensator $\boldsymbol{C}$ with a phase shift $\Delta = \pi/2$ and the fast axis parallel to the x axis. Then, it passes through a Faraday modulator $\boldsymbol{F}$ with modulation angle $\theta_F = \theta_0 \sin(\Omega t)$. Finally, the plane wave travels through an analyzer $\boldsymbol{A}$, which is a linear polarizer at an angle ϕ with respect to the x-axis, and impinges onto a detector measuring the intensity. The detector is not able to resolve the fast oscillation with angular frequency ω of the plane wave but the slower oscillation Ω of the modulation.

(a) Calculate the time dependence of the intensity, measured at the detector, and the Fourier component with frequency Ω. How can the ellipticity be determined from this?

(b) What happens to the measurement if the position of Soleil-Babinet compensator and Faraday modulator are interchanged? Are we still able to determine the ellipticity?

4.6 Solutions

4.6.1 Circular polarizer

Let us start with recalling the Jones vectors for the polarization states needed to treat this problem. A linear polarizer involves two prominent polarization states: linearly polarized parallel to the polarization axis, where light with this polarization is fully transmitted, and linearly polarized perpendicular to the polarization axis, where light with this polarization is completely blocked. Assuming that the polarization axis is along the x-axis, we get two prominent Jones vectors, a fully transmitted one for x-polarized light,

$$\begin{pmatrix} 1 \\ 0 \end{pmatrix}, \tag{4.74}$$

and a blocked one for y-polarized light,

$$\begin{pmatrix} 0 \\ 1 \end{pmatrix}, \tag{4.75}$$

The circular polarizer, that we want to construct, has two prominent polarization states as well: rcp, where light with this polarization is fully transmitted,

$$\frac{1}{\sqrt{2}}\begin{pmatrix} 1 \\ \mathrm{i} \end{pmatrix}, \tag{4.76}$$

and lcp, where light with this polarization is fully blocked,

$$\frac{1}{\sqrt{2}}\begin{pmatrix} 1 \\ -\mathrm{i} \end{pmatrix}. \tag{4.77}$$

Of course, we could as well interchange the two polarization directions, rcp and lcp. This would also produce a circular polarizer. This would be a polarizer fully transmitting lcp light and completely blocking rcp light.

(a) Now, we can start to construct our polarizer. There are two approaches. We can describe the Jones matrix in a circular basis, where it is easy to describe because it must be a diagonal matrix $\boldsymbol{P}_{\pm}$, transmitting rcp light and blocking lcp light, or we can assume a general 2×2 matrix, $\boldsymbol{P}_{\mathbf{c}}$, and calculate the four unknown matrix elements from the polarization behavior. Let us start with the second approach. We assume a general 2×2 matrix, $\boldsymbol{P}_{\mathbf{c}}$:

$$\mathbf{P_c} = \begin{pmatrix} a & b \\ c & d \end{pmatrix} \tag{4.78}$$

The transmission behavior is as follows: rcp light must be fully transmitted, leading to the following equation:

$$\frac{1}{\sqrt{2}}\begin{pmatrix}1\\ \mathrm{i}\end{pmatrix}=\mathbf{P_c}\frac{1}{\sqrt{2}}\begin{pmatrix}1\\ \mathrm{i}\end{pmatrix}. \tag{4.79}$$

On the other hand, lcp light must be completely blocked:

$$\begin{pmatrix}0\\ 0\end{pmatrix}=\mathbf{P_c}\frac{1}{\sqrt{2}}\begin{pmatrix}1\\ -\mathrm{i}\end{pmatrix}. \tag{4.80}$$

This yields four equations from the components of the Jones vectors:

$$1=a+\mathrm{i}b, \tag{4.81}$$

$$\mathrm{i}=c+\mathrm{i}d, \tag{4.82}$$

$$0=a-\mathrm{i}b, \tag{4.83}$$

$$0=c-\mathrm{i}d. \tag{4.84}$$

This is easy to solve. Adding eqs. (4.83) and (4.81) yields $a=\frac{1}{2}$. Adding eqs. (4.84) and (4.82), amounts to $c=\frac{1}{2}\mathrm{i}$. Entering the values for a and c in eqs. (4.83) and (4.84), respectively, yields $b=-\frac{1}{2}\mathrm{i}$ and $d=\frac{1}{2}$. So, the Jones matrix of a circular polarizer is:

$$\mathbf{P_c}=\frac{1}{2}\begin{pmatrix}1 & -\mathrm{i}\\ \mathrm{i} & 1\end{pmatrix}. \tag{4.85}$$

The alternative approach, describing the Jones matrix $\boldsymbol{P}_{\pm}$ in the rcp/lcp basis, where rcp light is fully transmitted and lcp light fully blocked, yields:

$$\mathbf{P}_{\pm}=\begin{pmatrix}1 & 0\\ 0 & 0\end{pmatrix}. \tag{4.86}$$

To get the Jones matrix, $\boldsymbol{P}_{\mathbf{c}}$, in the xy basis, we must transform $\boldsymbol{P}_{\pm}$ back according to the transformation rules given by linear algebra [see eq. (3.28)]:

$$\mathbf{P_c}=\mathbf{T}\,\mathbf{P}_{\pm}\,\mathbf{T}^{-1} \tag{4.87}$$

with $\boldsymbol{T}$ being the transformation matrix from circular to linear basis, as defined in eq. (3.29), but reduced to the x and y dimensions.

(b) We know the following optical elements: a compensator, a rotator, and a linear polarizer. The only polarizing element is the linear polarizer. So, we must in step (I) transform the state from rcp to linearly polarized. This can be achieved by a compensator with a phase shift of $\pm\pi/2$. Then, we must in step (II) align the polarizer with the direction of the linear polarization to get the light polarized. In step (III), we transform the polarization state back to the original circular state using another compensator. For step (I), we need a compensator according to eq. (4.29) with phase shift $\Delta = \pi/2$ and fast axis parallel to the x-axis ($\phi = 0°$):

$$\mathbf{C}\ (\phi = 0°,\ \Delta = \pi/2) = \begin{pmatrix} \mathrm{i} & 0 \\ 0 & 1 \end{pmatrix}. \tag{4.88}$$

This will yield linearly polarized light with the polarization axis at 45° with respect to the x-axis as can be seen when multiplying $\boldsymbol{C}$ with the Jones vector for rcp light as defined in eq. (4.76). In the next step we set the polarizer, eq. (4.16), at $\phi = 45°$, so that it transmits rcp light:

$$\mathbf{P}_{45°} = \frac{1}{2}\begin{pmatrix} 1 & 1 \\ 1 & 1 \end{pmatrix} \tag{4.89}$$

In the last step, we must transform the polarization state back to the original one. For this we need the inverse action, $\boldsymbol{C}^{-1}$, of the compensator $\boldsymbol{C}$. This is achieved by setting the phase difference $\Delta = -\pi/2$ instead of $\pi/2$:

$$\mathbf{C}^{-1}(\phi = 0°,\ \Delta = -\pi/2) = \begin{pmatrix} -\mathrm{i} & 0 \\ 0 & 1 \end{pmatrix}. \tag{4.90}$$

To check the validity, we can calculate the Jones matrix of the total action by simply multiplying the individual matrices:

$$\begin{aligned} \mathbf{P}_\mathrm{c} = \mathbf{C}^{-1}\mathbf{P}_{45°}\mathbf{C} &= \frac{1}{2}\begin{pmatrix} -\mathrm{i} & 0 \\ 0 & 1 \end{pmatrix}\begin{pmatrix} 1 & 1 \\ 1 & 1 \end{pmatrix}\begin{pmatrix} \mathrm{i} & 0 \\ 0 & 1 \end{pmatrix} \\ &= \frac{1}{2}\begin{pmatrix} -\mathrm{i} & 0 \\ 0 & 1 \end{pmatrix}\begin{pmatrix} \mathrm{i} & 1 \\ \mathrm{i} & 1 \end{pmatrix} = \frac{1}{2}\begin{pmatrix} 1 & -\mathrm{i} \\ \mathrm{i} & 1 \end{pmatrix}. \end{aligned} \tag{4.91}$$

This yields, as expected, the same matrix $\boldsymbol{P}_\mathrm{c}$ as in eq. (4.85).

4.6.2 Faraday modulator

(a) To calculate the intensity on the detector, we must first determine the electric-field vector on the detector by calculating the changes to the incoming electric-field vector while passing successively through all the optical elements involved. We start with the incident wave which is x-polarized after having passed the polarizer:

$$\vec{E}_0(0) = E_0 \begin{pmatrix} 1 \\ 0 \end{pmatrix}. \tag{4.92}$$

In the next step, the light passes the Faraday modulator, ***F***, and gets modulated by $\theta_F(t) = \theta_0 \sin(\Omega t)$:

$$\mathbf{F} = \begin{pmatrix} \cos[\theta_F(t)] & -\sin[\theta_F(t)] \\ \sin[\theta_F(t)] & \cos[\theta_F(t)] \end{pmatrix}. \tag{4.93}$$

This produces the following electric-field vector:

$$\vec{E}_0(t) = \mathbf{F}\, E_0 \begin{pmatrix} 1 \\ 0 \end{pmatrix} = E_0 \begin{pmatrix} \cos[\theta_F(t)] \\ \sin[\theta_F(t)] \end{pmatrix}. \tag{4.94}$$

Finally, the light passes through the analyzer,

$$\mathbf{A} = \begin{pmatrix} \cos^2\phi & \sin\phi\cos\phi \\ \sin\phi\cos\phi & \sin^2\phi \end{pmatrix}, \tag{4.95}$$

which yields the electric-field vector incident on the detector:

$$\begin{aligned} \vec{E}(t) &= \mathbf{A}\vec{E}_0(t) = E_0\,[\cos\phi\cos[\theta_F(t)] + \sin\phi\sin[\theta_F(t)]] \begin{pmatrix} \cos\phi \\ \sin\phi \end{pmatrix} \\ &= E_0 \cos[\phi - \theta_F(t)] \begin{pmatrix} \cos\phi \\ \sin\phi \end{pmatrix} \end{aligned} \tag{4.96}$$

The detector can only measure the light intensity and does not resolve the fast oscillation frequency ω of the light wave. For simplicity, we consider in the following again a reduced intensity by dividing the intensity by all the constant prefactors so that the reduced intensity is equal to the time-averaged square of the ***real part*** of the electric-field vector. The reduced intensity, time-averaged over ω, is obtained by performing the dot product with the conjugate complex vector as discussed in Section 4.2.1:

$$I(t) = \vec{E}(t)\cdot\vec{E}^*(t) = E_0^2\, \cos^2[\phi - \theta_F(t)] = \frac{1}{2}E_0^2\, (1 + \cos[2\,[\phi - \theta_F(t)]]). \tag{4.97}$$

Separating the trigonometric functions with combination of angles into functions with just one angle leads for the reduced intensity to

$$I(t) = \frac{1}{2}E_0^2\, [1 + \cos(2\phi)\cos(2\theta_F(t)) + \sin(2\phi)\sin(2\theta_F(t))]. \tag{4.98}$$

Because the modulation angle θ_F is itself a harmonic function, eq. (4.98) involves a nesting of trigonometric functions. This type of function can be developed in a

Fourier series according to eq. (4.71) and (4.72). Up to the second harmonic of the modulation frequency Ω, this results for the reduced intensity in

$$I(t)=\frac{1}{2}E_0^2\,[1+\cos(2\phi)\,[J_0(2\theta_0)+2J_2(2\theta_0)\cos(2\Omega t)+\cdots]\\ +\sin(2\phi)\,[2J_1(2\theta_0)\sin(\Omega t)+\cdots]], \tag{4.99}$$

where J_m is the mth Bessel function. Arranging the order of the harmonics of Ω yields the reduced intensity on the detector expressed as a Fourier series up to the second harmonic:

$$I(t)=E_0^2\left[\frac{1}{2}[1+\cos(2\phi)J_0(2\theta_0)]+\sin(2\phi)J_1(2\theta_0)\sin(\Omega t)\right.\\ \left.+\cos(2\phi)2J_2(2\theta_0)\cos(2\Omega t)+\cdots\right]. \tag{4.100}$$

(b) Before we start to discuss the Fourier components, we should note that we can obtain from the reduced intensity, eq. (4.100), the angle of the incoming polarization similar to eq. (4.47), with the difference that here the Faraday rotation is $\theta_F = 0$, representing the incoming x-polarized light. Furthermore, the first nonvanishing Fourier component is here the first harmonic.

As we can see from eq. (4.100), the first nonvanishing Fourier component I_Ω is prop. to Ωt:

$$I_\Omega(t)=E_0^2\sin(2\phi)J_1(2\theta_0)\sin(\Omega t). \tag{4.101}$$

I_Ω is zero if $\phi = m(\pi/2)$ with $m \in \mathbb{Z}$. For $m = 0$, we obtain $\phi = 0$, corresponding to the polarization direction of the incoming x-polarized light. In this case, analyzer and polarizer are aligned parallel. Entering this in eq. (4.100) yields the total reduced intensity on the detector at the zero crossing:

$$I(t)=\frac{1}{2}E_0^2\,[1+J_0(2\theta_0)+\mathcal{O}(2\Omega t)], \tag{4.102}$$

However, there is a second zero crossing at $m = 1$: $\phi = \pi/2$. Now, analyzer and polarizer are crossed. The reduced intensity is:

$$I(t)=\frac{1}{2}E_0^2\,[1-J_0(2\theta_0)+\mathcal{O}(2\Omega t)]. \tag{4.103}$$

In this case, the total reduced intensity on the detector is much smaller than in eq. (4.102) because of the negative second term on the right. The smaller total reduced intensity on the detector results in a lower signal load on the detector and, consequently, in a higher signal-to-noise ratio for this measurement configuration. So, this is the preferred setting for the relative position of analyzer and polarizer.

We conclude this problem with a numerical example: Let the modulation be $\theta_0 = 14.3°$. Then, $2\theta_0 = 0.5$ rad and the Bessel functions amount to: $J_0(0.5) = 0.938$, $J_1(0.5) = 0.242$, and $J_2(0.5) = 0.0031$. This yields a total reduced intensity on the detector (up to the third harmonic) for parallel alignment, $\phi = 0$: $I(t)/E_0^2 = \underline{\mathbf{0.969}} + 0.031 \cos(2\Omega t) + \mathcal{O}(3\Omega t)$. The total reduced intensity for crossed analyzer and polarizer, $\phi = \pi/2$, however, is only: $I(t)/E_0^2 = \underline{\mathbf{0.031}} - 0.031 \cos(2\Omega t) + \mathcal{O}(3\Omega t)$. So, we see that the DC load on the detector is more than thirty times smaller for crossed than for parallel alignment producing a much higher signal-to-noise ratio for crossed alignment.

4.6.3 Ellipticity measurement

(a) We start with an elliptically polarized plane wave with ellipticity η and with its semimajor axis parallel to the x-axis:

$$\vec{E} = E_0 \begin{pmatrix} 1 \\ \mathrm{i} \tan\eta \end{pmatrix} e^{\mathrm{i}v}, \tag{4.104}$$

where the phase v is defined according to eq. (4.3). The light is first transmitted through a compensator with fast axis parallel to the x-axis ($\phi = 0$) and phase shift $\Delta = \pi/2$:

$$\mathbf{C}\left(\Delta = \frac{\pi}{2}, \phi = 0\right) = \begin{pmatrix} \mathrm{i} & 0 \\ 0 & 1 \end{pmatrix}. \tag{4.105}$$

Thereafter, the light passes through a Faraday modulator,

$$\mathbf{F} = \begin{pmatrix} \cos[\theta_F(t)] & -\sin[\theta_F(t)] \\ \sin[\theta_F(t)] & \cos[\theta_F(t)] \end{pmatrix}, \tag{4.106}$$

and is modulated with the modulation angle $\theta_F = \theta_0 \sin(\Omega t)$. Before hitting the detector, the light is analyzed by a polarizer with polarization axis at an angle ϕ with respect to the x-axis:

$$\mathbf{A} = \begin{pmatrix} \cos^2\phi & \sin\phi\cos\phi \\ \sin\phi\cos\phi & \sin^2\phi \end{pmatrix}. \tag{4.107}$$

The electric field arriving on the detector is determined by multiplying, in the order of traversing, the Jones matrices of all optical elements involved. After the light having passed $\boldsymbol{C}$ and $\boldsymbol{F}$, this leads to the following electric-field vector:

$$\vec{E}''' = \mathbf{AFC}\,\vec{E} = \mathbf{AF}\,E_0\,\mathrm{i}\begin{pmatrix} 1 \\ \tan\eta \end{pmatrix} e^{\mathrm{i}\,v} = \mathbf{A}\,E_0\,\mathrm{i}\begin{pmatrix} \cos[\theta_{\mathrm{F}}(t)] - \sin[\theta_{\mathrm{F}}(t)]\tan\eta \\ \sin[\theta_{\mathrm{F}}(t)] + \cos[\theta_{\mathrm{F}}(t)]\tan\eta \end{pmatrix} e^{\mathrm{i}v}. \tag{4.108}$$

This equation can be simplified by combining the angles θ_{F} and η:

$$\begin{aligned} \vec{E}''' &= \mathbf{A}\,E_0\,\frac{\mathrm{i}}{\cos\eta}\begin{pmatrix} \cos[\theta_{\mathrm{F}}(t)]\cos\eta - \sin[\theta_{\mathrm{F}}(t)]\sin\eta \\ \sin[\theta_{\mathrm{F}}(t)]\cos\eta + \cos[\theta_{\mathrm{F}}(t)]\sin\eta \end{pmatrix} e^{\mathrm{i}v} \\ &= \mathbf{A}\,E_0\,\frac{\mathrm{i}}{\cos\eta}\begin{pmatrix} \cos[\theta_{\mathrm{F}}(t)+\eta] \\ \sin[\theta_{\mathrm{F}}(t)+\eta] \end{pmatrix} e^{\mathrm{i}v}. \end{aligned} \tag{4.109}$$

Finally, after passing the analyzer and before arriving at the detector, the electric-field vector becomes:

$$\vec{E}''' = E_0\,\frac{\mathrm{i}}{\cos\eta} e^{\mathrm{i}v}\,[\cos\phi\cos[\theta_{\mathrm{F}}(t)+\eta] + \sin\phi\sin[\theta_{\mathrm{F}}(t)+\eta]]\begin{pmatrix} \cos\phi \\ \sin\phi \end{pmatrix}. \tag{4.110}$$

Here, we have extracted a common factor from the vector part, leaving a simple unit vector pointing along the direction of the polarization axis, ϕ. This decomposition into a common factor and a simple unit vector must always be possible because the polarizer transmits only nonzero electric-field components along its polarization axis. The factor in front of the unit vector can be further simplified by combining the angles involved:

$$\vec{E}''' = E_0\,\frac{\mathrm{i}}{\cos\eta} e^{\mathrm{i}v}\,\cos[\phi - \theta_{\mathrm{F}}(t) - \eta]\begin{pmatrix} \cos\phi \\ \sin\phi \end{pmatrix}. \tag{4.111}$$

This is the electric field that arrives at the detector. As we know, the detector will measure only the light intensity and it is not sensitive to the fast angular frequency ω of the plane wave. For simplicity, we consider in the following again a reduced intensity by dividing the intensity by all the constant prefactors so that the reduced intensity is equal to the time-averaged square of the ***real part*** of the electric-field vector. The reduced intensity, averaged over ω, is obtained by performing the complex dot product between the electric-field vector and the conjugate complex vector:

$$I(t) = \vec{E}'''(t)\cdot\vec{E}'''^{*}(t) = \frac{E_0^2}{\cos^2\eta}\cos^2[\phi - \theta_{\mathrm{F}}(t) - \eta] = \frac{E_0^2}{\cos^2\eta}\frac{1}{2}\,(1 + \cos[2\phi - 2\theta_{\mathrm{F}}(t) - 2\eta]). \tag{4.112}$$

To be able to expand into a Fourier series, the terms involving θ_F must be separated from the other angles using the well-known sum and difference formulas for trigonometric functions:

$$I(t) = \frac{E_0^2}{2\cos^2\eta}\,(1 + \cos[2(\phi - \eta)]\cos[2\theta_F(t)] + \sin 2(\phi - \eta)\sin[2\theta_F(t)]). \qquad (4.113)$$

Applying eqs. (4.71) and (4.72), the reduced intensity can be expressed in terms of higher harmonics of the modulation frequency Ω. Up to the second harmonic, we obtain

$$\begin{aligned} I(t) = \frac{E_0^2}{2\cos^2\eta}\,&(1 + \cos[2(\phi - \eta)]\,[J_0(2\theta_0) + 2J_2(2\theta_0)\cos(2\Omega t) + \cdots]. \\ &+ \sin[2(\phi - \eta)]\,[2J_1(2\theta_0)\sin(\Omega t) + \cdots]). \end{aligned} \qquad (4.114)$$

The most useful term is the Fourier component proportional to Ωt as it contains the sine function of the difference between the ellipticity η and the polarizer angle ϕ:

$$I_\Omega = \frac{E_0^2}{\cos^2\eta} J_1(2\theta_0)\sin[2(\phi - \eta)]\;\sin(\Omega t). \qquad (4.115)$$

I_Ω exhibits a zero crossing for analyzer position $\phi = \eta$. This enables an exact determination of the ellipticity η. Note that there is a second zero crossing for $\phi = \eta + \pi/2$. Here, the second, constant term, proportional to $\cos[2[\phi - \eta)$, becomes negative and lowers the overall DC component incident on the detector yielding a lower signal load and, therefore, better signal-to-noise ratio as already discussed in the solution of the problem in Section 4.5.2.

(b) Now, we want to study what happens to the measurement if we put the Faraday modulator in a different place. We could first modulate the plane wave and then apply a phase shift. This looks like a minor and innocuous change but let us see what the effect on the measurement is. We must now evaluate the following chain of optical components acting on the electric field of the incoming elliptically polarized plane wave according to eq. (4.104):

$$\vec{E}''' = \mathbf{ACF}\,\vec{E} = \mathbf{AC}\,E_0 \begin{pmatrix} \cos[\theta_F(t)] - \mathrm{i}\sin[\theta_F(t)]\tan\eta \\ \sin[\theta_F(t)] + \mathrm{i}\cos[\theta_F(t)]\tan\eta \end{pmatrix} e^{\mathrm{i}\nu}. \qquad (4.116)$$

After having passed the Faraday modulator and the compensator, the electric-field vector is:

$$\vec{E}''' = \mathbf{A}\, E_0 \begin{pmatrix} \mathrm{i}\cos[\theta_F(t)] + \sin[\theta_F(t)]\tan\eta \\ \sin[\theta_F(t)] + \mathrm{i}\cos[\theta_F(t)]\tan\eta \end{pmatrix} e^{\mathrm{i}v}. \tag{4.117}$$

By multiplying with the matrix of the analyzer, the electric-field vector arriving at the detector has the following shape:

$$\begin{aligned}\vec{E}''' = E_0\, e^{\mathrm{i}v}\, [\cos\phi\, [\mathrm{i}\cos[\theta_F(t)] + \sin[\theta_F(t)]\tan\eta] + \sin\phi\, [\sin[\theta_F(t)] \\ + \mathrm{i}\cos[\theta_F(t)]\tan\eta]] \begin{pmatrix} \cos\phi \\ \sin\phi \end{pmatrix}.\end{aligned} \tag{4.118}$$

Again, we have extracted a common factor, leaving just the unit vector in the direction of the polarization axis of the analyzer. Next, we bundle real and imaginary terms:

$$\begin{aligned}\vec{E}''' = \frac{E_0\, e^{\mathrm{i}v}}{\cos\eta} & (\cos\phi\sin[\theta_F(t)]\sin\eta + \sin\phi\sin[\theta_F(t)]\cos\eta \\ & + \mathrm{i}\, [\cos\phi\cos[\theta_F(t)]\cos\eta + \sin\phi\cos[\theta_F(t)]\sin\eta]) \begin{pmatrix} \cos\phi \\ \sin\phi \end{pmatrix}\end{aligned} \tag{4.119}$$

We can further simplify the equation by combining the angles in the factor in front of the unit vector:

$$\vec{E}''' = \frac{E_0\, e^{\mathrm{i}v}}{\cos\eta} (\sin(\phi+\eta)\sin[\theta_F(t)] + \mathrm{i}\cos(\phi-\eta)\cos[\theta_F(t)]) \begin{pmatrix} \cos\phi \\ \sin\phi \end{pmatrix}. \tag{4.120}$$

Next, we determine the reduced intensity and get rid of the $\sin\theta_F$ functions by converting them to cosine functions:

$$I(t) = \vec{E}'''(t) \cdot \vec{E}'''^{*}(t) = \frac{E_0^2}{\cos^2\eta} \left(\sin^2(\phi+\eta) \underbrace{\sin^2[\theta_F(t)]}_{1-\cos^2[\theta_F(t)]} + \cos^2(\phi-\eta)\cos^2[\theta_F(t)] \right). \tag{4.121}$$

Then, we eliminate the squares of sine and cosine functions in the square bracket by converting them to functions involving the double angle:

$$I(t) = \frac{E_0^2}{\cos^2\eta} \left(\sin^2(\phi+\eta) + \cos^2[\theta_F(t)] \left[\underbrace{\cos^2(\phi-\eta)}_{\frac{1}{2}[1+\cos(2\phi-2\eta)]} - \underbrace{\sin^2(\phi+\eta)}_{\frac{1}{2}[1-\cos(2\phi+2\eta)]} \right] \right). \tag{4.122}$$

This yields the following expression for the reduced intensity on the detector:

$$I(t) = \frac{E_0^2}{\cos^2\eta}\left(\sin^2(\phi+\eta) + \cos^2[\theta_F(t)]\left[\frac{1}{2}\cos(2\phi-2\eta) + \frac{1}{2}\cos(2\phi+2\eta)\right]\right). \quad (4.123)$$

This can be further simplified by applying the sum and difference formulas to the sine and cosine function in the square bracket:

$$I(t) = \frac{E_0^2}{\cos^2\eta}\left(\sin^2(\phi+\eta) + \frac{1}{2}\cos^2[\theta_F(t)]\,[\cos(2\phi)\cos(2\eta) + \sin(2\phi)\sin(2\eta) + \cos(2\phi)\cos(2\eta) - \sin(2\phi)\sin(2\eta)]\right). \quad (4.124)$$

Leading to the following result:

$$I(t) = \frac{E_0^2}{\cos^2\eta}\left(\sin^2(\phi+\eta) + \cos^2[\theta_F(t)]\,\cos(2\phi)\cos(2\eta)\right). \quad (4.125)$$

To be able to apply the Fourier series for the modulation angle, we convert $\cos^2\theta_F$ to $\cos(2\theta_F)$:

$$I(t) = \frac{E_0^2}{\cos^2\eta}\left(\sin^2(\phi+\eta) + \frac{1}{2}(1 + \cos[2\theta_F(t)])\,\cos(2\phi)\cos(2\eta)\right), \quad (4.126)$$

and collect the constant terms in front of the equation:

$$I(t) = \frac{E_0^2}{\cos^2\eta}\left(\sin^2(\phi+\eta) + \frac{1}{2}\cos(2\phi)\cos(2\eta) + \frac{1}{2}\cos(2\phi)\cos(2\eta)\cos[2\theta_F(t)]\right). \quad (4.127)$$

The most important term for the measurement in eq. (4.127) is the last term on the right as it is the amplitude of the modulation term $\cos[2\theta_F(t)]$. We immediately see that a change of the analyzer position ϕ has no influence on the ellipticity η because the two angles are now separated from each other. Therefore, a zeroing procedure is not possible and the unknown ellipticity η cannot be determined by rotating the analyzer. So, the position of the modulator in the sequence of the optical components is highly relevant for the success of the measurement.

5 Microscopic model

5.1 Becquerel equation and its relation to Zeeman splitting

Before we discuss the quantum-mechanical derivation of the diagonal and off-diagonal elements of the optical conductivity tensor, we will briefly present a further macroscopic model, which leads to the generalized Becquerel formula. This model will allow us later to make a connection to variables derived in quantum mechanics.

In Section 3.2, we have determined for an electron in a homogeneous magnetic field in eq. (3.26) that the shift of the resonance frequency $\omega_{0\pm}$ depends on the circular polarization:

$$\omega_{0\pm} - \omega_0 = \pm\frac{1}{2}\omega_c = \pm\,\omega_L, \tag{5.1}$$

where the Larmor frequency $\omega_L < 0$ for electrons. To establish the connection with the Faraday rotation, we recall eq. (2.135):

$$\theta_F = \frac{\omega d}{2c}(n_+ - n_-). \tag{5.2}$$

Let us assume that the line shape of $n(\omega)$ does not change when the resonance frequency is shifted. $n(\omega)$ just displaces along the frequency axis. Equation (5.2) can then be written as

$$\theta_F = \frac{\omega d}{2c}\left[n(\omega - \omega_L) - n(\omega + \omega_L)\right]. \tag{5.3}$$

For $|\omega_L| \ll \omega$, eq. (5.3) is equivalent to the first derivative with respect to the angular frequency. This is the ***Becquerel equation***:

$$\theta_F = -\,\omega_L\frac{\omega d}{c}\frac{\partial n}{\partial\omega}. \tag{5.4}$$

Of course, eq. (5.4) can only represent a crude approximation. The assumption that the line shape of $n(\omega)$ does not change when a shift in the resonance frequency occurs is certainly an oversimplification. For example, consider the influence of the Zeeman splitting on the line shape of $\theta_F = \frac{\omega d}{2c}(n_+ - n_-)$. In Fig. 5.1, on the left, we have plotted the dipole-allowed $S \to P$ transitions for the simple case of total angular momentum $J = 1/2$. The optical transitions occurs from the $^2S_{1/2}$ to the $^2P_{1/2}$ atomic levels, obeying the selection rules $\Delta L = \pm 1$ and $\Delta m_J = \pm 1$. Suppose now that the oscillator strength for the right circularly polarized (rcp, $\Delta m_J = -1$) and left circularly polarized (lcp, $\Delta m_J = +1$) light transition is not equal and, thus, n_+ and n_- are different, then the frequency dependence of the Faraday rotation θ_F will resemble the dispersion $n(\omega)$ itself. This is visualized in Fig. 5.1 on the right. If the two transitions have about equal oscillator strength,

https://doi.org/10.1515/9783110635300-005

then the Faraday rotation will have an absorptive shape, showing a pronounced dip. It follows that the density of states as well as the transition probability will influence the line shape of the Faraday rotation. Nevertheless, the simple Becquerel formula can be useful to estimate the shape of the Faraday rotation spectra. In the problem in Section 5.7.1, it is used to calculate the sign of the Faraday rotation of quartz.

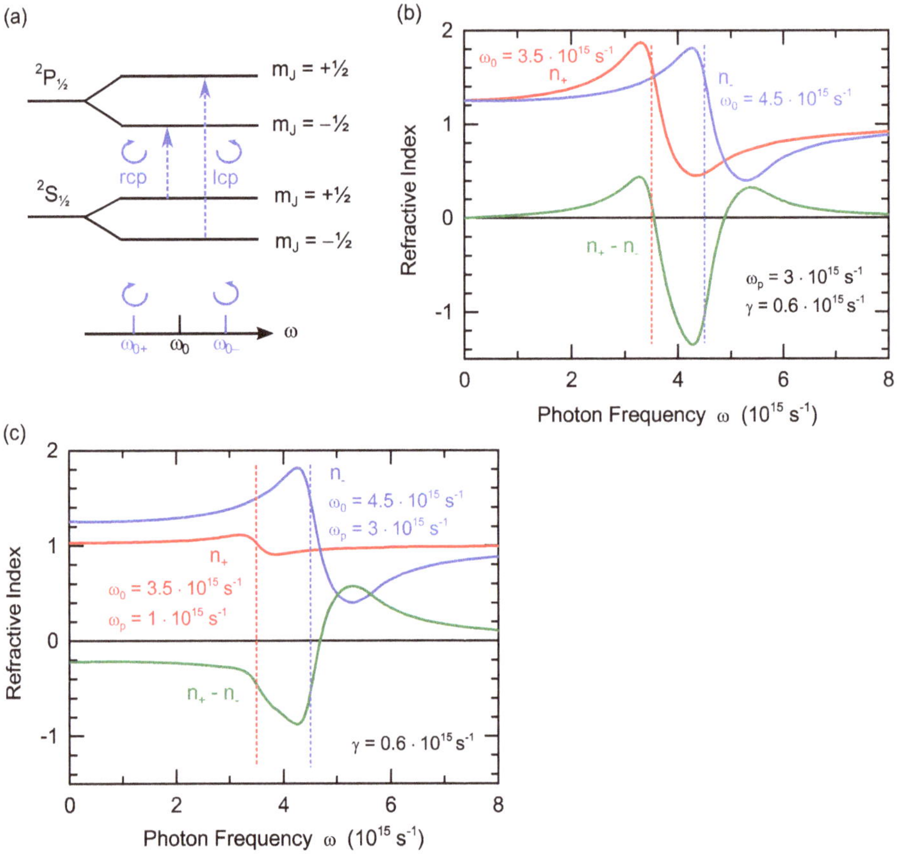

Fig. 5.1: Influence of different intensities on the Faraday rotation $\theta_F = (\omega d/2c)(n_+ - n_-)$ of an ${}^2S_{1/2} \rightarrow {}^2P_{1/2}$ transition in the Zeeman effect. (a) Energy-level scheme of the ${}^2S_{1/2} \rightarrow {}^2P_{1/2}$ transition. (b) and (c) Difference between the two refractive indices in circular basis for equal and unequal intensities, respectively.

A more general equation is obtained if Becquerel's equation is not only differentiated with respect to the photon energy ω but also to the population density N and to the oscillator strength f:

$$\theta_F = -\frac{\omega^2 d}{2\omega_0 c}\left\{\frac{\partial n}{\partial \omega_0}\delta\omega_0 + \frac{\partial n}{\partial f}\delta f + 2\frac{\partial n}{\partial N}\delta N\right\}. \tag{5.5}$$

The prefactors in eq. (5.5) result from a quantum-mechanical derivation, which can be found in Ref. (100). The physical interpretation of the three terms in eq. (5.5) is as follows:

Term 1: The Zeeman splitting leads to a ***diamagnetic rotation***. The term "diamagnetic" means temperature independent. The following relation applies:

$$\delta\omega_0 = -2a\omega_L, \text{ where } a \cong 1. \tag{5.6}$$

Term 2: This term describes the disturbance of the intensities in a strong magnetic field, which is known as the Paschen–Back effect. It occurs when the Zeeman splitting becomes comparable to the multiplet splitting. This term is called ***Darwin term***. The following equation applies:

$$\frac{\partial f}{f} = 2b\frac{\omega_L}{\Delta\omega_{\text{s.o.}}}, \text{ where } \Delta\omega_{\text{s.o.}} = \text{multiplet splitting } (\text{L}\cdot\text{S}). \tag{5.7}$$

Term 3: This term describes the different population of the energy levels according to the Boltzmann distribution. It leads to a temperature dependence that is inversely proportional as in a paramagnet. Therefore, it is called ***paramagnetic rotation***. We have

$$\frac{\delta N}{N} = -\frac{\mu_B B}{kT} = -\frac{\hbar\omega_L}{kT}. \tag{5.8}$$

For vanishingly small absorption ($\gamma = 0$) and magnetic field $B = 0$, the real part of the dielectric function is, according to eqs. (3.4) and (3.9), given by

$$\varepsilon_1 - 1 = n^2 - 1 = \frac{Ne^2}{\varepsilon_0 m}\frac{f}{\omega_0^2 - \omega^2}, \tag{5.9}$$

where f is the oscillator strength. After performing the differentiation in eq. (5.5), we obtain the ***generalized Becquerel equation***:

$$\theta_F = -\frac{Ne^2}{\varepsilon_0 mnc} f\omega_L d\left\{\frac{a_1\omega^2}{\left(\omega_0^2 - \omega^2\right)^2} + \frac{a_2\omega^2}{2\omega_0\left(\omega_0^2 - \omega^2\right)\Delta\omega_{\text{s.o.}}} - \frac{a_3\omega^2\hbar}{2\omega_0\left(\omega_0^2 - \omega^2\right)kT}\right\}. \tag{5.10}$$

Details of the derivation can be found in problem (a) in Section 5.7.2. Now, we explicitly see that the name "diamagnetic" for the first and "paramagnetic" for the last term on the right refers to the temperature dependence. Since the three constants have values $a_1, a_2, a_3 \cong 1$, the ratio of the magnitude of the three terms is as follows:

$$\frac{2}{\hbar\left(\omega_0 - \frac{\omega^2}{\omega_0}\right)} : \frac{1}{\hbar\Delta\omega_{\text{s.o.}}} : \frac{1}{kT}. \tag{5.11}$$

Using the Lorentz formula, we calculate the constant a_1 of the first term for an atomic $^2S_{1/2} \rightarrow {}^2P_{1/2}$ transition in problem (b) and the constant a_1 for the basic Becquerel formula in problem (c) in Section 5.7.2. Both calculations corroborate the statement from above that $a_1 \cong 1$.

5.2 Absorptive/dispersive quantities and Kramers-Kronig transformation

Before we calculate the optical conductivity tensor with the help of quantum mechanics, we want to make some general statements. We distinguish between ***absorptive*** and ***dispersive*** quantities, which are usually combined into a single complex quantity. In the diagonal element $\tilde{\sigma}_{xx}$ of the optical conductivity tensor, the real part σ_{1xx} constitutes the absorptive quantity and the imaginary part σ_{2xx} constitutes the dispersive quantity. Considering the off-diagonal element $\tilde{\sigma}_{xy}$ of the optical conductivity tensor, the situation is reversed: The absorptive part is σ_{2xy} and the dispersive part is σ_{1xy}. The origin for this commutation is found in eq. (3.36), where the optical conductivity in the circular basis, which is the physically relevant basis for discussing the optical transitions, is expressed as a linear combination of diagonal element and off-diagonal element multiplied with the imaginary unit. It is important to know that the absorptive and dispersive parts of a complex physical quantity are not independent of each other if we presume conservation of energy and the principle of causality (88). In general, we only need to derive by a theory or in experiment one part of a complex quantity. This is usually the absorptive part, because it is easier to derive or to measure. The dispersive part is then calculated from the absorptive part with the help of an integral equation: the ***Kramers–Kronig transformation***. In case of the optical conductivity, it is defined as

$$\sigma_{1ij}(\omega) = \frac{2}{\pi} P \int_0^\infty \frac{\omega' \sigma_{2ij}(\omega')}{\omega'^2 - \omega^2} d\omega', \tag{5.12}$$

$$\sigma_{2ij}(\omega) = -\frac{2\omega}{\pi} P \int_0^\infty \frac{\sigma_{1ij}(\omega')}{\omega'^2 - \omega^2} d\omega', \tag{5.13}$$

where “P” denotes the principal value of the integral, which is the integral evaluated along the integration path omitting the poles. The foundation of the Kramers–Kronig transformation is Cauchy’s integral formula and the residue theorem in complex analysis.

In the next section, we will show briefly where the concept of spin–orbit energy is originating from. Then, we will move on to the quantum-mechanical treatment of the optical conductivity.

5.3 Origin of spin–orbit interaction

To get an idea of the spin–orbit energy, we first consider the motion of an electron of charge $q = -e$ with speed $\vec{v}$ in a magnetic field $\vec{B}$. In addition to the force due to the electric field, the Lorentz force due to the magnetic field is acting:

$$\vec{F} = -e\left[\vec{E} + \left(\vec{v} \times \vec{B}\right)\right]. \tag{5.14}$$

The term $\vec{v} \times \vec{B}$ can be looked at as an effective electric field, generated by the magnetic field, acting on the electron. Analogously, the torque on a magnet (or spin), moving in a magnetic and electric field, is not only produced by the magnetic field $\vec{B}$ alone but also by an effective magnetic field which is proportional to $\left(\vec{v} \times \vec{E}\right)$:

$$\vec{M} = \vec{\mu} \times \left[\vec{B} - \frac{1}{c^2}\left(\vec{v} \times \vec{E}\right)\right]. \tag{5.15}$$

The magnetic moment $\vec{\mu}$ of the electron is defined as

$$\vec{\mu} = -\frac{e}{2m} g \hbar \vec{s}, \tag{5.16}$$

with electron spin $s = \pm \frac{1}{2}$ and g-factor $g \cong 2$ for electrons. Because the potential energy is the dot product between magnetic moment and total magnetic field, the spin of the electron should have a term in its potential energy, which is defined by the product of the magnetic moment $\vec{\mu}$ with the effective magnetic field generated by the movement in the electric field:

$$E'_{\text{s.o.}} = \vec{\mu} \cdot \frac{1}{c^2}\left(\vec{v} \times \vec{E}\right). \tag{5.17}$$

This is not the correct equation. A relativistic correction must be considered, which originates from the acceleration due to the curvature of the electron orbital. This leads to the so-called Thomas precession (101), which is exactly half the size of the classically derived spin–orbit energy $E'_{\text{s.o.}}$ but opposite in sign. With the correction, the spin–orbit energy is finally obtained as

$$E_{\text{s.o.}} = -\frac{e}{2m^2c^2}(g-1)\,\hbar\vec{s} \cdot \left(\vec{p} \times \vec{E}\right), \tag{5.18}$$

where we have set $\vec{p} = m\vec{v}$. Now, eq. (5.18) is in accordance with the relativistic Dirac theory. For bound electrons, the electric field is due to the central potential U, which is generated by the nucleus and the core electrons. The electric field can be expressed as the gradient of the central potential:

$$\vec{E} = \frac{1}{e}\frac{\vec{r}}{r}\frac{\partial U}{\partial r}. \tag{5.19}$$

Substituting eq. (5.19) into (5.18) yields the spin–orbit energy:

$$E_{\text{s.o.}} = -\frac{(g-1)\hbar}{2m^2c^2r}\frac{\partial U}{\partial r}\vec{s}\cdot(\vec{p}\times\vec{r}). \tag{5.20}$$

Because of the relation $\vec{r}\times\vec{p} = \hbar\vec{L}$, where $\vec{L}$ is the orbital angular momentum of the electron, $E_{\text{s.o.}}$ is expressed as follows:

$$E_{\text{s.o.}} = \frac{(g-1)\hbar^2}{2m^2c^2r}\frac{\partial U}{\partial r}\left(\vec{s}\cdot\vec{L}\right). \tag{5.21}$$

Equation (5.21) contains the dot product of $\vec{s}$ and $\vec{L}$, explaining the name spin–orbit energy. Summarizing, the spin–orbit energy is a consequence of the movement of an electric charge in the Coulomb field that binds the electron to the atomic nucleus. It generates an extra torque as if an effective magnetic field would be present.

5.4 Quantum-mechanical calculation of the optical conductivity tensor elements

If we introduce the Pauli spin operator $\vec{\sigma} = 2\vec{s}$, assume that $g = 2$, and use for the central potential the expression $\vec{\nabla}U = (\vec{r}/r)\partial U/\partial r$, then eq. (5.20) can be rewritten as

$$E_{\text{s.o.}} = \frac{\hbar}{4m^2c^2}\vec{p}\cdot\left(\vec{\sigma}\times\vec{\nabla}U\right). \tag{5.22}$$

In order to compose the ground-state Hamiltonian $\mathscr{H}_0$, we include, besides a static magnetic field, the spin–orbit interaction $E_{\text{s.o.}}$, and apply the Bloch approximation, that is, that $\mathscr{H}_0$ can be written as the sum of one-electron Hamiltonian $\mathscr{H}_{0i}$ for electron i at position $\vec{r}_i$ with momentum $\vec{p}_i$. Now, we can express $\mathscr{H}_0$ without radiation field but including a static external magnetic field $\vec{B}$ as

$$\mathscr{H}_0 = \sum_i\left\{\frac{1}{2m}\left[\vec{p}_i + e\vec{A}_{\text{M}}(\vec{r}_i)\right]^2 + \frac{\hbar e}{2m}\vec{\sigma}_i\cdot\vec{B} + U(\vec{r}_i) + \frac{\hbar}{4m^2c^2}\left[\vec{p}_i + e\vec{A}_{\text{M}}(\vec{r}_i)\right]\cdot\left[\vec{\sigma}_i\times\vec{\nabla}U(\vec{r}_i)\right]\right\}. \tag{5.23}$$

In eq. (5.23), $\vec{A}_{\text{M}}$ is the vector potential of the applied magnetic field, $\vec{B} = \vec{\nabla}\times\vec{A}_{\text{M}}$, and $\vec{p}_i$ and $\vec{p}_i + e\vec{A}_{\text{M}}(\vec{r}_i)$ are the canonical and the modified momentum operators, respectively. The first term on the right represents the kinetic energy, the second term the Zeeman energy, the third term represents the central potential, and the fourth term the spin–orbit energy. Note that the canonical momentum operator must be replaced by the modified momentum operator everywhere in the Hamiltonian to take correctly into account the vector potential of the static magnetic field.

This is a consequence of the Euler–Lagrange equations of motion. The commutator of $\vec{r}_i$ with $\mathcal{H}_0$ satisfies the relation:

$$[\vec{r}_i, \mathcal{H}_0] = \frac{i\hbar}{m}\vec{\pi}_i, \tag{5.24}$$

defining the kinetic momentum operator $\vec{\pi}_i$ of electron i including spin–orbit interaction:

$$\vec{\pi}_i = \vec{p}_i + e\vec{A}_{\mathrm{M}}(\vec{r}_i) + \frac{\hbar}{4mc^2}\left[\vec{\sigma}_i \times \vec{\nabla}U(\vec{r}_i)\right]. \tag{5.25}$$

Switching on the electromagnetic field, it will be described by the vector potential $\vec{A}_{\mathrm{L}}(\vec{r}_i)$ of the oscillating electric field of the light. The magnetic field of the light can be neglected as it is much weaker than the electric field. The Hamiltonian $\mathcal{H}$ can now be written according to first-order perturbation theory as the sum of the ground-state Hamiltonian $\mathcal{H}_0$ and the interaction Hamiltonian $\mathcal{H}_{\mathrm{int}}$:

$$\mathcal{H} = \mathcal{H}_0 + \mathcal{H}_{\mathrm{int}}. \tag{5.26}$$

The interaction term $\mathcal{H}_{\mathrm{int}}$,

$$\mathcal{H}_{\mathrm{int}} = \sum_i \frac{e}{m}\vec{\pi}_i \cdot \vec{A}_{\mathrm{L}}(\vec{r}_i), \tag{5.27}$$

represents in a solid-state material the interaction of light with electron i in a static magnetic field, whereupon the electron is excited from the initial state $|\alpha\rangle$ into the final state $|\beta\rangle$.

In time-dependent perturbation theory, the eigenvalues E_{in} of the wave function u_{in} of a non-perturbed Hamiltonian $\mathcal{H}_{0i}$ are defined as

$$\mathcal{H}_{0i}\, u_{in} = E_{in}\, u_{in}, \quad n \in \mathbb{N}. \tag{5.28}$$

We can apply Fermi's golden rule to derive the transition rate, that is, the transition probability per unit time:

$$W_{\alpha\beta} = \frac{2\pi}{\hbar}\left|\langle\beta|\mathcal{H}_{\mathrm{int}}|\alpha\rangle\right|^2 D(E)|_{E=E_\beta - E_\alpha}, \tag{5.29}$$

where $D(E)$ is the density of states evaluated at the transition energy:

$$D(E)|_{E=E_\beta - E_\alpha} \cong \frac{1}{\hbar}\left[\delta\left(\omega_{\alpha\beta} - \omega\right) + \delta\left(\omega_{\alpha\beta} + \omega\right)\right], \tag{5.30}$$

where $\hbar\omega_{\alpha\beta}$ is the transition energy and $\delta(\omega)$ is the (Dirac) delta function. We apply Fermi's golden rule to calculate the transition rate for an electron that is excited from the initial state $|\alpha\rangle$ into the final state $|\beta\rangle$ by a plane wave which is linearly polarized in the x direction:

$$W_{\alpha\beta x} = \frac{\pi}{2}\left(\frac{Ee}{m\hbar\omega}\right)^2 |\langle\beta|\pi_x|\alpha\rangle|^2 \left[\delta\left(\omega_{\alpha\beta}-\omega\right)+\delta\left(\omega_{\alpha\beta}+\omega\right)\right]. \tag{5.31}$$

If the electron is excited by rcp (+) and lcp (−) light, we get the transition rate in the circular basis:

$$W_{\alpha\beta\pm} = \frac{\pi}{4}\left(\frac{Ee}{m\hbar\omega}\right)^2 |\langle\beta|\pi_\pm|\alpha\rangle|^2 \left[\delta\left(\omega_{\alpha\beta\pm}-\omega\right)+\delta\left(\omega_{\alpha\beta\pm}+\omega\right)\right]. \tag{5.32}$$

Now, the question is, how do we connect the quantum world, which is represented by the transition rate, to the macroscopic world, where we want to measure a quantity in experiment? The transition rate describes how many electrons will be excited in the material by absorbing the photons of the plane wave. From the point of view of the plane wave, light gets absorbed by the material and loses, therefore, a certain amount of energy per unit time. So, the quantity that connects both worlds is the absorbed power. In the macroscopic optic world, the absorbed power in a volume V of the solid-state material is according to eq. (2.48):

$$P=\left\langle\int \mathrm{Re}\left\{\vec{\tilde{j}}\right\}\cdot\mathrm{Re}\left\{\vec{E}\right\}dV\right\rangle_t. \tag{5.33}$$

For the integration over the volume, we may neglect the small spatial variation of the electric field of the light in comparison with atomic distances and assume it to be constant. Ohm's law then yields:

$$P=V\left\langle\mathrm{Re}\left\{\overleftrightarrow{\tilde{\sigma}}\vec{E}\right\}\cdot\mathrm{Re}\left\{\vec{E}\right\}\right\rangle_t. \tag{5.34}$$

According to eq. (3.43), we get for a linearly polarized wave in x direction:

$$P_x=\frac{1}{2}V\sigma_{1xx}E^2. \tag{5.35}$$

Using the definition of the optical conductivity in a circular basis from eq. (3.36), we obtain for rcp and lcp light, accordingly:

$$\begin{aligned}P_\pm &= V\left\langle\mathrm{Re}\left\{\tilde{\sigma}_\pm\vec{E}_\pm\right\}\cdot\mathrm{Re}\left\{\vec{E}_\pm\right\}\right\rangle_t = V\left\langle\mathrm{Re}\left\{\left(\tilde{\sigma}_{xx}\pm\mathrm{i}\tilde{\sigma}_{xy}\right)\vec{E}_\pm\right\}\cdot\mathrm{Re}\left\{\vec{E}_\pm\right\}\right\rangle_t\\ &= V\left\langle\mathrm{Re}\left\{\left((\sigma_{1xx}+\mathrm{i}\sigma_{2xx})\pm\mathrm{i}(\sigma_{1xy}+\mathrm{i}\sigma_{2xy})\right)\vec{E}_\pm\right\}\cdot\mathrm{Re}\left\{\vec{E}_\pm\right\}\right\rangle_t\\ &= V\left\langle\mathrm{Re}\left\{\left((\sigma_{1xx}\mp\sigma_{2xy})+\mathrm{i}(\sigma_{2xx}\pm\sigma_{1xy})\right)\vec{E}_\pm\right\}\cdot\mathrm{Re}\left\{\vec{E}_\pm\right\}\right\rangle_t.\end{aligned} \tag{5.36}$$

In the last part of eq. (5.36), the time average of the real part of the first two terms of the optical conductivity yields a factor 1/2, whereas the time average of the real part of the last two terms of the optical conductivity vanishes, as explained in eq. (3.43). So, we finally get:

$$P_{\pm} = \frac{1}{2} VE^2 \left(\sigma_{1xx} \mp \sigma_{2xy}\right). \tag{5.37}$$

From eq. (5.35), we obtain the absorptive part of the diagonal element:

$$\sigma_{1xx} = \frac{2P_x}{VE^2}. \tag{5.38}$$

For the absorptive part of the off-diagonal element, it follows from eq. (5.37) that

$$\sigma_{2xy} = \frac{P_- - P_+}{VE^2}. \tag{5.39}$$

To connect to the quantum world, we must consider how the absorbed power relates to the transition rates. For each electron which is excited per unit time, a photon of transition energy $\hbar\omega_{\alpha\beta}$ is absorbed. Hence, the absorbed power is simply the photon energy times the number of excitations per unit time: $P = \hbar\omega_{\alpha\beta} \cdot W_{\alpha\beta}$. But we must be careful when we equate the two equations for the power with each other because we are then relating the quantum world with the macroscopic optic world. The problem is hidden in different time dependences used in both worlds for describing plane waves, as mentioned in Section 4.1. In quantum mechanics, a time dependence $e^{-\mathrm{i}\omega t}$ is chosen in place of $e^{+\mathrm{i}\omega t}$, which is common in optics. This will interchange the definitions for rcp and lcp light as defined by the Jones vectors. To take this into account, we must swap rcp by lcp and vice versa in the quantum world. This will only affect the result for quantities in the circular basis:

$$P_{\pm} = \hbar\omega_{\alpha\beta\pm} W_{\alpha\beta\mp}. \tag{5.40}$$

Finally, we can connect the absorptive quantities of the optical conductivity tensor to the quantum-mechanically derived transition rates:

$$\sigma_{1xx} = \frac{2\hbar\omega}{VE^2} W_{\alpha\beta x} = \frac{\pi e^2}{\hbar\omega m^2 V} \sum_{\alpha\beta} |\langle\beta|\pi_x|\alpha\rangle|^2 \left[\delta\left(\omega_{\alpha\beta} - \omega\right) + \delta\left(\omega_{\alpha\beta} + \omega\right)\right], \tag{5.41}$$

$$\sigma_{2xy} = \frac{\hbar\omega}{VE^2} \left(W_{\alpha\beta+} - W_{\alpha\beta-}\right) = \frac{\pi e^2}{4\hbar\omega m^2 V} \sum_{\alpha\beta} \left[|\langle\beta|\pi_+|\alpha\rangle|^2 - |\langle\beta|\pi_-|\alpha\rangle|^2\right] \left[\delta\left(\omega_{\alpha\beta} - \omega\right) + \delta\left(\omega_{\alpha\beta} + \omega\right)\right], \tag{5.42}$$

where, a little bit sloppy, we have approximated $\omega_{\alpha\beta\pm} \cong \omega_{\alpha\beta}$. This is a good approximation as long as the splitting of the two resonance frequencies, which is (according to eq. (3.26)) equal to the cyclotron resonance frequency ω_c is much smaller than the resonance frequency $\omega_{\alpha\beta}$. Equation (5.42) yields a very simple interpretation of magneto-optic effects in terms of a quantum-mechanical explanation. The imaginary part of the off-diagonal optical conductivity element, which directly relates to the Faraday and polar Kerr effect, is proportional to the difference of the transition rates for rcp and lcp light. So, the larger the difference in the transition rates or, equivalently, in the square of the absolute value of the transition-matrix element of rcp and lcp light, the larger are the magneto-optic effects.

With the help of the Kramers–Kronig transformation (eqs. (5.12) and (5.13)), the dispersive part of the optical conductivity tensor can be computed:

$$\sigma_{2xx} = -\frac{2\omega e^2}{\hbar m^2 V}\sum_{\alpha\beta}\frac{|\langle\beta|\pi_x|\alpha\rangle|^2}{|\omega_{\alpha\beta}|\left(\omega_{\alpha\beta}^2-\omega^2\right)}, \tag{5.43}$$

$$\sigma_{1xy} = \frac{e^2}{2\hbar m^2 V}\sum_{\alpha\beta}\frac{|\langle\beta|\pi_+|\alpha\rangle|^2-|\langle\beta|\pi_-|\alpha\rangle|^2}{\omega_{\alpha\beta}^2-\omega^2}. \tag{5.44}$$

In general, for non-ferromagnetic materials, a series expansion of eqs. (5.43) and (5.44) can be performed as a function of the magnetic field. Exceptions occur when the magnetic splitting of the terms is comparable to or larger than the band gap. So, if we expand, for example, σ_{1xy} in powers of B (and do not approximate $\omega_{\alpha\beta\pm} \cong \omega_{\alpha\beta}$), then we obtain up to first order in B:

$$\begin{aligned}\sigma_{1xy}^{(1)} = -\frac{e^2}{2\hbar m^2 V}\sum_{\alpha\beta}\Bigg[&\frac{2\omega_{\alpha\beta+}\,|\langle\beta|\pi_+|\alpha\rangle|^2}{\left(\omega_{\alpha\beta+}^2-\omega^2\right)^2}\frac{\delta\omega_{\alpha\beta+}}{\delta B}-\frac{1}{\omega_{\alpha\beta+}^2-\omega^2}\frac{\delta|\langle\beta|\pi_+|\alpha\rangle|^2}{\delta B}\\ &-\frac{2\omega_{\alpha\beta-}\,|\langle\beta|\pi_-|\alpha\rangle|^2}{\left(\omega_{\alpha\beta-}^2-\omega^2\right)^2}\frac{\delta\omega_{\alpha\beta-}}{\delta B}+\frac{1}{\omega_{\alpha\beta-}^2-\omega^2}\frac{\delta|\langle\beta|\pi_-|\alpha\rangle|^2}{\delta B}\Bigg].\end{aligned} \tag{5.45}$$

We notice from eq. (5.45) that $\sigma_{1xy}^{(1)}$ and, thus, the Faraday rotation arises from two effects: First, from the Zeeman splitting $\delta\omega_{\alpha\beta\pm}/\delta B$ and, second, from the change of the square of the absolute value of the matrix element. The first and third terms are important near the resonance frequency due to the denominator that contains the square of the energy difference. It corresponds to the diamagnetic rotation in eq. (5.10). The second and fourth terms give a contribution to the Darwin term in eq. (5.10). For large frequencies $\omega \gg \omega_{\alpha\beta\pm}$ the second and the fourth term are dominant.

To demonstrate the level of accuracy that modern ab initio calculations attain, we plot a comparison of the measured polar Kerr rotation and ellipticity of the noble metals gold, silver, and copper with the calculated spectra in Fig. 5.2 (102). The ab initio calculations were made using a spin-polarized fully relativistic linear-muffin-tin-orbital method in the atomic sphere approximation with combined correction term taken into account (102). The experimental spectra were obtained by the three-lock-in-amplifier method described in Section 4.2.3. The polar Kerr

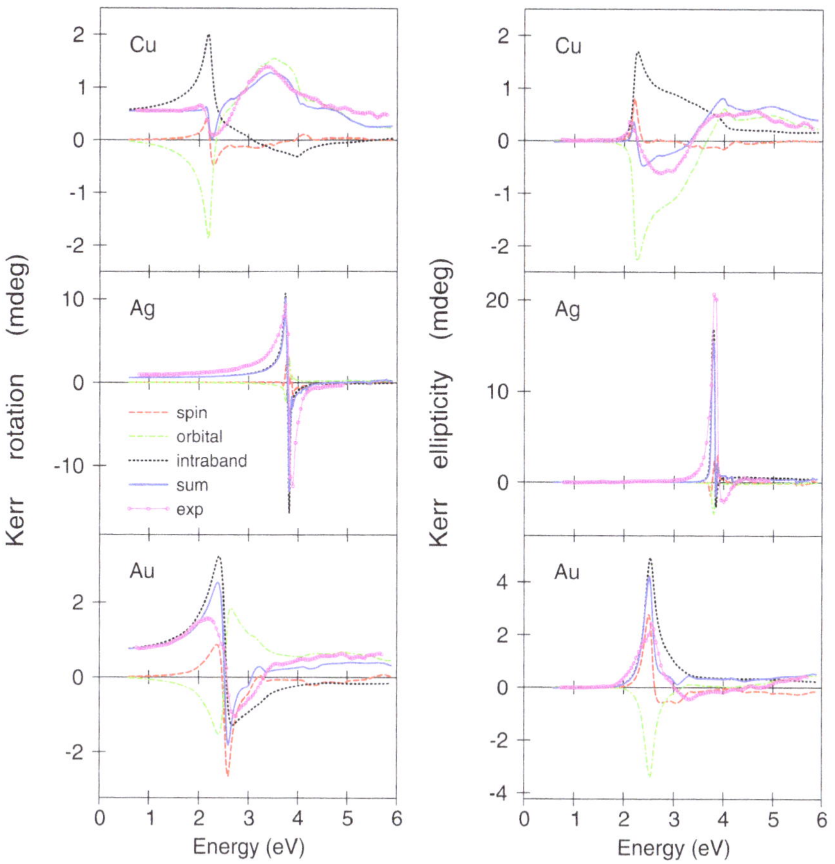

Fig. 5.2: Decomposition of the theoretically calculated polar Kerr rotation (left column) and polar Kerr ellipticity (right column) spectra of noble metals into spin (red dashed lines) and orbital (green dashed-dotted lines) contributions of the Zeeman term. Full blue lines present the theoretically calculated magneto-optic (MO) spectra with the orbital contribution reduced by 45%, 75%, and 45% in Cu, Ag, and Au, respectively. The experimentally measured MO spectra in the applied field of 1.5 T are presented by pink circles. The intraband contributions are shown by black dotted lines. Figure from Ref. (102).

rotation of the noble metals is known to be very small, reaching only a few thousandths of a degree, and therefore, not easy to measure. All the more, it will be difficult to accurately calculate these effects. Nevertheless, the calculated polar Kerr spectra (full blue lines) are for all three noble metals in excellent agreement with the experimental curves (pink circles). To get the best agreement between the calculated and the experimental polar Kerr rotation and ellipticity spectra in absolute values, the contribution of the orbital Zeeman term was decreased by 45%, 75%, and 45% in Cu, Ag, and Au, respectively (102).

5.5 Relation between off-diagonal conductivity and transition rates

In this last section, we want to discuss the conditions under which a magneto-optic effect occurs. This is equivalent to the statement that there exists a finite off-diagonal conductivity $\sigma_{2xy}(\omega) \neq 0$. The basic model had been proposed by Erskine and Stern (103) and was refined by Schoenes (104). To consider the energy bands of solid-state materials, as opposed to discrete energy levels of atoms, we assume broad transitions with width ΔE_{B} instead of line spectra as shown in Fig. 5.3. However, the total absorption should be preserved due to sum rules. Let us consider a fully polarized initial state which has only one spin direction, say, spin up (↑). The spin–orbit splitting causes an energy shift between the absorption bands, which are described by the transition rate $W^{\uparrow}_{\alpha\beta\pm}$ of rcp and lcp light as plotted in Fig. 5.3(a). The difference between the two transition rates $\Delta W^{\uparrow} = W^{\uparrow}_{\alpha\beta+} - W^{\uparrow}_{\alpha\beta-}$, according to eq. (5.42), is proportional to the absorptive part $\sigma^{\uparrow}_{2xy}(\omega)$ of the off-diagonal conductivity which causes the magnetic circular dichroism. As shown in Fig. 5.3(b), this difference has the shape of a procumbent "S". For unsophisticated density of states, the signal is proportional to the ratio of spin–orbit splitting $\Delta E_{\mathrm{s.o.}}$ to bandwidth ΔE_{B}.

For the other spin direction, spin down (↓), we can argue the same way. However, $\sigma^{\downarrow}_{2xy}(\omega)$ has now the ***opposite*** sign due to its direct proportionality to the magnetic field as depicted in Fig. 5.3(c). Note that the magnetic field is here generated by the magnetization, which is itself created by the spin. Thus, if the initial state is unpolarized, meaning the two spin directions equally populate the initial state (and the same number of final states are available), the two contributions $\Delta W^{\uparrow}$ and $\Delta W^{\downarrow}$ to the off-diagonal element cancel each other exactly, as shown in Fig. 5.3(d), yielding zero contribution to the off-diagonal optical conductivity: $\sigma^{\mathrm{unpol}}_{2xy} = \sigma^{\uparrow}_{2xy} + \sigma^{\downarrow}_{2xy} = 0$. Thus, an ***unpolarized initial state*** will provide ***no contribution*** to the magneto-optic signal. However, this is only true as long as the ***final state*** is ***degenerate*** for the two spin directions. This is demonstrated in Fig. 5.3(e), where the degeneracy is lifted and the two spin contributions to the off-diagonal optical conductivity are plotted by displacing along the photon energy axis. In this case, as a first approximation, the total signal $\sigma^{\mathrm{ex}}_{2xy}$ is proportional to the ratio of exchange splitting ΔE_{ex} to bandwidth ΔE_{B}. However, the line shape of $\sigma^{\mathrm{ex}}_{2xy}$, as shown in Fig. 5.3(f), is now ***clearly different*** from the case of the completely polarized initial state $\sigma^{\mathrm{pol}}_{2xy}$ of Fig. 5.3(b), which has the shape of a procumbent "S". This simple model is very useful in the assignment of magneto-optic structures to specific electronic transitions as we will see in the next chapter.

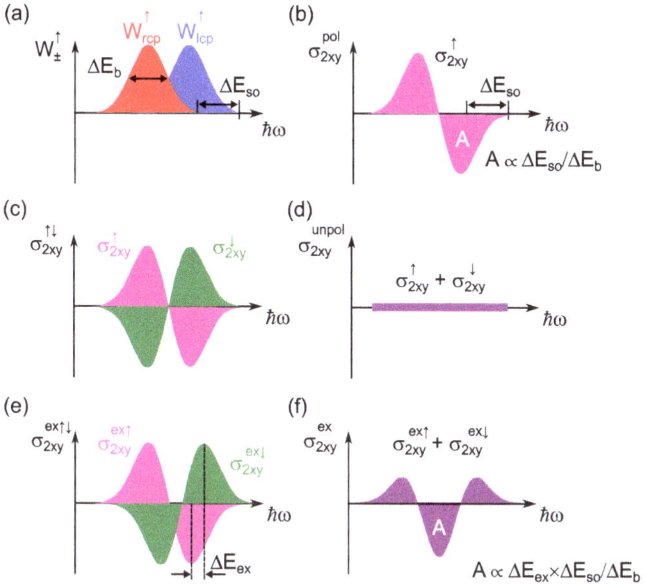

Fig. 5.3: Model to explain the relation between the transition rates $W_\pm$ and the absorptive part $\sigma_{2xy}(\omega)$ of the off-diagonal conductivity. ΔE_b is the width of the absorption band, ΔE_{so} is the spin–orbit splitting, and ΔE_{ex} is the exchange splitting of the final state.

5.6 Comprehension questions

5.6.1 Becquerel equation

(a) Which assumption must be made for the derivation of the Becquerel equation?
(b) What is the basic statement of the Becquerel equation? How can it be generalized?
(c) What is the physical meaning of the three terms of the generalized Becquerel equation?

5.6.2 Dispersive and absorptive quantities: Kramers–Kronig transformation

(a) Explain the difference between ***absorptive*** and ***dispersive*** physical quantities. Which one is easier to derive theoretically and why?
(b) We very often deal with complex physical quantities in optics. Can real and imaginary part be of the same type, that is, both absorptive or both dispersive?
(c) Is there a relation between the real part and the imaginary part?

(d) In order to know the complete complex optical function, do we have to know in general the real part as well as the imaginary part or is it possible to calculate one out of the other?

5.6.3 Spin–orbit interaction

(a) In the framework of classical mechanics, how can you explain the origin of the spin–orbit energy?
(b) How can this be generalized to a quantum-mechanical picture?

5.6.4 Quantum-mechanical calculation of the optical conductivity tensor

(a) How is usually an electromagnetic field incorporated into a Hamiltonian? Which operator must be substituted for that?
(b) In order to explain magneto-optic effects in quantum mechanics, we need to apply perturbation theory. Which terms must be considered for the ground-state Hamiltonian and how does the interaction Hamiltonian look like?
(c) Which quantity can be calculated using (time-dependent) perturbation theory? Which famous rule is used for that? How does this quantity relate to a physical quantity that can be derived from experiment?
(d) What is the general conclusion of the quantum-mechanical treatment of magneto-optics? In other words, how do the magneto-optic properties depend on the transition rates?
(e) Explain by using a simple graphic model under which circumstances a magneto-optic response will be present and which line shapes can be found for magneto-optically active electronic transitions.

5.7 Problems

5.7.1 Faraday rotation of quartz

Determine the sign of the Faraday rotation of quartz in the range of normal dispersion, that is, $\partial n(\omega)/\partial\omega > 0$.

Hint: Use Faraday's definition of the sign: a positive rotation corresponds to the direction of the positive current producing the corresponding magnetic field in a coil. Further, assume that the magnetic field is along the direction of light propagation. Remember how the resonance frequency $\omega_{0\pm}$ shifts for rcp and lcp light in the Lorentz model [*eq.* (3.26)] *and assume that* $n_\pm(\omega)$ *is derived from* $n(\omega)$ *by a simple shift along the frequency axis without changing the line shape* (*Becquerel equations* (5.3) and (5.4)).

5.7.2 Faraday rotation of an $S \rightarrow P$ transition

(a) Derive the generalized Becquerel equation (5.10):

$$\theta_F = -\frac{Ne^2}{\varepsilon_0 mnc} f\,\omega_L d \left\{ \frac{a_1\,\omega^2}{\left(\omega_0^2-\omega^2\right)^2} + \frac{a_2\,\omega^2}{2\omega_0\left(\omega_0^2-\omega^2\right)\Delta\omega_{\text{s.o.}}} - \frac{a_3\,\omega^2\hbar}{2\omega_0\left(\omega_0^2-\omega^2\right)kT} \right\} \tag{5.46}$$

by differentiating $n(\omega) = \sqrt{\varepsilon_1(\omega)}$ in the context of the Lorentz model, assuming an oscillator strength $f \neq 0$, no damping ($\gamma \cong 0$), and no absorption ($k \cong 0$).

Hint: Use eq. (5.9) *for* ε_1 *and from eqs.* (5.6)–(5.8) *the relations*

$$\delta\omega_0 = -2a_1\,\omega_L, \tag{5.47}$$

$$\frac{\delta f}{f} = 2a_2\frac{\omega_L}{\Delta\omega_{\text{s.o.}}}, \tag{5.48}$$

$$\frac{\delta N}{N} = -a_3\frac{\mu_B B}{kT} = -a_3\frac{\hbar\omega_L}{kT}. \tag{5.49}$$

Furthermore, use the extended Becquerel equation (5.5):

$$\theta_F = -\frac{\omega^2 d}{2\omega_0 c}\left\{\frac{\partial n}{\partial\omega_0}\delta\omega_0 + \frac{\partial n}{\partial f}\delta f + 2\frac{\partial n}{\partial N}\delta N\right\}. \tag{5.50}$$

(b) Determine the value of the constant a_1 for an atomic $^2S_{1/2} \rightarrow {}^2P_{1/2}$ transition.

Hint: Consider Fig. 5.1. *You see the two allowed electronic transitions for rcp and lcp light between the* $^2S_{1/2}$ *ground state and the* $^2P_{1/2}$ *excited state. Consider the Zeeman splitting,* $\Delta E_Z = \mu_B\, g(j,l,s)\, B$*, where* $g(j,l,s)$ *is the Landé g factor.*

(c) Compare the diamagnetic term (= first term) of the explicit form of the generalized Becquerel equation (5.46) with the basic Becquerel equation (5.4), applied to $n(\omega)$ from the Lorentz model. For which value of a_1 are they identical?

5.8 Solutions

5.8.1 Faraday rotation of quartz

For rcp and lcp light, we have a shift of the resonance frequency ω_0 when a magnetic field is switched on:

$$\omega_{0\pm} = \omega_0 \pm \frac{1}{2}\omega_c \text{ with } \omega_c = \frac{q B_0}{m_e} = 2\omega_L < 0 \quad \text{for } q = -e. \tag{5.51}$$

Assuming that the line shape of $n(\omega)$ does not change but only shifts when the magnetic field is switched on, we get:

$$n_\pm(\omega) = n(\omega \mp \omega_L). \tag{5.52}$$

For normal dispersion, we have $dn/d\omega > 0$ [i.e., $n(\omega)$ is increasing with increasing ω]. Furthermore, a Larmor frequency $\omega_L < 0$ yields, according to the Lorentz model, $n_+(\omega) > n_-(\omega)$; therefore,

$$\theta_F^{\text{quartz}} = \frac{\omega d}{2c}(n_+ - n_-) \cong -\omega_L \frac{\omega d}{c}\frac{\partial n}{\partial \omega} > 0. \tag{5.53}$$

So, quartz has a positive sign of the Faraday rotation in the range of normal dispersion.

The sign can be derived using an alternative approach. Using the definition of normal dispersion,

$$\frac{\partial n}{\partial \omega} > 0, \tag{5.54}$$

and the definition of the cyclotron resonance frequency,

$$\omega_c = \frac{q B_0}{m_e} = 2\omega_L < 0 \quad \text{for } q = -e, \tag{5.55}$$

we can evaluate the sign of the Faraday rotation by directly inserting eqs. (5.54) and (5.55) in the basic Becquerel formula (5.4),

$$\theta_F^{\text{quartz}} \cong -\underbrace{\omega_L}_{<0}\frac{\omega d}{c}\underbrace{\frac{\partial n}{\partial \omega}}_{>0} > 0, \tag{5.56}$$

yielding once more a positive sign of the Faraday rotation for quartz.

5.8.2 Faraday rotation of an $S \rightarrow P$ transition

(a) We start with plotting the frequency dependence of the refractive index, $n(\omega)$, as derived from Lorentz theory, eq. (3.9), for one transition, with the assumptions $k = 0$ and $\gamma = 0$:

$$n(\omega) = \sqrt{\varepsilon_1(\omega)} = \sqrt{1 + \frac{Ne^2}{\varepsilon_0 m} f \frac{1}{\omega_0^2 - \omega^2}}. \tag{5.57}$$

The extended Becquerel equation is defined as

$$\theta_{\mathrm{F}} = -\frac{\omega^2 d}{2\omega_0 c}\left\{\frac{\partial n}{\partial \omega_0}\delta\omega_0 + \frac{\partial n}{\partial f}\delta f + 2\frac{\partial n}{\partial N}\delta N\right\}. \tag{5.58}$$

Now, we can execute explicitly the individual differentiations of eq. (5.58). We start with the first one:

$$\frac{\partial n(\omega)}{\partial \omega_0} = \frac{1}{2n(\omega)}\frac{Ne^2}{\varepsilon_0 m} f \frac{-2\omega_0}{\left(\omega_0^2 - \omega^2\right)^2}. \tag{5.59}$$

Note the factor $1/n(\omega)$ that arises from the derivative of the square root of $\varepsilon_1(\omega)$. The second term yields

$$\frac{\partial n(\omega)}{\partial f} = \frac{1}{2n(\omega)}\frac{Ne^2}{\varepsilon_0 m}\frac{1}{\omega_0^2 - \omega^2}. \tag{5.60}$$

By evaluating the third term, we obtain

$$\frac{\partial n(\omega)}{\partial N} = \frac{1}{2n(\omega)}\frac{e^2 f}{\varepsilon_0 m}\frac{1}{\omega_0^2 - \omega^2}. \tag{5.61}$$

Inserting the three derivatives in eq. (5.58) results in

$$\theta_{\mathrm{F}} = -\frac{\omega^2 d}{2\omega_0 c}\frac{e^2}{\varepsilon_0 m}\left\{\frac{Nf}{2n}\frac{-2\omega_0}{\left(\omega_0^2 - \omega^2\right)^2}\delta\omega_0 + \frac{N}{2n}\frac{1}{\omega_0^2 - \omega^2}\delta f + 2\frac{f}{2n}\frac{1}{\omega_0^2 - \omega^2}\delta N\right\}. \tag{5.62}$$

Next, we enter the three relations from eqs. (5.6) to (5.8),

$$\delta\omega_0 = -2a_1\,\omega_{\mathrm{L}}, \tag{5.63}$$

$$\frac{\delta f}{f} = 2a_2\frac{\omega_{\mathrm{L}}}{\Delta\omega_{\mathrm{s.o.}}}, \tag{5.64}$$

$$\frac{\delta N}{N} = -a_3\frac{\mu_{\mathrm{B}} B}{kT} = -a_3\frac{\hbar\omega_{\mathrm{L}}}{kT}, \tag{5.65}$$

and obtain

$$\theta_{\mathrm{F}} = -\frac{\omega^2 d}{2\omega_0 c}\frac{e^2}{\varepsilon_0 m}\left\{\frac{Nf}{2n}\frac{4\omega_0 a_1\,\omega_{\mathrm{L}}}{\left(\omega_0^2 - \omega^2\right)^2} + \frac{Nf}{2n}\frac{2a_2\,\omega_{\mathrm{L}}}{\left(\omega_0^2 - \omega^2\right)\Delta\omega_{\mathrm{s.o.}}} - 2\frac{Nf}{2n}\frac{a_3\,\hbar\omega_{\mathrm{L}}}{\left(\omega_0^2 - \omega^2\right)kT}\right\}. \tag{5.66}$$

This can now be brought in the desired generalized form by extracting the appropriate factor:

$$\theta_{\mathrm{F}} = -\frac{Ne^2}{\varepsilon_0 mnc} f\,\omega_{\mathrm{L}} d\left\{\frac{a_1\,\omega^2}{\left(\omega_0^2 - \omega^2\right)^2} + \frac{a_2\,\omega^2}{2\omega_0\left(\omega_0^2 - \omega^2\right)\Delta\omega_{\mathrm{s.o.}}} - \frac{a_3\,\omega^2\hbar}{2\omega_0\left(\omega_0^2 - \omega^2\right)kT}\right\}. \tag{5.67}$$

(b) An atomic $S_{1/2} \rightarrow P_{1/2}$ transition is plotted in Fig. 5.1(a). Considering the selection rules for dipole-allowed (spin-conserving) transitions, two allowed electronic transitions exist (one for rcp and one for lcp light), between the $S_{1/2}$ ground state and the $P_{1/2}$ excited state. The levels split due to the Zeeman splitting,

$$\Delta E_Z = \mu_B \, g(j, l, s) \, B, \tag{5.68}$$

where $g(j,l,s)$ is the Landé g-factor,

$$g(j\,l\,s) = 1 + \frac{j\,(j+1) + s\,(s+1) - l\,(l+1)}{2j\,(j+1)}, \tag{5.69}$$

and μ_B is the Bohr magneton,

$$\mu_B = \frac{e\hbar}{2mc}. \tag{5.70}$$

First, we calculate the Landé g factor. For the ${}^2S_{1/2}$ ground state, we have $l = 0$, $j = s = 1/2$. According to eq. (5.69), this yields $g({}^2S_{1/2}) = 2$. For the excited state, ${}^2P_{1/2}$, the corresponding values are: $l = 1$, $j = s = 1/2$. This results in $g({}^2P_{1/2}) = 2/3$. According to Fig. 5.1(a), the difference $\hbar\delta\omega_0$ of the transition energies $\hbar\omega_{0+}$ for rcp and $\hbar\delta\omega_{0-}$ for lcp light is just the sum of the Zeeman splitting of the two levels. This amounts to

$$\hbar\delta\omega_0 = \hbar\omega_{0-} - \hbar\omega_{0+} = \mu_B B \left[g({}^2S_{1/2}) + g({}^2P_{1/2}) \right] = \frac{8}{3}\mu_B B = -\frac{8}{3}\hbar\,\omega_L. \tag{5.71}$$

Inserting the value for $\delta\omega_0$ from eq. (5.71) into eq. (5.63) results in

$$\delta\omega_0 = -\frac{8}{3}\,\omega_L = -2\,a_1\,\omega_L. \tag{5.72}$$

From this we obtain $a_1 = 4/3$ for an atomic $S_{1/2} \rightarrow P_{1/2}$ transition.

(c) With the help of the Lorentz theory, we obtain the first derivative of the refractive index, $n(\omega)$, with respect to ω. Inserting this into the basic Becquerel equation (5.4), it can be brought into the form of the first term of the generalized Becquerel equation (5.46):

$$\theta_F \cong -\omega_L \frac{\omega d}{c}\frac{\partial n}{\partial \omega} = -\omega_L \frac{\omega d}{c}\frac{1}{2n(\omega)}\frac{Ne^2}{\varepsilon_0 m} f \frac{2\omega}{\left(\omega_0^2 - \omega^2\right)^2}. \tag{5.73}$$

Note that we have added the oscillator strength, f. By rearranging the terms, we get

$$\theta_F \cong -\frac{Ne^2}{\varepsilon_0 mnc} f\,\omega_L\,d \frac{\omega^2}{\left(\omega_0^2 - \omega^2\right)^2}. \tag{5.74}$$

This is equal to the first (diamagnetic) term in eq. (5.46) if $a_1 = 1$.

6 Magneto-optic properties of magnetic semiconductors

6.1 General properties of magnetic semiconductors

In this chapter, we would like to show, on the basis of a model material system, how magneto-optics can contribute to the understanding of the electronic structure and of the nature of magnetic ordering of solid-state materials. The purpose of the chapter is not to give a thorough and up-to-date review on a specific material class. Rather, we want to follow the history of the discoveries in the 1970s and focus on specific highlights to point out the power of magneto-optic investigations.

The model material system, we want to illustrate, is that of the magnetic semiconductors. Surprisingly, they show magnetic ordering although not being conductive. They generally exhibit a large magnetization, or at least a large temperature dependence of the magnetization. The semiconducting character, in turn, provides a spectral separation of electronic transitions. In the midinfrared to infrared region, transitions within the conduction band, so-called ***intraband*** transitions (< 1 eV) occur. In the visible and the ultraviolet region, ***interband*** transitions (> 1 eV), that is, transitions from the valance band or energetically lower lying bands into the conduction band, occur so that a quantitative analysis for both transitions will be possible.

Among magnetic semiconductors, the group of the europium chalcogenides constitutes an interesting model system. There are four materials in this group: EuO, EuS, EuSe, and EuTe. They all crystallize in the simple, cubic NaCl structure as shown in Fig. 6.1. The lattice sites of the Eu ions are all equivalent and bear a localized magnetic moment. The electronic configuration of a neutral Eu atom is [Xe] $4f^7 5d^0 6s^2$. In the chalcogenides, the two 6*s* electrons form a more or less ionic bond with the chalcogen ion while the 4*f* electrons practically do not participate in the binding. Therefore, in these compounds, a clear separation exists between the electrons responsible for chemical bonding and the electrons responsible for the magnetic ground state.

As said before, the europium atom will give away both 6*s* electrons to form an ionic bond in the chalcogenides and remains as a doubly charged ion Eu^{2+}. Divalent Eu^{2+} with an electronic configuration [Xe] $4f^7$ has a half-filled 4*f* shell. According to Hund's rules, the ground state is $^8S_{7/2}$. This is a fully polarized ground state with a magnetic moment of $7\mu_B$. It is a pure *S* state. Thus, no orbital angular momentum *L* exists and the total angular momentum *J* equals the spin angular momentum *S*. The 4*f* electrons are spatially close to the nucleus as shown in Fig. 6.2. As a result, the 4*f* states behave almost like atomic levels and do not form a band. In addition, the crystal-field splitting of the 4*f* states is generally small and the spherical symmetry of the *S* state in Eu^{2+} leads practically to its disappearance (Eu^{2+} in CaF_2: 10 Dq = 0.2 cm^{-1}). Therefore, the ground state in the solid state can still be described with the free-ion configuration.

https://doi.org/10.1515/9783110635300-006

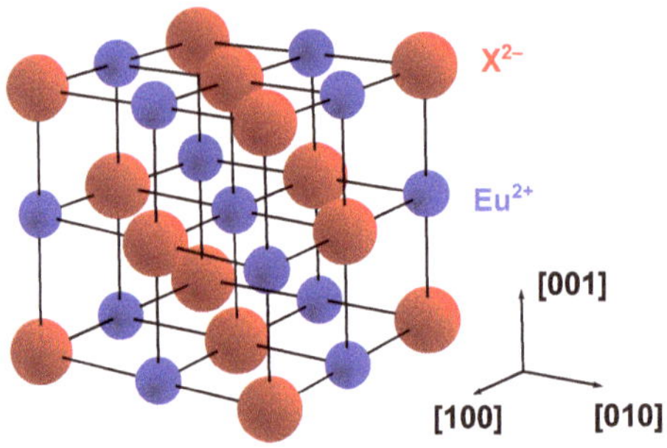

Fig. 6.1: Crystal structure of $Eu^{2+}X^{2-}$, $X = O$, S, Se, Te.
The NaCl structure consists of two interlaced face-centered cubic lattices, displaced along the <100> direction by half a lattice constant. Pictured is one unit cell.

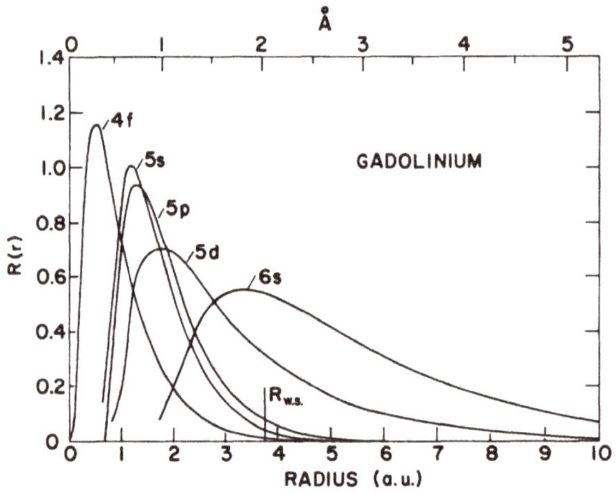

Fig. 6.2: Radial part of the modified Hartree–Fock wave function for a neutral gadolinium atom with electron configuration $4f^7 5d^1 6s^2$. Figure from Ref. (103).

Tab. 6.1: Lattice constant a, magnetic structure, ordering temperature, exchange parameters for europium chalcogenides, and gap energies at 300 K. Table from Ref. (63).

Material	a (Å)	Magnetic structure	$T_{N,C}$ (K)	J_1/k_B (K)	J_2/k_B (K)	E_G (eV) at 300 K
EuO	5.141 (105)	F	69.33 (106)	0.55 (107)	0.15 (107)	1.12 (108)
EuS	5.968 (105)	F	16.57 (109)	0.22 (110)	−0.10 (110)	1.65 (108)
EuSe	6.195 (105)	AF-IIA Ferri AF-II	4.60 (105) 2.8 (111) 1.8 (111)	0.11 (107)	−0.09 (107)	1.80 (108)
EuTe	6.598 (105)	AF-II	9.58 (105)	0.06 (107)	−0.20 (107)	2.00 (108)

The europium chalcogenides represent a model system for the localized Heisenberg exchange interaction. The corresponding Hamiltonian, which is usually approximated by nearest and next nearest neighbor interactions, is:

$$\mathcal{H}_{\text{ex}} = -\sum_{ij} J_{ij}\left(\vec{S}_i \cdot \vec{S}_j\right) \cong -\sum_{\text{nn}} J_1\left(\vec{S}_0 \cdot \vec{S}_{\text{nn}}\right) - \sum_{\text{nnn}} J_2\left(\vec{S}_0 \cdot \vec{S}_{\text{nnn}}\right), \tag{6.1}$$

where $\vec{S}_{\text{nn}}$ and $\vec{S}_{\text{nnn}}$ denominate the nearest and next-nearest-neighbor spins, respectively, and J_{ij} and $J_{1,2}$ are the exchange constants, which are a measure of the strength of the magnetic coupling. The type of coupling depends on the sign and relative magnitude of the exchange constants J_1 and J_2. As shown in Tab. 6.1, we find for positive exchange constants with magnitude $J_1 > J_2 > 0$ ferromagnetism in EuO. It shows the highest ordering temperature $T_C \cong 69$ K. For negative but small J_2, satisfying $J_1 > 0 > J_2$ with $J_1 > |J_2|$, we still have ferromagnetic ordering in EuS but with a very low ordering temperature of $T_C \cong 16.5$ K. In EuTe, the exchange constant J_2 is negative but larger in absolute value than J_1. Therefore, it is dominating the coupling: $J_1 > 0 > J_2$ and $J_1 < |J_2|$. As a result, we find antiferromagnetic ordering in EuTe. The chalcogenide in the middle of the group, EuSe, is a very special case. Here, the two exchange constants have approximately equal size but opposite sign: $J_1 > 0 > J_2$ and $J_1 \cong |J_2|$. This leads to a so-called metamagnetic behavior showing several magnetically ordered phases as a function of temperature, yielding a complicated magnetic phase diagram for EuSe (see Fig. 7.4).

6.2 Absorption of unpolarized light in Eu chalcogenides

Next, we will discuss the optical properties of the Eu chalcogenides. All compounds are semiconductors. Thus, one expects a band gap in the absorption spectrum in the near infrared. As an example, the absorption spectrum of EuS is shown in Fig. 6.3, measured on a thin film of about 100 nm thickness at various temperatures. To exclude scattering on magnetic domains, a magnetic field was applied for alignment at low temperatures. From the room-temperature measurement at $T = 300$ K, we derive from the absorption edge an energy gap of approximately 1.7 eV. As this is an optical measurement, the energy gap must represent a dipole-allowed transition. Possible allowed atomic transitions in this energy range are intra-Eu $4f \rightarrow 5d$ and $3p(\text{S}) \rightarrow 5d(\text{Eu})$ transitions. In a solid, the atomic selection rules cannot be strictly applied, because the symmetries of the individual atomic states are mixing when forming energy bands. Nevertheless, we want to make use of these terms to have a simple naming. At temperatures near the onset of magnetic ordering, which takes place at $T_C = 16.5$ K for EuS, strong changes in the optical absorption occur. Most prominently, we detect a strong redshift of the absorption edge. The redshift was first discovered by Busch, Junod, and Wachter in 1964 with the help of reemission spectroscopy on powders (112). The redshift of the energy gap can be found in both EuO and EuS. It is larger in EuO, where it reaches nearly 0.3 eV, which is huge in comparison with the energy gap of 1.2 eV.

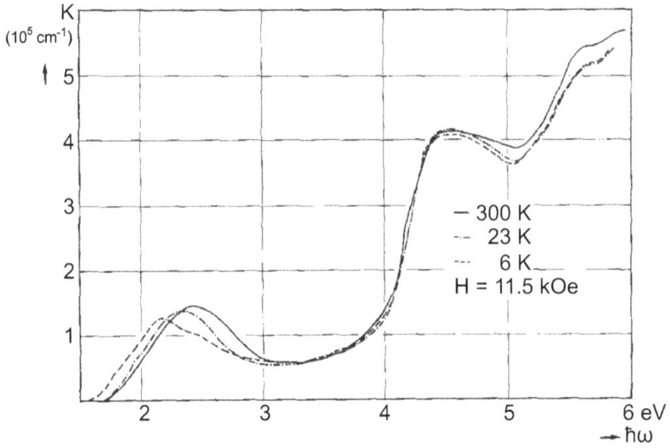

Fig. 6.3: Absorption coefficient of EuS below and above the ordering temperature of $T_C = 16.5$ K. Figure from Ref. (104).

The strong redshift, observed when crossing the ordering temperature, points to an optical transition involving the electrons that are responsible for the magnetic ordering. Obviously, these are the *4f* electrons of europium. The redshift, therefore, gives high evidence for an intra-Eu $4f \rightarrow 5d$ transition. Additional information may be obtained if diluted Eu is incorporated into a matrix which shows no absorption over a sufficiently large energy range (0–4 eV). Such measurements on Eu in KCl and in KBr are shown in Fig. 6.4 (113). We notice that in the energy gap of the matrix substances, an additional absorption occurs in the same energy range as in EuS. This fact further corroborates the assignment of the low-energy absorption maximum in EuS to an intra-Eu $4f \rightarrow 5d$ transition. In addition, a fine structure can be found in the absorption structure associated with Eu. The existence of a staircase-like fine structure hints to a multiplet splitting due to the crystal field of the host matrix, which would be expected for an intra-Eu $4f \rightarrow 5d$ transition.

More light on the origin of the redshift can be shed when the redshift of the absorption edge is plotted as a function of temperature, as shown in Fig. 6.5 for undoped and Gd-doped EuO (114). The redshift of the absorption edge already starts at a temperature of 150 K, well above the ordering temperature T_C. It is, thus, not proportional to the magnetization, which would be zero for temperatures larger than T_C in zero magnetic field. The onset of the shift at temperatures well above T_C is typical for a local-ordering effect. Consequently, the redshift of the absorption edge can be

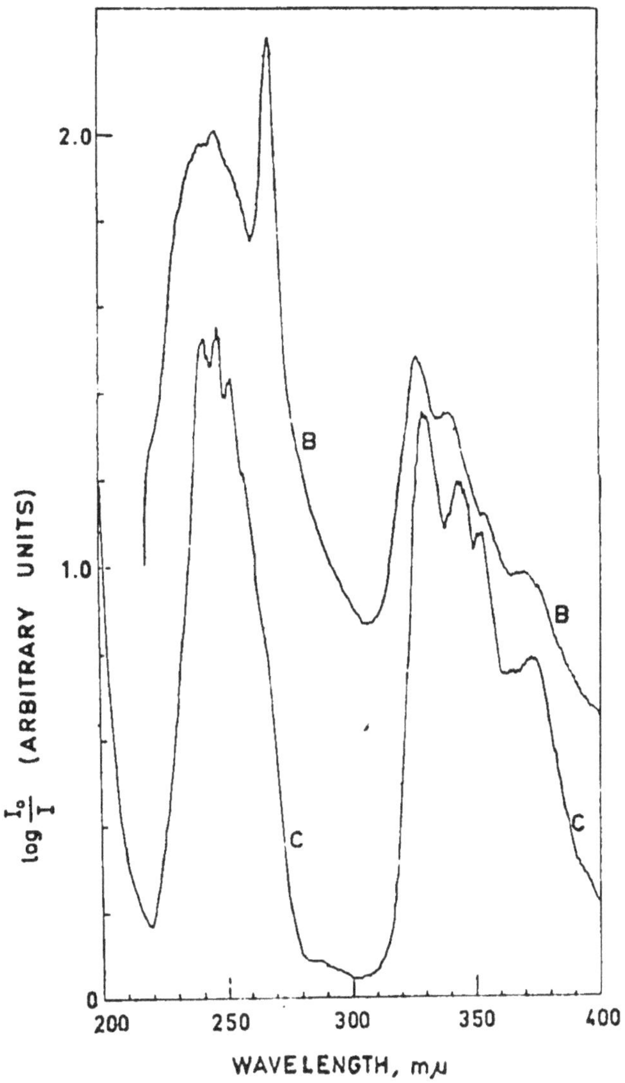

Fig. 6.4: Absorption spectra of Eu^{2+} in alkali halides at liquid-air temperature. (B) KCl:Eu (from mixture). (C) KBr:Eu (from melt). Reprinted with permission from Ref. (113) © The Optical Society.

fitted nicely with a spin-correlation function. While the magnetization is defined by the expectation value $\langle S_z \rangle$ of the spins, the two-particle spin-correlation function is an expression of the form $\langle \vec{S}_1 \cdot \vec{S}_2 \rangle$.

We conclude from this discussion that the redshift involves two different spin systems. If our assignment that the absorption edge is defined by the intra-Eu $4f \rightarrow 5d$

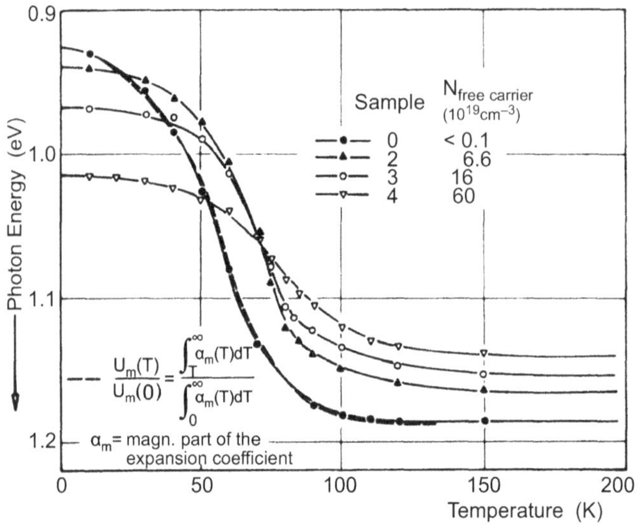

Fig. 6.5: Temperature shift of the absorption edge of undoped and Gd-doped EuO. Dashed curve represents the spin correlation function. Figure from Ref. (114).

transition is correct, then besides the 4*f* electrons, the 5*d* electrons will be spin-polarized as well. The 4*f* electrons are responsible for the ground state $^8S_{7/2}$. Hence, it is obvious that they are spin-polarized. The fact that the 5*d* electrons are spin-polarized, on the other hand, is not obvious at all. The conduction band usually splits due to the crystal field. For a splitting of the 5*d* band of magnetic origin, we must reflect on a coupling to the 4*f* electrons. The solution was presented by Peter Wachter (115). A schematic representation of his model, explaining the redshift of the absorption edge, is shown in Fig. 6.6. The main idea is that an electron, which is optically excited from the 4*f* state to the 5*d* conduction band, couples ferromagnetically to the remaining 4*f* electrons. So, parallel spin alignment is the preferred orientation that lowers the energy of the 5*d* electron. Antiparallel alignment, on the other hand, will increase the energy. Above T_C, the spin of the 5*d* electron is already feeling the exchange coupling to the spins of the 4*f* electrons. This correlation will lead energetically to a slight decrease of the 5*d* band with the same spin orientation as the 4*f* states. Below T_C, the ferromagnetic exchange interaction turns on and the 5*d* conduction band splits into two spin-polarized subbands. The subband with the same spin orientation as the 4*f* states shifts to lower energies. The 5*d* subband with antiparallel spin orientation will shift to higher energies.

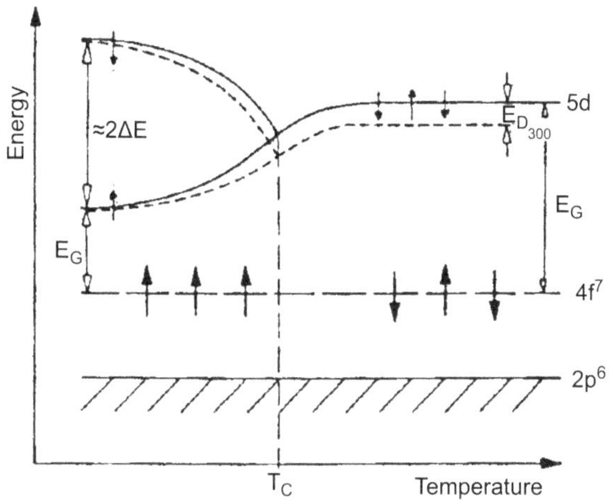

Fig. 6.6: Model of the 5*d* conduction band splitting of EuO. The conduction band contains a small number of excited charge carriers. Figure from Ref. (115).

6.3 Magnetic circular birefringence and dichroism

The simple model depicted in Fig. 6.6 is based on the assignment of the lowest energy peak which defines the absorption edge to a $4f \rightarrow 5d$ transition. This is indeed suggested by the strong magnetization dependence of the absorption edge but not yet proven. The interpretation can be much deeper if the Faraday rotation θ_F is measured and analyzed in detail. In Fig. 6.7, a measurement of θ_F is shown in the range of the low energy absorption maximum, as obtained with the dynamical measurement method mentioned in Section 4.2.3 (104). If the spectra in the ferromagnetically ordered state below $T_C \cong 16.5$ K are compared with those at $T = 40$ K, we see that the shift is not equal at all energies but there are changes in its structure. All these observations, which are possible due to the greater spectral resolution of θ_F, indicate that the picture shown in Fig. 6.6 is too simple. Let us have a closer look. The negative structure at the lowest energy is shifted from 1.6 to 1.4 eV. So, it experiences a redshift, as expected. This redshift is ultimately responsible for the redshift of the absorption edge. Looking at the spectrum at 5 K, the prominent positive peak at 2.1 eV and the shoulder at 1.9 eV experience a redshift as well. However, the weak shoulder at 2.5 eV is blueshifted, which is not what we would have expected. To explain this behavior, we need a more sophisticated model than the one proposed in Fig. 6.6.

For an improved understanding, which can explain the unexpected blueshift, we must consider how the electronic states split into various subbands. Because the 4*f* states are local and the conduction band is predominantly derived from the Eu 5*d* states, the $4f \rightarrow 5d$ transition takes place at the Eu site and can be looked at as a local atomic transition. Therefore, we should get a good approximation by describing the

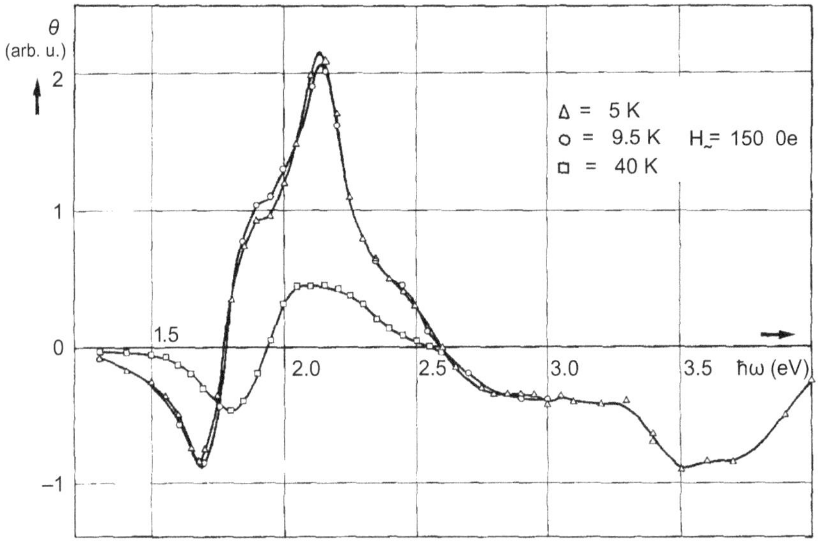

Fig. 6.7: Dynamic Faraday rotation of EuS below and above the ordering temperature $T_C \cong 16.5$ K. Figure from Ref. (104).

electronic states by an atomic model. Such a model has been proposed by Tadao Kasuya (116). We will now explain it in more detail. The resulting energy-level scheme is shown in Fig. 6.8.

To develop an energy-level scheme in an atomic model, the hierarchy of the energy splittings must be discussed. The following interactions and their relative strength must be considered: (1) crystal field splitting of the 5*d* final state due to the Coulomb potential of the crystal lattice, (2) spin–orbit splitting due to the interaction of electron spins with the central potential of the atomic nucleus, and (3) exchange splitting due to a direct spin–spin interaction between electrons in different electronic states. In the case considered here, without doubt, the crystal-field splitting of the 5*d* states is the largest. It amounts to approximately 2 eV and it splits the 5*d* states into two subbands, the $5dt_{2g}$ subband at lower energy and the $5de_g$ subband at higher energy. This is indicated in Fig. 6.8.

The redshift of 0.2 eV of the absorption edge in EuS gives a clue for the next interaction to be considered. As the redshift sets in near the ordering temperature, it must be of magnetic origin. Hence, it must relate to an exchange interaction. Following the model proposed by Wachter, the 5*d* conduction band splits into two spin-polarized subbands. Because the redshift is the energy difference between the unpolarized and the energetically lower spin-polarized subband, the splitting of the two spin-polarized subbands is twice the value of the redshift, amounting to 0.4 eV. However, the 5*d* band is only thermally occupied and the occupation numbers are very low at temperatures below the EuS ordering temperature of 16.5 K. Therefore, Kasuya considered a coupling

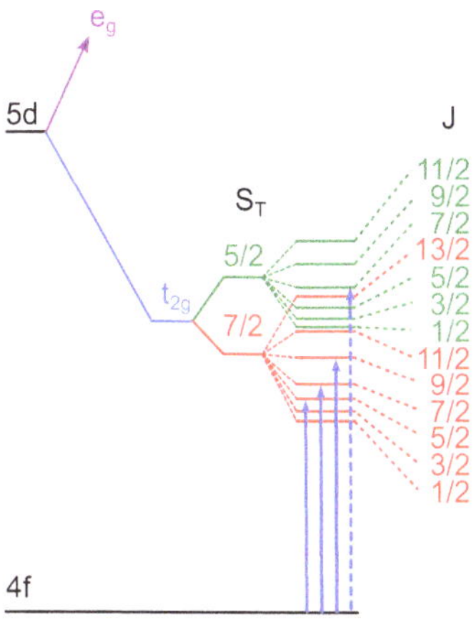

Fig. 6.8: Energy-coupling scheme of the $^7F(4f^6)5d$ configuration in the Eu chalcogenides. Figure from Ref. (116).

between the $5d$ electron that was optically excited from the $4f$ state with the remaining $4f$ electrons. The excited electron has spin $S_{5d} = \frac{1}{2}$, and the six remaining $4f$ electrons have spin $S_{4f} = 3$ because of the fully polarized ground state according to Hund's rules. This yields a total spin $S_T = S_{4f} \pm S_{5d}$, which amounts to $S_T = \frac{7}{2}$ for the energetically favorable parallel coupling between the $5d$ and the $4f$ spins and $S_T = \frac{5}{2}$ for antiparallel coupling, which is energetically unfavorable. That the parallel coupling is the energetically preferred coupling is obvious from the observed redshift in experiment. The absorption edge is defined by the dipole-allowed optical transition $4f \longrightarrow 5dt_{2g}$, which is spin conserving by nature. The redshift tells us that the optically excited electron is excited in the energetically lower subband. Because the electron conserves the spin during transition, it must be parallelly aligned to the remaining $4f$ electrons. So, the $5d$ subband with spin direction parallel to the $4f$ states must be the one lower in energy.

The last term, we want to consider, is the spin–orbit splitting. The remaining six $4f$ electrons are left in a high angular state, $L_{4f} = 3$. The excited $5d$ electron has only a spin $S_{5d} = \frac{1}{2}$, which we already used to establish the exchange coupling. This yielded the total spin S_T with two spin states, $S_T = \frac{7}{2}$ and $S_T = \frac{5}{2}$. In Kasuya's model, this total spin is coupled to the orbital moment of the remaining $4f$ electrons to calculate the total angular momentum $\vec{J} = \vec{S}_T + \vec{L}_{4f}$ for the spin–orbit coupling. This provides two multiplets, namely:

$$\text{for } S_T = \frac{7}{2}: J = \frac{1}{2}, \frac{3}{2}, \frac{5}{2}, \frac{7}{2}, \frac{9}{2}, \frac{11}{2}, \frac{13}{2} \tag{6.2}$$

$$\text{and for } S_T = \frac{5}{2}: J = \frac{1}{2}, \frac{3}{2}, \frac{5}{2}, \frac{7}{2}, \frac{9}{2}, \frac{11}{2}. \tag{6.3}$$

The large number of spectral lines and the high J values result in an overlap of the two configurations and, thus, in a mixture of S and J terms.

Now, we are in the position to discuss the possible optical transitions within these energy levels and compare them with experiment. Note that each of the J levels is

$2J + 1$ times degenerate but the degeneracy is lifted in a local magnetic field due to the Zeeman effect. The local magnetic field is generated by an applied magnetic field and, below T_C, in addition by the spontaneous magnetization of the ferromagnetic state. This yields another multiplet with $2J + 1$ sublevels, $-J \le m_J \le J$, for each J level. Considering first non-overlapping terms, the following selection rules apply for spin-preserving transitions:

$$\Delta S = 0; \quad \Delta L = \pm 1; \quad \Delta J = 0, \ \pm 1; \quad \Delta m_J = \pm 1. \tag{6.4}$$

From the $4f^7$ ground state with configuration $^8S_{7/2}$, three dipole-allowed transitions are possible to the following levels: $J = \frac{5}{2}, \frac{7}{2}$, and $\frac{9}{2}$. Thus, we can assign in the dynamic Faraday rotation spectrum (Fig. 6.7), the three peaks lowest in energy to these three states: at 1.7, 1.9, and 2.1 eV into $J = \frac{5}{2}, \frac{7}{2}$, and $\frac{9}{2}$, respectively. The different signs of the Faraday rotation of the structures can be explained as well. They are a consequence of the Zeeman splitting of the m_J levels and the different selection rules for rcp and lcp light: $\Delta m_J = +1$ for lcp and $\Delta m_J = -1$ the for rcp light (according to the definition of rcp and lcp used here, as discussed in Section 5.4). It can be shown that, for lcp light, the probability is in fact the largest for a transition to smaller J values, while for rcp light, a larger transition probability exists for transitions to greater J values. For a detailed discussion, see the problem in Section 6.5.1. Because the incoming light is linearly polarized and contains, therefore, equal amount of rcp and lcp light, it follows that the transitions to $J = \frac{5}{2}$ must have a negative and to $J = \frac{9}{2}$ a positive Faraday rotation as is found indeed in the experiment. The transition to $J = \frac{7}{2}$ is roughly equal for the two polarization states, showing only a small rotation as evidenced in the experiment.

Due to the intermixing of S and J terms by spin–orbit coupling in Kasuya's model, only J^2 and J_z remain good quantum numbers. Consequently, spin–flip transitions are no longer strictly forbidden. In Russell–Saunders coupling (i.e., L–S coupling), the following selection rules then apply:

$$\Delta S_T = \pm 1; \quad \Delta L = 0; \ \pm 1; \quad \Delta J = 0; \quad \Delta m_J = 0. \tag{6.5}$$

We, therefore, expect a further transition to $S = \frac{5}{2}$, $J = \frac{7}{2}$ which is a spin–flip transition and, hence, rather weak. In addition, it must be the energetically highest transition of the four transitions discussed so far because the electron is excited into the upper spin branch with $S = \frac{5}{2}$. Looking at the dynamical Faraday rotation spectrum in Fig. 6.7 again, we can assign the structure at 2.5 eV to the spin–flip transition. Because it excites into the upper spin branch, we expect, according to Wachter's model (Fig. 6.6), a blueshift when the temperature falls below T_C. This is indeed what the experiment shows, and we have found a satisfying explanation for the unexpected blueshift at 2.5 eV. The three transitions to lower energy must display a redshift as the electron is excited into the lower spin branch with $S = \frac{7}{2}$. This is again in accordance with the experiment.

Now, we move on to the high-energy range of the Faraday rotation spectrum, up to 6 eV, which is depicted for EuS in Fig. 6.9. We find a second structure which looks very similar to the one at approximately 2 eV. This suggests that it could be assigned to the $4f \rightarrow 5de_g$ transition. This is consistent with the assumption that the $5d$ final state splits up into the t_{2g} and e_g subbands due to the crystal field as supposed in Kasuya's model. Furthermore, we observe smaller structures between 3 and 4 eV and beyond 5 eV.

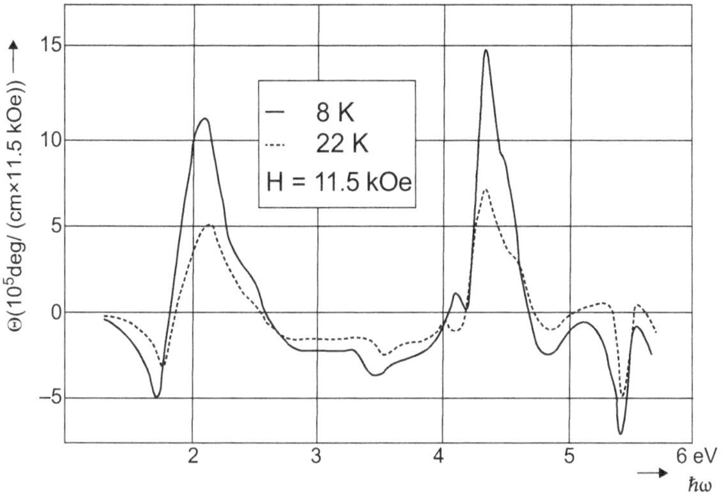

Fig. 6.9: Faraday rotation of EuS below and above the ordering temperature $T_C = 16.5$ K in the energy range up to 6 eV. Figure from Ref. (104).

Up to now, we have discussed the Faraday rotation spectra to assign the optical transitions. However, this is not accurate. We have learnt in Section 3.1 that the correct energy of an optical transition is displayed in the optical conductivity. Therefore, we must determine the energies of magneto-optically active transitions from the absorptive part σ_{2xy} of the off-diagonal element of the optical conductivity tensor. In addition, we want to quantitatively compare the expressions for the optical conductivity, as derived in Section 5.4, with experimental data. Because it is still very difficult to calculate the spin-polarized band structure due to the interaction of localized (Eu $4f$) and delocalized (Eu $5d$ and $6s$) bands (117, 118), which would allow to calculate $\sigma_{2xy}(\omega)$, a simple model will be used here as presented in Section 5.5. We cannot determine $\sigma_{2xy}(\omega)$ from the Faraday rotation alone. We need, in addition, a measurement of the Faraday ellipticity η_F. We know that the Faraday ellipticity combines with the Faraday rotation to the complex Faraday rotation $\tilde{\theta}_F = \theta_F - i\eta_F$ as shown in eq. (3.64). Therefore, the simplest way to get the ellipticity is by means of the Kramers–Kronig transformation. In the Kramers–Kronig transformation, the integral extends at higher energies to infinity and at lower energies to zero, both of which are impossible to measure. Fortunately, one does not make a large mistake when integrating only over a finite energy or frequency range because the Faraday rotation and ellipticity have about an equal number of regions with negative and positive sign. So, the integration at higher or lower energies, than the known spectral region, yields each negative and positive contributions, which nearly cancel each other. Therefore, the total contribution of the unknown spectral regions to the integral is rather small.

To be on the safe side and to prove if the arguments given above for the Kramers–Kronig transformation are correct, the Faraday ellipticity can also be derived from measuring the magnetic circular dichroism (MCD). According to eqs. (2.133) and (2.136), the ellipticity is directly proportional to the difference of the absorption coefficients $K_\pm$ for rcp (+) and lcp (–) light. The MCD, on the other hand, is defined as the normalized difference of the absorbance in a circular basis: $\Delta A_{\mathrm{MCD}} = (A_- - A_+)/(A_- + A_+)$, where the absorbance is $A = -\log(I/I_0)$ and I, I_0 are the transmitted and incoming intensity, respectively. Thus, we expect that the MCD and the Faraday-ellipticity spectra must almost look alike. In Fig. 6.10, the MCD spectra of EuS (top) are plotted for various temperatures and (bottom) the Faraday ellipticity η_{F}, as obtained from the Kramers–Kronig transformation. We see immediately that both spectra are in excellent agreement with each other. This corroborates that for Eu chalcogenides, the usage of the Kramers–Kronig transformation yields a good approximation of the Faraday ellipticity.

For the calculation of the off-diagonal element from the Faraday effect, we also need the optical functions $n(\omega)$ and $k(\omega)$, as can be seen from eqs. (3.69) and (3.70). They must be determined by an additional measurement such as reflectivity or absorption, covering at least the same photon-energy range as the measurements of the Faraday effect. From these four quantities, the real σ_{1xy} and imaginary part σ_{2xy} of the off-diagonal conductivity can be computed. The result is shown in Fig. 6.11. We notice an opulent structure which we would like to assign to distinct optical transitions. For this, we look back to Section 5.5. In Fig. 5.3(b), we had shown that $\sigma_{2xy}(\omega)$ has the shape of a procumbent "S" for a 100% spin-polarized initial state. According to Hund's rules, this is exactly the case for the initial $4f$ state of the $4f \rightarrow 5d$ transition. We could therefore assign the two dominant structures around 2 eV and 4.3 eV to the two $4f \rightarrow 5d$ transition as we have already done before on account of the Faraday effect. However, there is an inconsistency concerning the expected line shape. It is not a symmetric procumbent "S" but rather a dominant peak with a small undershoot.

This problem can be solved by remembering that the Faraday effect can take on both signs. Consequently, structures can be partially canceled by overlapping neighboring structures with opposite sign. Therefore, the two dominant peaks could indeed be symmetric and the high-energy part partially canceled by an overlapping broad structure extending to higher photon energies. Following these ideas, the value of the integral over the two dominant positive structures at 2 and 4.3 eV is calculated. Then, the negative part on their high-energy side is complemented in such a way that the structures exhibit the procumbent and symmetric "S" shape with equal area for the positive and negative part as expected for a $4f \rightarrow 5dt_{2g}$ and a $4f \rightarrow 5de_{g}$ transition. The negative area, that has been added to complement the two main structures, must

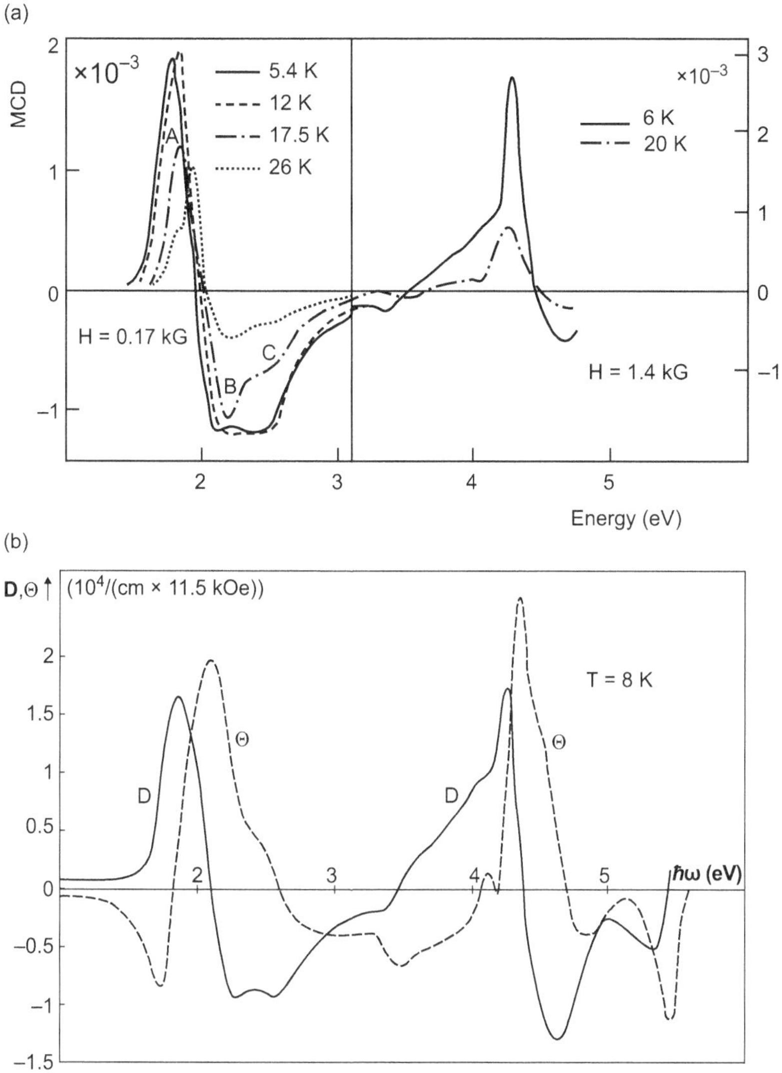

Fig. 6.10: Top: Directly measured magnetic circular dichroism. Figure from Ref. (119). Bottom: Faraday ellipticity η_F (solid line at bottom) of EuS. Figure from Ref. (104). Note the excellent agreement!

now be added as a positive area to the remaining structures to preserve the shape of the total spectrum. This will result in a decomposition of σ_{2xy} into four structures as shown in Fig. 6.12.

Let us now have a closer look at the two structures that we have created through the complement of the $4f \rightarrow 5d$ transitions. Both structures look alike but their shape differs a lot from the shape of the $4f \rightarrow 5d$ transitions. It has the form of two procumbent "S" that are arranged head to head. Referring again to our simple model from Section 5.5, we

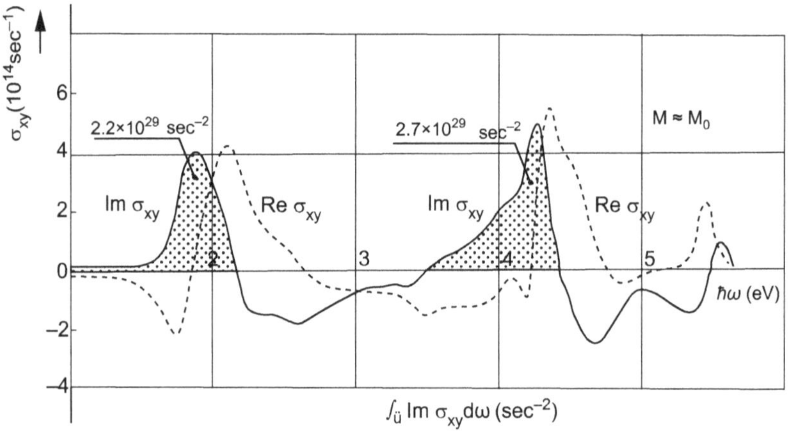

Fig. 6.11: Real and imaginary part of the off-diagonal optical conductivity $\tilde{\sigma}_{xy}$ of EuS. The numbers refer to the positive integrals of the structures in σ_{2xy}. Figure from Ref. (104).

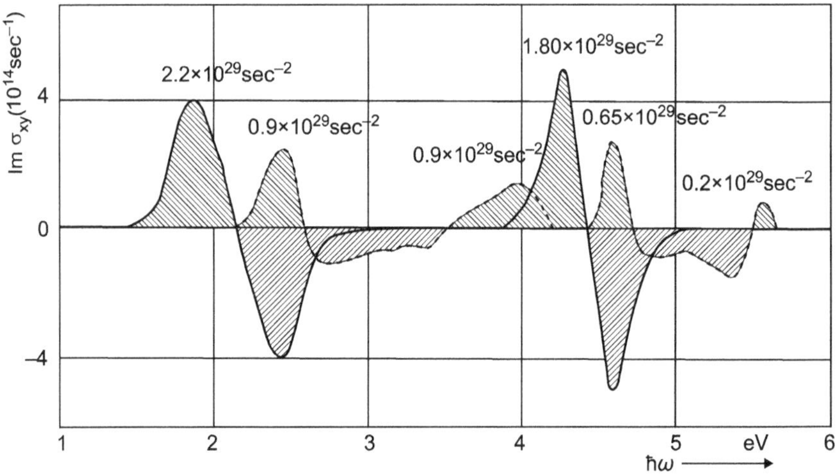

Fig. 6.12: Decomposition of $\tilde{\sigma}_{xy}$ in two procumbent "S" shaped curves, as expected for a $4f_{\uparrow} \rightarrow 5d_{\uparrow}$ transition. Figure from Ref. (104).

notice that the shape is quite like the case shown in Fig. 5.3(f). This shape is expected for a transition from an unpolarized initial state to a final state which is split due to an exchange coupling. The exchange-split final state can only be the $5dt_{2g}$ or $5de_g$ state but what is the unpolarized initial state? It must deliver a dipole-allowed transition. Therefore, besides the $4f$ states only the $3p$ states from the sulfur ion fulfill the correct selection rules. In addition, the $3p$ states are expected to be nearly unpolarized because the $3p$ shell is full. In conclusion, we can assign the two remaining structures to $3p \rightarrow 5dt_{2g}$ and $3p \rightarrow 5de_g$ transitions. Since the spin–orbit splitting of the $4f$ electrons of europium is an order of magnitude greater than that of the p electrons of sulfur, the transitions of the "magnetic" f electrons should dominate as opposed to the transitions of the "nonmagnetic" p electrons. This is exactly what is found in the decomposition.

We shall now discuss how the expected MO signals can be calculated in the context of an atomic model and compared with measurement. We start from eq. (5.42), and sum over all $4f$ initial states α and over all $5dt_{2g}$ and $5de_g$ final states β:

$$\langle \sigma_{2xy} \rangle = \int_{\alpha,\beta} |\sigma_{2xy}(\omega)| \, d\omega. \tag{6.6}$$

The spin–orbit coupling disturbs the initial-state wave function and splits degenerate levels. Only this contribution is to first-order dependent on the spin–orbit energy, in contrast to the contributions derived from the kinetic operator π, which is defined in eq. (5.25). The latter can, therefore, be simplified to first order to

$$\pi_{\pm} = p_{\pm} = \mathrm{i}\,\omega\, m\,(x \pm \mathrm{i}y). \tag{6.7}$$

Let n be the number of oscillators per unit volume, then it follows for eq. (6.6), without approximating $\omega_{\alpha\beta\pm} \cong \omega_{\alpha\beta}$ as in eq. (5.42):

$$\langle \sigma_{2xy} \rangle = \frac{\pi n e^2}{4\hbar} \sum_{\alpha\beta} \left| \omega_{\alpha\beta+} \,|\langle \beta | x + \mathrm{i}y | \alpha \rangle|^2 - \omega_{\alpha\beta-} \,|\langle \beta | x - \mathrm{i}y | \alpha \rangle|^2 \right|, \tag{6.8}$$

with the position operator in the circular basis,

$$x \pm \mathrm{i}y = \pm 2 \sqrt{\frac{2\pi}{3}}\, r\, Y_1^{\pm 1}(\theta, \phi). \tag{6.9}$$

We can approximate the initial and final states by hydrogen-like wave functions,

$$|\alpha\rangle, |\beta\rangle = R_{l,n}(r)\, Y_l^{m_l}(\theta, \phi), \tag{6.10}$$

where $Y_l^{m_l}(\theta, \phi)$ are spherical harmonics with angular-momentum quantum number l and magnetic quantum number m_l. Furthermore, it has been assumed for the decomposition of the wave function into a radial $R_{l,n}(r)$ and an angular part $Y_l^{m_l}(\theta, \phi)$ that the wave functions $|\alpha\rangle$ and $|\beta\rangle$ are based on a central field. Thus, it is possible to estimate the oscillator strength for the $4f \to 5dt_{2g}$ transition on the basis of an atomic model and compare it with experiment. The solid state can be considered by multiplying the oscillator strength with the ratio of spin–orbit energy to bandwidth (103). For EuS, the following values for $\langle \sigma_{2xy} \rangle$ were obtained (120): $2.6 \times 10^{-29}\ \mathrm{s}^{-2}$ and $1.3 \times 10^{-29}\ \mathrm{s}^{-2}$ for the $4f \to 5dt_{2g}$ and $3p \to 5dt_{2g}$ transition, respectively. The experimental values from Fig. 6.12 are $2.2 \times 10^{-29}\ \mathrm{s}^{-2}$ and $0.9 \times 10^{-29}\ \mathrm{s}^{-2}$ for the $4f \to 5dt_{2g}$ and $3p \to 5dt_{2g}$ transition, respectively. The agreement is very good, justifying neglecting the spin–orbit interaction in the momentum operator. In the problem in Section 6.5.1, we calculate $\langle \sigma_{2xy} \rangle$ for a $4f \to 5d$ using a purely atomic model.

6.4 Comprehension questions

6.4.1 Eu chalcogenides

(a) What is the chemical composition of the Eu chalcogenides? What are the structural and electronic properties (crystal lattice, band structure, and ground state)?
(b) What are the magnetic properties (magnetic ordering)? Which quantum mechanical model is used to describe the magnetic ordering? How can the model describe the different types of magnetic ordering in the Eu chalcogenides?
(c) What are the optical and magneto-optic properties? How can they be explained?
(d) How does the energy-level scheme of Kasuya look like? Which energy terms does it contain?
(e) Why does the polar Kerr rotation spectrum of EuS show a second structure around 4 eV with similar shape as the structure originating from the energy gap around 2 eV? Which quantity can be estimated from the separation of the two structures?
(f) Which spectra are needed to calculate the imaginary part of the off-diagonal element $\sigma_{2xy}(\omega)$? In which two transitions can $\sigma_{2xy}(\omega)$ of EuS be decomposed?

6.5 Problems

6.5.1 Faraday rotation of a $4f \rightarrow 5d$ transition

(a) Draw an energy-level diagram of an $4f \rightarrow 5d$ transition for a single electron. Consider only spin–orbit splitting of the orbital moment by the Zeeman effect. Determine all dipole-allowed optical transitions for rcp ($\Delta m = -1$) and lcp ($\Delta m = +1$) light.

(b) Calculate the imaginary part $\sigma_{2xy}(\omega)$ of the off-diagonal optical conductivity element for a $4f \rightarrow 5d$ transition. Determine the line shape of σ_{2xy} as a function of ω and calculate the integrated value $\langle \sigma_{2xy} \rangle$ according to eq. (6.8). Compare with the $4f \rightarrow 5dt_{2g}$ transition of EuS plotted in Fig. 6.12.

Hint: Use atomic hydrogen wave functions for the 4f and 5d electrons and take into account only spin–orbit interaction on the orbital moment,

$$\Delta E_{\text{s.o.}} = -\zeta \left(\vec{L} \cdot \vec{S}\right) = -\zeta\, m_l\, m_s, \tag{6.11}$$

for the splitting of the electronic states. Assume $\zeta = 0.116$ eV and 0.05 eV for 4f and 5d electrons, respectively. Consider only spin-conserving transitions of a single electron with a spin polarization of 100% in the 4f initial state.
Assume dipole approximation:

$$p_{\text{kin},\pm} = \frac{m}{i\hbar}\left[r_\pm, \mathcal{H}\right] \cong -\, i m\, \omega_{4f\rightarrow 5d}\, r_\pm, \tag{6.12}$$

where $r_\pm = x \pm iy$ is the position operator in circular basis and $\hbar\omega_{4f\to 5d}$ is the average transition energy of the $4f \to 5d$ transition. Calculate the square of the absolute value $|\boldsymbol{M}_\pm|^2$ of the matrix elements $\boldsymbol{M}_\pm = \langle 5d|(x \pm iy)|4f\rangle$ by replacing x and y by spherical functions and by separating the wave function into radial and orbital part.

The radial contribution to the overlap integral, $R^2 = \left|\int R_{5d}(r)\, r\, R_{4f}(r)\, d^3r\right|^2$, *can be evaluated numerically and yields* $1.4 \times 10^{-17}\ \text{cm}^2$.

To solve the integral of the orbital part, use the connection between spherical harmonics and Clebsch–Gordan coefficients:

$$\int_0^\pi \sin\theta\, d\theta \int_0^{2\pi} d\phi\, Y_{l_1}^{m_1 *}(\theta,\phi)\, Y_{l_2}^{m_2 *}(\theta,\phi)\, Y_L^M(\theta,\phi) = \sqrt{\frac{(2l_1+1)(2l_2+1)}{4\pi\,(2L+1)}}\,\langle l_1\, l_2; 0\, 0|l_1\, l_2; L\, 0\rangle\,\langle l_1\, l_2; m_1\, m_2|l_1\, l_2; L\, M\rangle, \tag{6.13}$$

where "" means the conjugate complex function and*

$$\langle l_1\, l_2; m_1\, m_2|l_1\, l_2; L\, M\rangle \tag{6.14}$$

are the Clebsch–Gordan coefficients. In addition, we assume $L = l_1 + l_2$, $-l_{1,2} \le m_{1,2} \le l_{1,2}$, *and* $-L \le M \le L$. *The Clebsch–Gordan coefficients can be easily looked up on Wikipedia* (121). *We will provide a table in the solutions, as well.*

6.6 Solutions

6.6.1 Faraday rotation of a $4f \to 5d$ transition

(a) In Fig. 6.13, we plot an energy-level diagram of a $4f \to 5d$ transition for a single electron. We consider only the Zeeman effect acting on the orbital moment L. Note that the dipole selection rules are $\Delta m_l = +1$ for lcp and $\Delta m_l = -1$ for rcp light. The helicity is reversed due to the negative sign in the time dependence of the wave function in quantum theory as discussed in Sections 4.1 and 5.4. We see that due to the larger spin–orbit constant ζ the splitting of the $4f$ states is larger than the splitting of the $5d$ states. Therefore, transitions starting from energetically lower levels of the $4f$ multiplet have higher transition energies than transitions starting from energetically higher levels of the $4f$ multiplet. Hence, the transitions excited by lcp light take place on average at higher photon energies than for rcp light.

(b) For the calculation of the imaginary (absorptive) part σ_{2xy} of the off-diagonal element of the optical-conductivity tensor, we assume dipole approximation. Then, the kinetic momentum operator becomes proportional to the position operator $r_\pm$

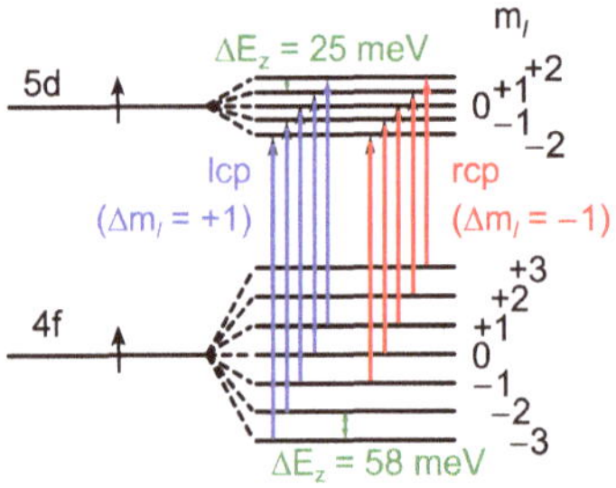

Fig. 6.13: Energy-level scheme of an atomic $4f_\uparrow \to 5d_\uparrow$ transition. Note that the selection rules request for lcp and rcp light $\Delta m_l = +1$ and -1, respectively.

in circular basis as explained in eq. (6.12). Thus, we must evaluate the matrix element in eq. (6.8):

$$\boldsymbol{M}_\pm = \langle \beta | x \pm \mathrm{i}y | \alpha \rangle. \tag{6.15}$$

The matrix elements include the position operator in circular basis, eq. (6.9), which can be expressed in terms of spherical functions:

$$x \pm \mathrm{i}y = \pm 2\sqrt{\frac{2\pi}{3}}\, r\, Y_1^{\pm 1}(\theta, \phi). \tag{6.16}$$

The $4f$ initial state $|\alpha\rangle$ and $5d$ final state $|\beta\rangle$ are approximated by hydrogen wave functions,

$$|\alpha\rangle, |\beta\rangle = R_{l,n}(r)\, Y_l^{m_l}(\theta, \phi), \tag{6.17}$$

where $Y_l^{m_l}(\theta, \phi)$ are spherical harmonics with angular-momentum quantum number l and magnetic quantum number m_l.

Now, we are ready to calculate σ_{2xy}. For this, we must evaluate the matrix element $\boldsymbol{M}_\pm$ by replacing x and y by spherical functions and by separating the wave function into radial and orbital part:

$$\boldsymbol{M}_\pm = \int_0^\infty d^3r \int_0^\pi \sin\theta\, d\theta \int_0^{2\pi} d\phi\, R_{5d}(r)\, r\, R_{4f}(r)\, Y_2^{m'*}(\theta, \phi)\, Y_1^{\pm 1}(\theta, \phi)\, Y_3^{m}(\theta, \phi). \tag{6.18}$$

To solve the double integral of the orbital part,

$$\int_0^\pi \sin\theta\, d\theta \int_0^{2\pi} d\phi\, Y_2^{m'*}(\theta, \phi)\, Y_1^{\pm 1}(\theta, \phi)\, Y_3^{m}(\theta, \phi), \tag{6.19}$$

we use eq. (6.13), the relation between spherical harmonics and Clebsch–Gordan coefficients, which are defined in eq. (6.14). The Clebsch–Gordan coefficients for $l_1 = 2$, $l_2 = 1$, $m_2 = \pm 1$, $L = 3$, and $m_1 - m_2 = M$ are listed in Tab. 6.2. Note that in eq. (6.19) the second spherical harmonic, which represents the dipole operator, is not in the conjugate complex form as in eq. (6.13) because we use for the matrix element the format $\langle 5d|r_{\pm}|4f\rangle$. Therefore, for transitions involving Y_1^{+1} we must take the tabulated Clebsch–Gordan coefficient for Y_1^{-1} and vice versa. In Tab. 6.2, we have already taken this into account.

Tab. 6.2: Clebsch–Gordan coefficients for $l_1 = 2$, $l_2 = 1$, $m_2 = \pm 1$, $L = 3$, and $m_1 - m_2 = M$.

m_2 \ m_1	−2	−1	0	1	2
+1	1	$\sqrt{\frac{2}{3}}$	$\sqrt{\frac{2}{5}}$	$\sqrt{\frac{1}{5}}$	$\sqrt{\frac{1}{15}}$
−1	$\sqrt{\frac{1}{15}}$	$\sqrt{\frac{1}{5}}$	$\sqrt{\frac{2}{5}}$	$\sqrt{\frac{2}{3}}$	1

The Clebsch–Gordan coefficient for $m_1 = m_2 = M = 0$ is $\langle 2\,1;\ 0\,0|\ 2\,1;\ 3\,0\rangle = \sqrt{3/5}$. The prefactor on the right side of eq. (6.13) amounts for $l_1 = 2$, $l_2 = 1$, and $L = 3$ to $\sqrt{(2l_1+1)\,(2l_2+1)/[4\pi\,(2L+1)]} = \sqrt{15/28\pi}$. Putting all the factors together, we get the values for the integral over the spherical harmonics, eq. (6.13). They are listed in Tab. 6.3.

Tab. 6.3: Integral of the orbital part of the matrix elements $\boldsymbol{M}_{\pm}$ for the $4f \to 5d$ transition ($l_1 = 2$, $l_2 = 1$, $m_2 = \pm 1$, $L = 3$, and $m_1 - m_2 = M$).

m_2 \ m_1	−2	−1	0	1	2
+1	$\sqrt{\frac{9}{28\pi}}$	$\sqrt{\frac{3}{14\pi}}$	$\sqrt{\frac{9}{70\pi}}$	$\sqrt{\frac{9}{140\pi}}$	$\sqrt{\frac{3}{140\pi}}$
−1	$\sqrt{\frac{3}{140\pi}}$	$\sqrt{\frac{9}{140\pi}}$	$\sqrt{\frac{9}{70\pi}}$	$\sqrt{\frac{3}{14\pi}}$	$\sqrt{\frac{9}{28\pi}}$

Of course, we could have done the integration over the product of spherical harmonics directly without using Clebsch–Gordan coefficients. The spherical harmonics are defined as

$$Y_l^m(\theta,\phi) = \sqrt{\frac{(l-|m|)!\,(2l+1)}{(l+|m|)!\,4\pi}}\, e^{im\phi} P_l^{|m|}(\cos\theta), \tag{6.20}$$

where $p_l^{|m|}(\cos\theta)$ are associated Legendre polynomials. They are simple power functions of $z = \cos\theta$ which makes the integration over a product of spherical harmonics straight forward. We will demonstrate this for $l_1 = 2$, $m_1 = -2$, $l_2 = 1$, $m_2 = +1$, $L = 3$, and $M = -3$. Substituting $z = \cos\theta$, the spherical harmonics involved are then:

$$Y_2^{-2}(z,\phi) = \sqrt{\frac{15}{32\pi}}\,(1-z^2)\,e^{-2i\phi}, \tag{6.21}$$

$$Y_1^1(z,\phi) = \sqrt{\frac{3}{8\pi}}\,\sqrt{1-z^2}\,e^{i\phi}, \tag{6.22}$$

$$Y_3^{-3}(z,\phi) = \sqrt{\frac{35}{64\pi}}\,(1-z^2)^{\frac{3}{2}}\,e^{-3i\phi}. \tag{6.23}$$

Now, we can evaluate eq. (6.13):

$$\begin{aligned}\int_{-1}^{1} dz \int_{0}^{2\pi} d\phi\, Y_2^{-2*}\, Y_1^1\, Y_3^{-3} &= \sqrt{\frac{15\times3\times35}{32\times8\times64\pi^3}} \int_{-1}^{1} (1-z^2)^3 dz \int_0^{2\pi} e^{i(2+1-3)\phi} d\phi \\ &= \sqrt{\frac{15\times3\times35}{32\times8\times64\pi^3}} \int_{-1}^{1} \left(1-3z^2+3z^4-z^6\right) dz \int_0^{2\pi} d\phi \\ &= 2\pi\sqrt{\frac{15\times3\times35}{32\times8\times64\pi^3}} \left[z - z^3 + \frac{3}{5}z^5 - \frac{1}{7}z^7\right]_{-1}^{1} \\ &= 2\pi\sqrt{\frac{15\times3\times35}{32\times8\times64\pi^3}}\,\frac{32}{35} = \sqrt{\frac{9}{28\pi}},\end{aligned} \tag{6.24}$$

which is exactly the same as the Clebsch–Gordan coefficient in Tab. 6.3.

Looking at Tab. 6.3, we see that the values increase/decrease for rcp/lcp transitions ($m_2 = -1/+1$) with increasing magnetic quantum number m_1. From Fig. 6.13, we know that rcp transitions are energetically lower than lcp transitions. Furthermore, the imaginary part σ_{2xy} of the optical conductivity element is according to eq. (5.42), proportional to the difference between the square of the absolute value of the matrix element for rcp and lcp light. So, the line shape of σ_{2xy} for a $4f \to 5d$ transition should be dominated at low photon energies by positive values due to large contributions of rcp transitions and at high photon energies by negative values (because of the subtraction) due to large contributions of lcp transitions leading to the form of a procumbent "S". This is indeed what we see in experiment shown in Fig. 6.12. So, our simple model cannot only explain the line shape, it even gives the correct sign of the $4f \to 5d$ transition.

The last property we can compute with our simple model is the integrated imaginary part $\langle\sigma_{2xy}\rangle$ of the off-diagonal optical conductivity element of the transition according to eq. (6.8) and compare with the experiment shown in Fig. 6.12. $\langle\sigma_{2xy}\rangle$ follows from integration of $|\sigma_{2xy}|$ over ω which is straightforward due to the delta functions. The radial contribution,

$$R^2 = \left|\int_0^\infty R_{5d}(r)\, r\, R_{4f}(r)\, d^3r\right|^2, \tag{6.25}$$

can be evaluated numerically and yields 1.4×10^{-17} cm^2 = 1.4×10^{-21} m^2. Next, we need to evaluate the prefactor in eq. (6.8). We get the density n_{Eu} of Eu^{2+} ions from the face-centered cubic crystal lattice. We have four Eu^{2+} ions per cubic unit cell, which has a lattice constant of $a_{\mathrm{EuS}} = 5.968$ Å according to Tab. 6.1. Therefore, the density is

$$n_{\mathrm{Eu}} = \frac{4}{a_{\mathrm{EuS}}^3}. \tag{6.26}$$

Inserting this, we get for the prefactor:

$$\frac{\pi\, n_{\mathrm{Eu}} e^2}{4\hbar} = \frac{\pi\, e^2}{\hbar\, a_{\mathrm{EuS}}^3} = 3.6 \times 10^{24} \frac{\mathrm{A}}{\mathrm{Vm}^3}, \tag{6.27}$$

The last value missing is the sum over the absolute value of the difference of the squares of the matrix elements for rcp and lcp transitions. The amplitudes of the matrix elements $\boldsymbol{M}_\pm$ are given in Tab. 6.3. By using the approximation, $\omega_{\alpha\beta+} \cong \omega_{\alpha\beta-} \cong \omega_{4f\to5d}$, we get for the sum of the absolute value of the difference of the squares of $\boldsymbol{M}_\pm$:

$$\begin{aligned}\sum_{\alpha\beta} \left| \omega_{\alpha\beta+} |\boldsymbol{M}_+|^2 - \omega_{\alpha\beta-} |\boldsymbol{M}_-|^2 \right| &\cong 2\left[1 + \frac{1}{15} + \frac{2}{3} + \frac{1}{5} + \frac{2}{5}\right] \\ &\times \frac{9}{28\pi} \times \omega_{4f\to5d} = \frac{3}{2\pi}\omega_{4f\to5d}.\end{aligned} \tag{6.28}$$

We can estimate the transition frequency $\omega_{4f\to5d}$ from Fig. 6.12. We find

$$\omega_{4f\to5d} = \frac{2.2\ \mathrm{eV}}{\hbar} = 2.2 \times \frac{1.6022 \times 10^{-19}\ \mathrm{AsV}}{1.055 \times 10^{-34}\ \mathrm{Js}} = 3.34 \times 10^{15}\mathrm{s}^{-1}. \tag{6.29}$$

Putting all values together, we finally get for $\langle \sigma_{2xy} \rangle$:

$$\langle \sigma_{2xy} \rangle = 0.48 \times 3.34 \times 10^{15}\mathrm{s}^{-1} \times 1.14 \times 10^{-21}\mathrm{m}^2 \times 3.6 \times 10^{24} \frac{\mathrm{A}}{\mathrm{V\,m}^3} = 6.58 \times 10^{18} \frac{\mathrm{A}}{\mathrm{V\,m\,s}}. \tag{6.30}$$

In Fig. 6.12, the experimental value is $\langle \sigma_{2xy} \rangle = 2.2 \times 10^{29}\mathrm{s}^{-2}$ in cgs units. So, we must convert eq. (6.30) to cgs units. The conversion factor can be taken from eq. (2.36), because the dielectric function is dimensionless in both unit systems:

$$\langle \sigma_{2xy} \rangle_{\mathrm{cgs}} = \frac{1}{4\pi\varepsilon_0} \langle \sigma_{2xy} \rangle_{\mathrm{SI}} = \frac{6.58 \times 10^{18} \frac{\mathrm{A}}{\mathrm{Vms}}}{4\pi \times 8.85 \times 10^{-12} \frac{\mathrm{As}}{\mathrm{Vm}}} = 5.9 \times 10^{28}\mathrm{s}^{-2}. \tag{6.31}$$

As we have integrated over all possible initial and final states, the calculated value must be divided by a factor 2 to get the integral of just one lobe of the procumbent "S" transition. The result is a factor of 7 smaller than the experimental value which is alright for this simple approach.

7 Relation between magnetism and magneto-optics

In this chapter, we would like to discuss the relation between magnetism and magneto-optics. We are mainly interested in the question how magnetic properties manifest themselves in magneto-optic measurements. We start with some basic considerations on magnetic fields. For instance, we have repeatedly claimed that the applied magnetic field $\vec{B}_0$ must be replaced by the magnetization $\vec{M}$ when considering magnetically ordering materials. This is illustrated in Fig. 7.1, where the specific Faraday rotation $\theta_F' = \theta_F/d$, where d is the film thickness, of a thin film of EuSe at a temperature of 2.5 K is plotted versus applied magnetic field $\vec{B}_0$. At this temperature, EuSe is in a ferrimagnetically ordered state. So, it behaves more or less like a ferromagnet. One can see a linear increase of θ_F'. Above a magnetic field of approximately $2T$, we notice saturation behavior. The measurement was performed on a thin layer with $\vec{M}$ perpendicular to the surface due to the applied field. Therefore, a demagnetization factor of $N = 1$ (in SI units) must be considered.

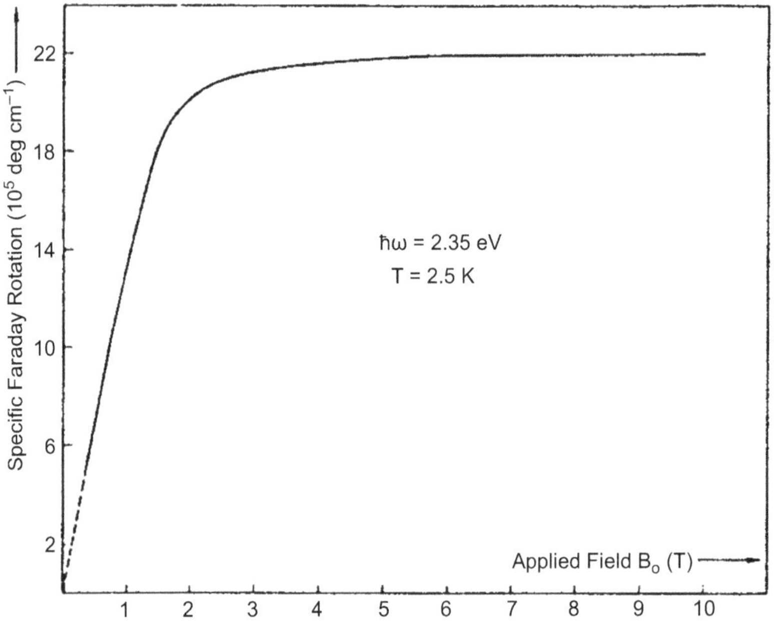

Fig. 7.1: Field dependence of the specific Faraday rotation θ_F' of EuSe at a photon energy of $\hbar\omega = 2.35$ eV and temperature of 2.5 K. Figure from Ref. (122).

https://doi.org/10.1515/9783110635300-007

The internal magnetic field $\vec{B}_\mathrm{i} = \mu_0 \vec{H}$ is not equal to the applied magnetic field $\vec{B}_0 = \mu_0 \vec{H}_0$. It must be corrected by the demagnetizing field:

$$\vec{B}_\mathrm{i} = \mu_0 \vec{H} = \vec{B}_0 + \vec{B}_\mathrm{s} = \vec{B}_0 - \mu_0 N\vec{M} = \mu_0 \left(\vec{H}_0 + \vec{H}_\mathrm{s}\right). \tag{7.1}$$

In eq. (7.1), $\vec{B}_\mathrm{s} = \mu_0 \vec{H}_\mathrm{s} = -\ \mu_0 N\vec{M}$ is the stray magnetic field, also called demagnetizing field, and N is the demagnetization factor, which depends on the geometry. Note that $\vec{B}_\mathrm{s}$ and $\vec{B}_0$ have opposite direction, explaining the name demagnetizing field. The origin of the linear behavior of the Faraday rotation θ_F which follows the magnetization, is due to the compensation of the applied magnetic field $\vec{B}_0$ by the demagnetizing field $\vec{B}_\mathrm{s}$. As long as the magnetization $\vec{M}$ is not yet saturated, the internal field $\vec{B}_\mathrm{i}$ will stay zero when increasing $\vec{B}_0$ as is evident from eq. (7.1). This leads to a direct proportionality between $\vec{B}_0$ and $\vec{M}$ and, consequently, θ_F. The saturation magnetization $\mu_0 M_\mathrm{sat}$ of EuSe is reached at approximately 1.4 T. To achieve the saturated state, the condition $B_\mathrm{i} \geq 0.2$ T must be fulfilled. Therefore, magnetic saturation is expected at $B_0 \geq 1.6$ T in good agreement with experiment.

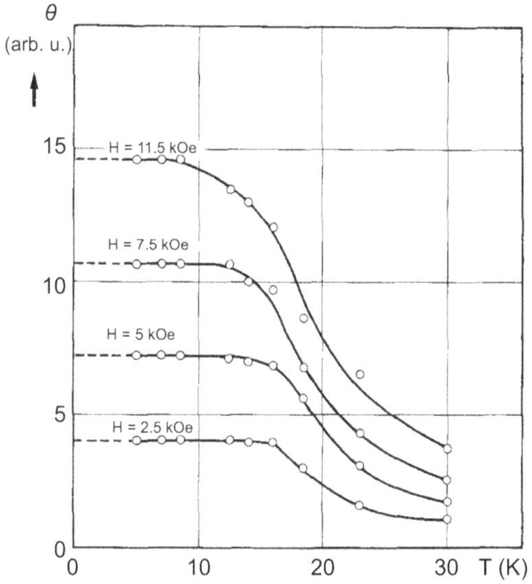

Fig. 7.2: Temperature dependence of the Faraday rotation θ_F of EuS in various fields measured in the photon-energy range of the maximum between 2.05 and 2.15 eV. Figure from Ref. (104).

Next, we want to discuss the temperature behavior of the Faraday rotation θ_F. The temperature dependence of θ_F is shown for a thin film of EuS at various fields in Fig. 7.2. For a temperature below the ordering temperature, $T < T_\mathrm{C} = 16.5$ K, and for small magnetizations far away from saturation, $M \ll M_\mathrm{sat}$, the applied field $\vec{B}_0$ solely creates a magnetization along its direction and must, at the same time, compensate the stray magnetic field $\vec{B}_S$. Thus, the internal field vanishes: $B_i \cong 0$ T. Assuming negligible hysteresis, it follows from eq. (7.1), as already discussed before, that

$$\vec{B}_0 = \mu_0 N \vec{M} = \mu_0 \vec{M}. \tag{7.2}$$

Hence, the magnetization is proportional to $\vec{B}_0$ and is independent of temperature below T_C. The ferromagnetic ordering temperature T_C can be determined in Fig. 7.2 from the kink in θ_F at vanishing external magnetic field. At higher magnetic fields, θ_F is no longer temperature independent below T_C. The curves become rounded and T_C can no longer be assigned accurately.

For temperatures larger than $T > T_C$, the material is in a paramagnetic phase. The magnetization follows the internal magnetic field and the proportionality constant is given by the magnetic susceptibility χ_m:

$$\vec{M} = \chi_m \vec{H} = \frac{\chi_m}{\mu_0} \vec{B}_i = \frac{\chi_m}{\mu_0}\left(\vec{B}_0 + \vec{B}_s\right), \tag{7.3}$$

By substituting eq. (7.1) into (7.3), we can take care of the fact that the stray magnetic field depends on the magnetization and solve the equation for $\vec{M}$:

$$\vec{M} = \frac{\chi_m}{\mu_0}\left(\vec{B}_0 + \vec{B}_s\right) = \frac{\chi_m}{\mu_0}\vec{B}_0 - \chi_m N \vec{M} = \frac{\chi_m}{\mu_0\,(1 + N\chi_m)}\vec{B}_0. \tag{7.4}$$

If the temperature is much larger than T_C, then $N\chi_m \ll 1$ and we get the well-known Curie–Weiss law:

$$\vec{M} = \frac{\chi_m}{\mu_0}\vec{B}_0, \tag{7.5}$$

with a temperature-dependent magnetic susceptibility,

$$\chi_m \propto \frac{1}{T - \theta_p}, \tag{7.6}$$

where θ_p is the paramagnetic Curie temperature, which is approximately equal to the ferromagnetic ordering temperature T_C. We suppose again that θ_F is proportional to the magnetization. To verify this assumption, $1/\theta_F$ is shown as a function of temperature in Fig. 7.3. We find indeed a straight line as expected from the Curie–Weiss law. From the intercept of $1/\theta_F$ with the temperature axis, we can determine a paramagnetic Curie temperature $\theta_p = 19$ K. Near T_C, which is near θ_p, deviations from a straight line occur due to local-ordering effects.

EuSe is a metamagnet as is evident from Table 6.1. It shows a complicated magnetic phase diagram with magnetic-field-induced transitions, which is described in detail in Fig. 7.4(a) (116). If we keep the temperature fixed at 1.7 K (dashed vertical line) and slowly increase the applied magnetic field, EuSe will change from an antiferromagnetic to a ferrimagnetic state at 0.03 T and to a ferromagnetic state at 0.25 T.

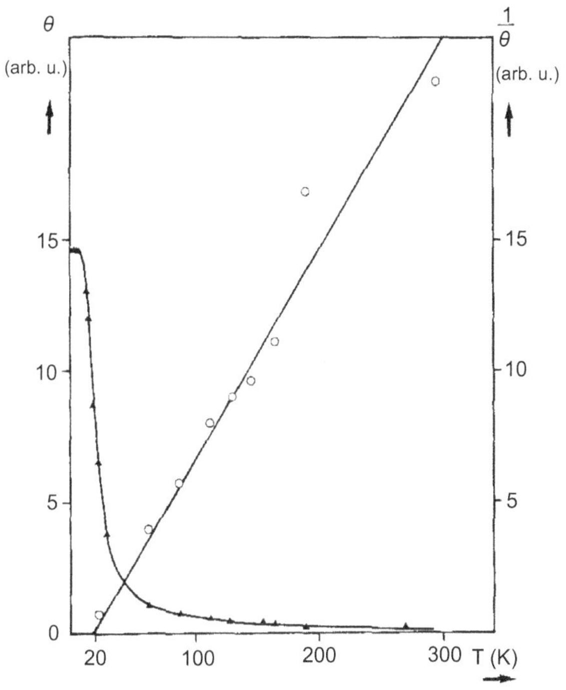

Fig. 7.3: Temperature dependence of the Faraday rotation θ_F of EuS in the photon-energy range of the maximum between 2.05 and 2.15 eV. Figure from Ref. (104)

Nevertheless, as depicted in Fig. 7.4(b), the specific Faraday rotation (red dots) is, apart from a narrow magnetic field range around 0.4 T, always proportional to the magnetization (solid line) across the magnetic phases (123).

The small deviation around 0.4 T indicates that the proportionality between magnetization and Faraday (or polar Kerr) rotation is not always given. One must act with caution, especially in antiferromagnetic or ferrimagnetic systems, where by nature of the magnetic ordering at least two spin sublattices exist with opposite spin direction. In case these spin sublattices have different photon energy (or – what is the same – wavelength or frequency) dependence of the magneto-optic properties, it can happen that at a certain photon energy, the magneto-optic property of one spin sublattice is predominant. Then, of course, a magneto-optic measurement would primarily be sensitive to the magnetization of one spin sublattice and not, as usually assumed, to the total magnetization. An instructive example is given in Fig. 7.5, where polar Kerr hysteresis curves of a Co/EuS multilayer are plotted at a temperature of 10 K at two different photon energies $\hbar\omega = 0.85$ and 3.6 eV (124). In this sample, a portion of each EuS layer is antiferromagnetically coupled to the Co layers and for the thickness chosen, the magnetization of the sum of the Co layers is larger than for all EuS layers. So, the total system behaves like a ferrimagnet. The Co

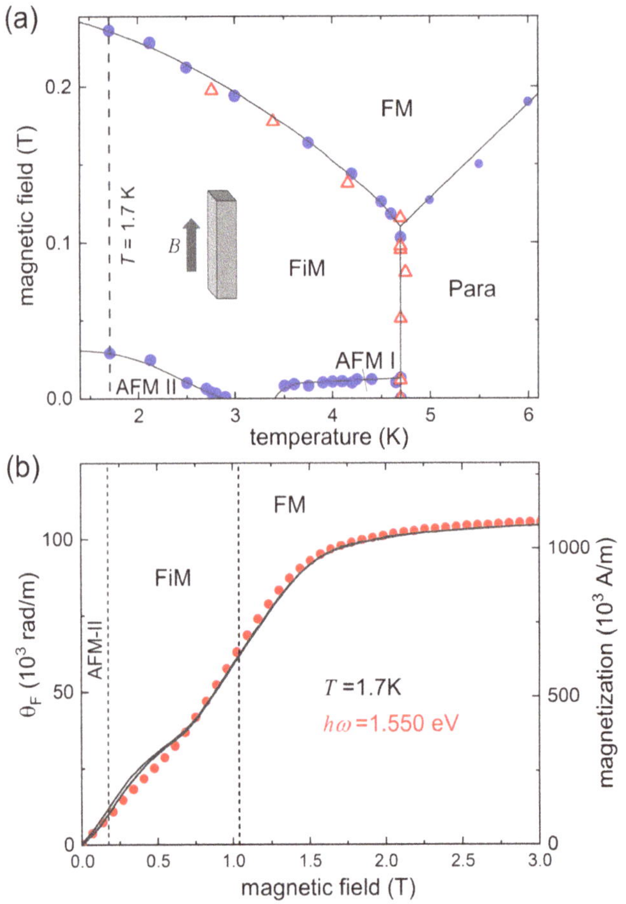

Fig. 7.4: (a) Magnetic phase diagram of the studied EuSe layer. The full and open dots represent data obtained from *M* versus *B* (T = const.) and *M* versus *T* (B = const.) traces, respectively. The magnetic field was applied parallel to the surface of the epitaxial sample. The solid lines are guides to the eye. The dashed line shows that at $T = 1.7$ K, a magnetic field drives EuSe through an antiferromagnetic (AFM II), a ferrimagnetic (FiM), and a ferromagnetic (FM) phase. (b) Magnetization (solid line) and specific Faraday rotation at $\hbar\omega = 1.55$ eV (red dots), as a function of magnetic field, at $T = 1.7$ K. The magnetic field was applied perpendicular to the surface of the epitaxial sample. When *B* is applied perpendicular to the layer, the AFM II –FiM and FiM – FM phase boundaries are shifted to $B = 0.17$ T and $B = 1.04$ T, respectively, due to the demagnetization effect. Vertical lines show the boundaries between the magnetic phases indicated. Figure from Ref. (123).

magnetization is dominant and will follow the applied magnetic field and a part of the EuS magnetization is aligned in opposite direction for not too large magnetic fields. At $\hbar\omega = 0.85$ eV (blue dots), the photon energy is within the gap of EuS and the polar Kerr effect is probing predominantly the Co magnetization. At $\hbar\omega = 3.6$ eV (red squares), due to an optical enhancement effect, the polar Kerr effect is mainly sensitive to the

EuS magnetization. Consequently, the polar Kerr hysteresis curves at the two photon energies look totally different. At 0.85 eV (blue dots), a square hysteresis curve is measured, indicating a ferromagnetically ordered spin system, which is expected for a magnetically dominating Co magnetization following the applied magnetic field. At 3.6 eV (red squares), however, we see the behavior of a magnetization which is antiferromagnetically coupled to Co. Starting at zero magnetic field, it first follows the Co magnetization and switches at 0.23 T, the coercive field of the Co hysteresis curve. When the magnetic field exceeds 0.5 T, the antiferromagnetic coupling breaks up. The magnetization of the EuS layers aligns with the applied magnetic field, being now parallel to the Co magnetization.

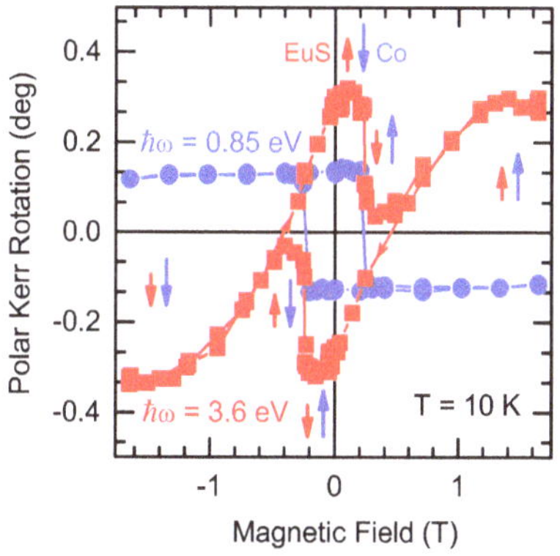

Fig. 7.5: Polar Kerr hysteresis curves at 10 K. At $\hbar\omega = 0.85$ eV (blue dots), EuS is transparent and the hysteresis loop only shows ferromagnetic Co characteristics. At $\hbar\omega = 3.6$ eV (red squares), the polar Kerr rotation is positive for small magnetic fields, that is, EuS moments are aligned antiparallel to the Co moments. The magnetic moment of EuS and Co is symbolized by the red and blue arrow, respectively. Increasing the magnetic field breaks up the antiferromagnetic coupling and aligns the EuS moments parallel to the applied field, changing the sign of the Kerr rotation. Figure from Ref. (124).

The fact that θ_F is typically proportional to the magnetization can not only be used for the determination of T_C or θ_p. In addition, statements can be made about the direction of magnetization, magnetic coupling, or phase transitions. This will be briefly illustrated in the following on three examples.

7.1 Determination of easy axis of magnetization

In the first example, we want to show how the direction of the easy axis can be determined using magneto-optics. The material, we want to look at, is a Co/Pt multilayer. It consists of several identical bilayers of Co/Pt deposited on a suitable substrate. The direction of the easy axis of magnetization depends strongly on the ratio of the thickness of the Pt and Co layer. With a suitable choice of layer thickness, a perpendicular magnetic anisotropy can be created in such a multilayer structure. The perpendicular magnetic anisotropy is then so strong that it overcomes the demagnetizing field, which wants to keep the magnetization within the plane of the thin film. Consequently, the easy direction of magnetization will be perpendicular to the film surface.

In Fig. 7.6, the hysteresis curves, measured using polar Kerr rotation, are recorded for a fixed wavelength of 530 nm (125). The Pt-layer thickness is kept constant at 17.4 ± 0.3 Å and the Co-layer thickness varies from 2.4 to 20.2 Å. The number of Co/Pt bilayers varies from 25 down to 14 to keep a constant overall thickness of roughly 500 Å. Starting with Fig. 7.6(a), we recognize an almost square hysteresis curve. This means that the magnetization reversal takes place in a very narrow field range and that the magnetization reaches a remanence of 100% at zero magnetic field, that is, the magnetization stays constant when decreasing the magnetic field and assumes at zero field still the saturation value. Such a square-loop behavior is typical for an easy magnetization axis pointing out of plane because the polar Kerr effect is only sensitive to the magnetization component perpendicular to the film surface. The sudden magnetization reversal indicates a fast switching between two magnetization directions parallel and antiparallel to the surface normal as expected for an easy axis of magnetization which is perpendicular to the film surface. The square hysteresis curve proves that for sufficiently thin Co layers, a perpendicular magnetic anisotropy exists. For thicker Co layers, the curve becomes more and more sheared. First, the 100% remanence is still maintained, as seen in Fig. 7.6(b), but the magnetization reversal is now less steep. Further increasing the Co-layer thickness, as plotted in Fig. 7.6(c)–(e), the remanence gets smaller and smaller and the magnetization reversal flattens out. This implies that the perpendicular magnetic anisotropy weakens and the easy axis rotates away from the normal direction into the plane of the layer. Finally, in Fig. 7.6(f), the hysteresis curve has changed to a typical hard-axis curve. Magnetization reversal starts already at high magnetic field and decreases linearly with almost no hysteresis. This behavior is expected when the easy axis of magnetization lies in the plane of the film and is dragged out of equilibrium position by the applied magnetic field, which is now along the hard axis of magnetization because it is perpendicular to the surface.

The creation of perpendicular magnetic anisotropy is very important for optimizing magneto-optic storage materials and for high-density magnetic recording in the form of perpendicular magnetic recording as it is used in some of today's hard-disk drives. In fact, Co/Pt multilayer structures were viewed as promising candidates for magneto-optic storage materials of the second generation using Blu-ray technology at a blue wavelength of 405 nm.

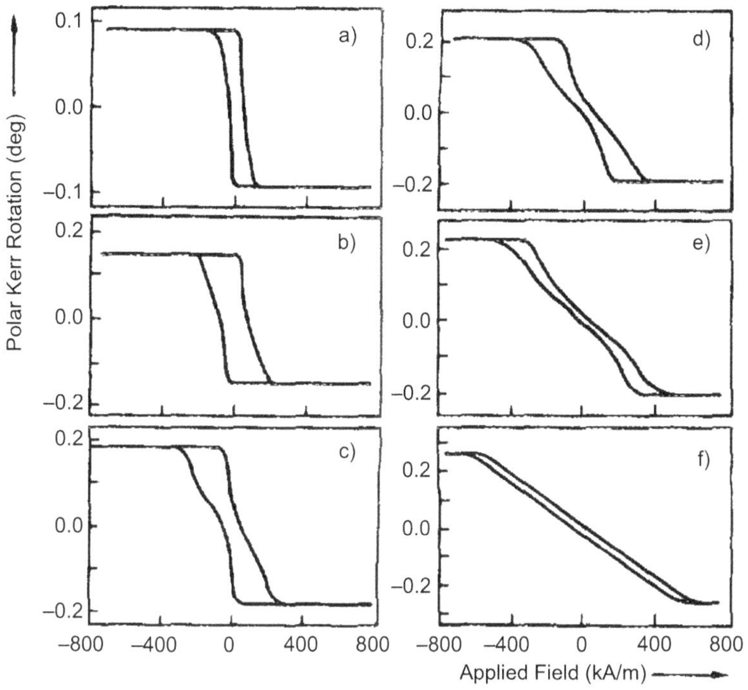

Fig. 7.6: Polar Kerr hysteresis curves of Co/Pt multilayers on Si. The curves have been measured at a wavelength of 530 nm. The Pt-layer thickness is kept constant at 17.4 ± 0.3 Å, the Co-layer thickness is (a) 2.4 Å [25], (b) 4.5 Å [22], (c) 7.0 Å [20], (d) 9.2 Å [19], (e) 14.2 Å [17], and (f) 20.2 Å [14]. The value in square brackets indicates the number of Co/Pt bilayers. Figure from Ref. (125).

7.2 Determination of exchange coupling

Another interesting phenomenon in multilayer structures is the oscillatory exchange coupling of magnetic layers which are separated by a nonmagnetic interlayer. Here, nonmagnetic means not magnetically ordering. So, the interlayer can be paramagnetic or diamagnetic. The magnetic layers couple alternately ferromagnetically and antiferromagnetically as a function of the thickness of the interlayer. This coupling can very nicely be explored with the help of magneto-optics.

The sample, we would like to discuss, is a Fe(100)/Mo(100)/Fe(100) wedge structure (126). The value (100) is the Miller index and denominates the crystal orientation of the surface. The paramagnetic Mo wedge between the 14-monolayer thick Fe layers allows for the measurement of different interlayer thicknesses on the same sample. This has the advantage that a wide thickness range of spacer layers can be produced under identical conditions. The disadvantage is that the sample cannot be measured as a whole but must be scanned. However, for an optical measurement using lasers, this is no problem. The interlayer thickness resolution is then limited by the lateral resolution of the scan.

In Fig. 7.7, on the left, the hysteresis curves measured with longitudinal Kerr effect, are shown for a Fe/Mo/Fe wedge structure (126). Going from very thin to thicker Mo spacer layer, one recognizes an oscillatory behavior. The first hysteresis loop on top with a Mo thickness of 3.8 monolayers is an almost perfect square hysteresis curve as already discussed in the previous example. Because the measurement has been done using longitudinal Kerr effect, which is sensitive to magnetization components parallel to the surface, the magnetization must be in the plane of the sample. One monolayer of Fe is approximately 0.15 nm. Then, 14 monolayers are only about 2 nm thick. According to eq. (2.47), the penetration depth of the light is deep enough to see both iron layers. A square hysteresis curve, therefore, indicates a ferromagnetic, that is, parallel, alignment of the in-plane magnetization of the two Fe layers. The contribution of the paramagnetic Mo interlayer should at most lead to small linear increase at high fields. If one looks very carefully, one can detect a tiny linear increase in the saturation region.

Increasing the interlayer by just about one and a half monolayer to 5.2 monolayers, the hysteresis curve changes completely. Now it shows two symmetric regions with hysteresis and a staircase-like behavior. We note that the longitudinal Kerr rotation is zero in the low-field region indicating zero total magnetization of the Fe layers. The simplest explanation for this behavior is an antiferromagnetic, that is, antiparallel, alignment of the two Fe layers. The appearance of the small hysteresis symmetrically on both sides of the magnetization loop indicates the breakup of the antiferromagnetic coupling between the two Fe layers. The layers will then switch to parallel alignment at high magnetic fields. The magnetic field, at which the hysteresis appears, is called switching field B_s. It is a measure of the antiferromagnetic coupling strength. It is determined at the center of the hysteresis.

By further increasing the Mo spacer-layer thickness, an oscillation between ferromagnetic and antiferromagnetic coupling is evident. The thicker the spacer layer gets, the closer the switching hystereses move together. Thus, the switching field strength decreases with increasing interlayer thickness. At 17-monolayer thickness of the Mo interlayer, the antiferromagnetic switching field is smaller than the coercive field of its hysteresis. Therefore, a complete switching does not take place anymore. Just the onset of switching is present as a small step at the flank of the hysteresis.

On the right side of Fig. 7.7, the oscillations of the coupling are illustrated by plotting the switching field $H_s = B_s/\mu_0$, which describes the field at which the antiferromagnetic coupling breaks up. The exchange-coupling strength per unit area J can be estimated from theory (127) as $J \cong M_{sat} B_s d$, were M_{sat} is the saturation magnetization and d the thickness of the Fe layer. Each peak determines the maximum antiferromagnetic coupling and the valleys in between the maximum ferromagnetic coupling. We see a sort of damped oscillation with periodicity of approximately three monolayers. Note that the first maximum around 3 monolayer is missing. The coupling is

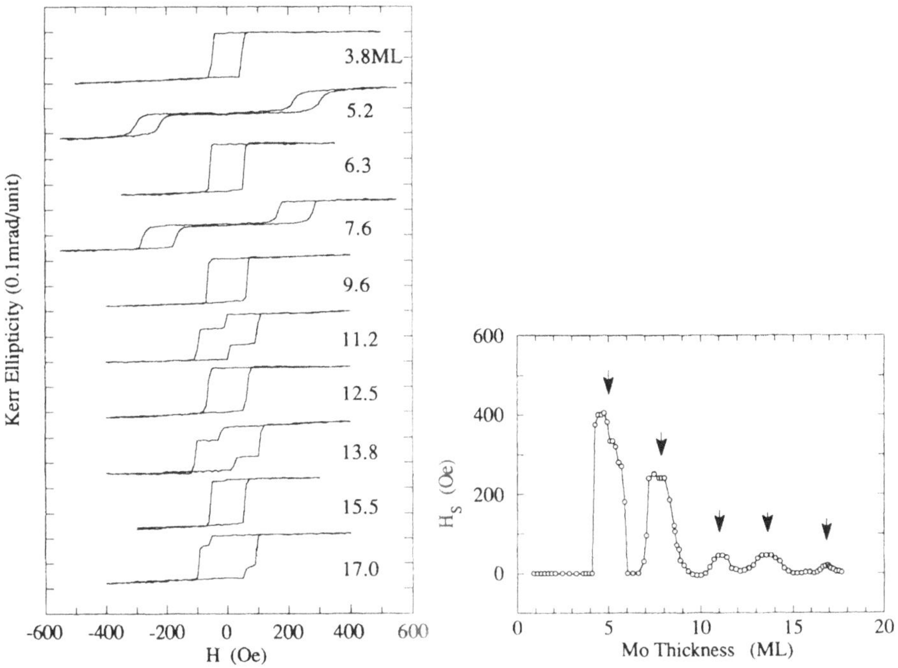

Fig. 7.7: Left: Longitudinal Kerr hysteresis curves of a Fe/Mo/Fe wedge structure for various Mo-layer thicknesses. The curves were measured at the wavelength of a HeNe Laser. Right: The switching field as a function of Mo-layer thickness. The oscillations can clearly be seen. Figure from Ref. (126).

thought to be due to Ruderman–Kittel–Kasuya–Yosida interaction (RKKY) (128). It is not a direct but an indirect exchange interaction which is mediated by the conduction electrons in the molybdenum layer. At the Fe/Mo interface, they get spin-polarized by the magnetic moments of the iron layer. According to the RKKY interaction, the spin-polarization oscillates with distance. Therefore, the coupling that is transferred to the next Fe/Mo interface oscillates with the distance the electrons must travel in the Mo layer.

7.3 Determination of phase separation

In the last example, we will see that one can use magneto-optics even to investigate structural phase separation by measuring the magnetization as a function of temperature. In a mixed material, we are able to study crystallographic phase transitions or phase separation as a function of concentration by using magneto-optic methods if the magneto-optic properties are different for the different phases. An interesting example is provided by thin layers of amorphous ferromagnetic uranium compounds.

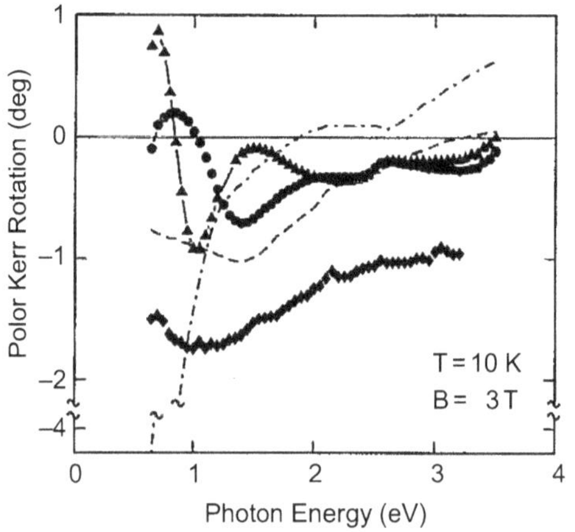

Fig. 7.8: Polar Kerr rotation in amorphous U_xAs_{100-x} layers for $x = 49$ (●), 51 (▲) and 62 (◆). Figure from Ref. (129). Also shown is the data for the crystalline compounds UAs (dashed line) and U_3As_4 (dot-dashed line) from Ref. (95).

In Fig. 7.8, the polar Kerr rotation of amorphous U_xAs_{100-x} layers are plotted at a temperature of 10 K and in a magnetic field of 3 T as a function of photon energy for three different uranium concentrations $x = 49$, 51, and 62 (129). In addition, the spectra of crystalline samples of UAs and U_3As_4 are shown for comparison. The magnetic field was not high enough to saturate the magnetization and, hence, the polar Kerr rotation. We clearly see that near a photon energy of 1 eV a positive structure appears for uranium concentrations $x < 55$ which is not present at concentrations $x > 55$. Furthermore, the negative peak, also present in samples with high uranium concentration, is narrowing.

We get more information about the magnetic state of the samples by measuring the Kerr spectra in remanence. Remanence measurements are performed by applying prior to starting the measurement of the polar Kerr rotation a saturating magnetic field to the samples and then switching it off. This procedure certifies that all magnetic domains are aligned prior to measuring in zero magnetic field. This is like starting a hysteresis curve and stopping at zero magnetic field where the sample resides in the remanent state. The difference is that a complete spectrum is taken. The result is shown in Fig. 7.9 for the two samples, $U_{51}As_{49}$ and $U_{62}As_{38}$. Usually, one would expect the same spectral dependence of the polar Kerr rotation as in the spectra measured in a magnetic field because the Kerr rotation is proportional to the magnetization and the magnetization is reduced to the value in remanence. Hence, the remanence spectra should be reduced by a factor equal to the ratio between remanent magnetization and the magnetization induced by the applied magnetic field. We find that for $x > 55$, the polar Kerr rotation spectrum has indeed the identical form as in high magnetic field. For $x < 55$, however, the remanence spectrum has

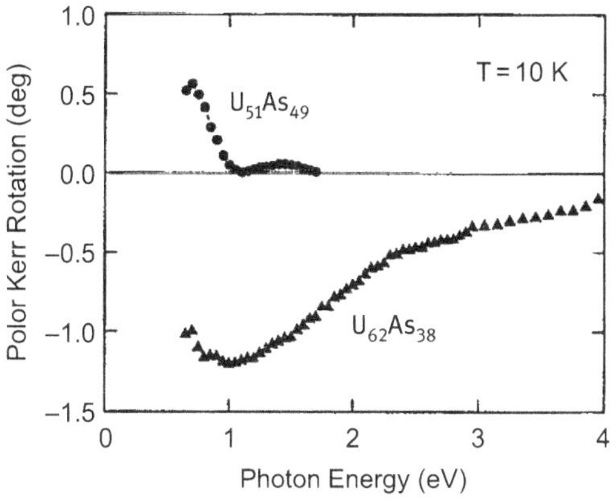

Fig. 7.9: The remanence polar Kerr rotation of the amorphous films $U_{51}As_{49}$ (●) and $U_{62}As_{38}$ (▲) at a temperature of 10 K. Figure from Ref. (129).

changed quite a lot. In comparison with the spectrum taken in high magnetic field, only the positive, low-energy peak exhibits a remanence. The negative structure at slightly higher photon energy has completely vanished.

This behavior can be further illuminated by measuring polar Kerr hysteresis curves in the same sample at different photon energies. We measure once at the photon energy of the positive structure at 0.7 and 0.85 eV for $x = 51$ and 49, respectively, showing a large remanence. The second measurement is taken at the photon energy of the negative structure at 1.05 and 1.4 eV for $x = 51$ and 49, respectively, where the samples exhibit a small remanence. The result for the two layers with concentration $x < 55$ is shown in Fig. 7.10. The positive low-energy peak exhibits for both samples a hysteresis with a finite remanence. The coercive field, which defines the width of the hysteresis, and the remanence is larger for the sample with the higher uranium concentration. The negative peak at higher photon energy, on the other hand, does not show a hysteresis. Consequently, it does neither show remanence. For a homogeneous material, such a behavior is not allowed. The magnetization is well defined and must be proportional to the polar Kerr rotation at all photon energies. Therefore, the material cannot be homogeneous, it must be inhomogeneous. We must assume at least two coexisting phases with different Kerr spectra and magnetic properties to explain the measurements.

The assumption of coexisting phases is further confirmed by resistivity measurements where a percolation behavior between a metallic and a semiconducting phase is observed. In conclusion, with magneto-optic investigations, it is possible to distinguish between different magnetic phases in inhomogeneous materials. It is, however, not possible to unambiguously denominate the coexisting magnetic phases.

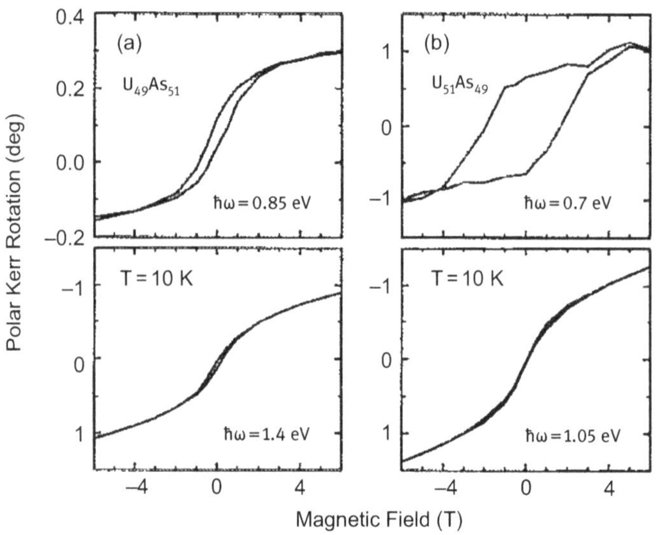

Fig. 7.10: Polar Kerr hysteresis curves in amorphous U_xAs_{100-x} layers with $x < 55$ at the photon energy of the positive and negative peaks. Note that the spectra at higher photon energies have been plotted with negative axis labels for convenience. Figure from Ref. (129).

7.4 Comprehension questions

7.4.1 Relation between magnetism and magneto-optics

(a) Why is a hard-axis magnetization curve $M(B)$ at small magnetic fields linear in the applied magnetic field B?
(b) How does the magnetization curve $M(T)$ as a function of temperature look like? How does it change when the applied magnetic field is increased?
(c) Which magnetic property can be investigated with MOKE in Co/Pt multilayers?
(d) Which magnetic property can be investigated with MOKE in a wedged Fe/Mo/Fe trilayer?
(e) Which property can be investigated with MOKE in U_xAs_{1-x} amorphous thin films?

7.5 Problems

7.5.1 Magnetic field dependence of Faraday and polar Kerr effect

Show that in terms of the Lorentz model, the Faraday effect and the polar Kerr effect are, up to second order, linear in the applied magnetic field.

Hint: Equations (3.68) and (3.76) define the relation of the complex Faraday effect and of the complex polar Kerr effect to the optical functions whose frequency dependence is given by the Lorentz model. By expanding the equations up to second order in the magnetic field, you get the magnetic field dependence. Note that $\omega_c \ll \omega_0, \gamma, \omega$.

7.5.2 Hysteresis curve

(a) The geometric demagnetization factor N is the cause that the internal magnetic field $\vec{B}_i$ inside a magnetic material is not equal to the applied magnetic field $\vec{B}_0$. To get the true dependence of the magnetization $\vec{M}$ on the inner magnetic field $\vec{B}_i$, the experimental magnetization curve $M(B_0)$ must be transformed (i.e., sheared back). Perform this so-called back shearing with the help of a diagram.

(b) What about the experimental Kerr hysteresis curves? Do we have to shear them back as well? And if yes, how can that be done?

7.6 Solutions

7.6.1 Magnetic field dependence of Faraday and polar Kerr effect

Let us first have a look at the two equations, eqs. (3.68) and (3.76), defining the relation between complex Faraday effect, complex polar Kerr effect, and optical conductivity:

$$\tilde{\theta}_F = \frac{\omega d}{2c}(\tilde{n}_+ - \tilde{n}_-) = \frac{d}{2\varepsilon_0 c}\frac{\tilde{\sigma}_{xy}}{\tilde{n}}, \tag{7.7}$$

$$\tilde{\theta}_K \cong -\frac{\tilde{\sigma}_{xy}}{\tilde{n}\tilde{\sigma}_{xx}} = \mathrm{i}\frac{1}{\varepsilon_0\omega}\frac{\tilde{\sigma}_{xy}}{\tilde{n}\left(\tilde{n}^2 - 1\right)} = \mathrm{i}\frac{1}{\varepsilon_0\omega}\frac{\tilde{\sigma}_{xy}}{A - \mathrm{i}B}. \tag{7.8}$$

Equation (7.7) shows that the complex Faraday effect $\tilde{\theta}_F$ is proportional to the off-diagonal element $\tilde{\sigma}_{xy}$ of the optical conductivity tensor and inversely proportional to the (non-magnetic) complex index of refraction $\tilde{n}$. However, as is obvious from eq. (3.67), we had approximated $(\tilde{n}_+ + \tilde{n}_-)$ by $2\tilde{n}$. Therefore, we must first examine if the sum $(\tilde{n}_+ + \tilde{n}_-)$ is constant up to first order in the magnetic field B or, which is equivalent, in the cyclotron resonance frequency $\omega_c = qB/m$, defined in eq. (3.16). Because $\omega_c \ll \omega_0, \gamma, \omega$, we can expand $\tilde{n}_\pm$ in a Taylor series of ω_c:

$$\tilde{n}_\pm = \sqrt{\tilde{\varepsilon}_\pm} = \sqrt{1 + \frac{\omega_\mathrm{p}^2}{\omega_0^2 - \omega^2 \pm \omega\omega_\mathrm{c} + \mathrm{i}\,\omega\gamma}} \cong \tilde{n}_\pm(\omega_\mathrm{c} = 0) + \left.\frac{d\tilde{n}_\pm}{d\omega_\mathrm{c}}\right|_{\omega_\mathrm{c}=0} \omega_\mathrm{c} + \left.\frac{1}{2}\frac{d^2\tilde{n}_\pm}{d\omega_\mathrm{c}{}^2}\right|_{\omega_\mathrm{c}=0} \omega_\mathrm{c}^2$$

$$\cong \tilde{n} \mp \frac{1}{2\tilde{n}} \frac{\omega_\mathrm{p}^2\omega}{\left(\omega_0^2 - \omega^2 + \mathrm{i}\,\omega\gamma\right)^2}\omega_\mathrm{c} + \left[-\frac{1}{8\tilde{n}^3}\frac{\omega_\mathrm{p}^4\omega^2}{\left(\omega_0^2 - \omega^2 + \mathrm{i}\,\omega\gamma\right)^4} + \frac{1}{2\tilde{n}}\frac{\omega_\mathrm{p}^2\omega^2}{\left(\omega_0^2 - \omega^2 + \mathrm{i}\,\omega\gamma\right)^3}\right]\omega_\mathrm{c}^2, \tag{7.9}$$

where ñ is the refractive index in the absence of a magnetic field B. So, it is constant with respect to B or ω_C. Using eq. (7.9) to expand the sum $(\tilde{n}_+ + \tilde{n}_-)$ in a Taylor series, we obtain

$$\tilde{n}_+ + \tilde{n}_- \cong 2\tilde{n} + \left[-\frac{1}{4\tilde{n}^3}\frac{\omega_\mathrm{p}^4\omega^2}{\left(\omega_0^2 - \omega^2 + \mathrm{i}\,\omega\gamma\right)^4} + \frac{1}{\tilde{n}}\frac{\omega_\mathrm{p}^2\omega^2}{\left(\omega_0^2 - \omega^2 + \mathrm{i}\omega\gamma\right)^3}\right]\omega_\mathrm{c}^2. \tag{7.10}$$

We see that the sum $(\tilde{n}_+ + \tilde{n}_-)$ is equal to the constant $2\tilde{n}$ up to first order in ω_C. Next, we expand $\tilde{\sigma}_{xy}$ in a power series of ω_c by combining eqs. (3.21), (3.37), and (3.38):

$$\begin{aligned}
\tilde{\sigma}_{xy} = \frac{1}{2\mathrm{i}}(\tilde{\sigma}_+ - \tilde{\sigma}_-) &= \frac{\varepsilon_0\omega_\mathrm{p}^2\omega}{2}\left(\frac{1}{\omega_0^2 - \omega^2 + \omega\omega_\mathrm{c} + \mathrm{i}\,\omega\gamma} - \frac{1}{\omega_0^2 - \omega^2 - \omega\omega_\mathrm{c} + \mathrm{i}\,\omega\gamma}\right)\\
&= -\varepsilon_0\frac{\omega_\mathrm{p}^2\omega^2\omega_\mathrm{c}}{\left(\omega_0^2 - \omega^2 + \mathrm{i}\,\omega\gamma\right)^2 - \omega^2\omega_\mathrm{c}^2}\\
&= -\varepsilon_0\frac{\omega_\mathrm{p}^2\omega^2\omega_\mathrm{c}}{\left(\omega_0^2 - \omega^2 + \mathrm{i}\,\omega\gamma\right)^2\left(1 - \frac{\omega^2\omega_\mathrm{c}^2}{\left(\omega_0^2 - \omega^2 + \mathrm{i}\,\omega\gamma\right)^2}\right)}\\
&\cong -\varepsilon_0\frac{\omega_\mathrm{p}^2\omega^2\omega_\mathrm{c}}{\left(\omega_0^2 - \omega^2 + \mathrm{i}\,\omega\gamma\right)^2}\left(1 + \frac{\omega^2\omega_\mathrm{c}^2}{\left(\omega_0^2 - \omega^2 + \mathrm{i}\,\omega\gamma\right)^2}\right).
\end{aligned} \tag{7.11}$$

The result shows that $\tilde{\sigma}_{xy}$ is proportional to ω_c up to second order. Consequently, the complex Faraday rotation $\tilde{\theta}_\mathrm{F}$ is proportional to ω_c up to second order.

In a similar way, we can now treat the complex polar Kerr rotation $\tilde{\theta}_\mathrm{K}$. To derive eq. (3.76),

$$\tilde{\theta}_\mathrm{K} \cong -\frac{\tilde{\sigma}_{xy}}{\tilde{n}\tilde{\sigma}_{xx}} = \mathrm{i}\frac{1}{\varepsilon_0\omega}\frac{\tilde{\sigma}_{xy}}{\tilde{n}\left(\tilde{n}^2 - 1\right)}, \tag{7.12}$$

we had used the approximations $(\tilde{n}_+ + \tilde{n}_-) \cong 2\tilde{n}$ and $\tilde{n}_+\tilde{n}_- \cong \tilde{n}^2$. We have already checked the sum. To evaluate the product, we can use eq. (7.9), neglecting the last term on the right as it only leads to second order and higher contributions. For the product, we then get

$$\tilde{n}_+\tilde{n}_- \cong \left(\tilde{n} - \frac{1}{2\tilde{n}}\frac{\omega_p^2\omega}{\left(\omega_0^2-\omega^2+i\omega\gamma\right)^2}\,\omega_c\right)\left(\tilde{n} + \frac{1}{2\tilde{n}}\frac{\omega_p^2\omega}{\left(\omega_0^2-\omega^2+i\,\omega\gamma\right)^2}\,\omega_c\right)$$
$$= \tilde{n}^2 - \frac{1}{4\tilde{n}^2}\frac{\omega_p^4\omega^2}{\left(\omega_0^2-\omega^2+i\,\omega\gamma\right)^4}\,\omega_c^2. \tag{7.13}$$

Indeed, the product is up to first order independent of ω_c. Therefore, the denominator in eq. (7.12) gives only contributions of higher order in ω_c. In conclusion, the magnetic field dependence of the complex polar Kerr effect is dominated by $\tilde{\sigma}_{xy}$ leading to a linear dependence on ω_c as for $\tilde{\theta}_F$.

7.6.2 Hysteresis curve

(a) The internal magnetic field $\vec{B}_i$ is defined in eq. (7.1):

$$\vec{B}_i = \mu_0\vec{H} = \vec{B}_0 + \vec{B}_s = \vec{B}_0 - \mu_0 N\vec{M} = \mu_0\left(\vec{H}_0 + \vec{H}_s\right). \tag{7.14}$$

The applied magnetic field $\vec{B}_0$ is reduced by the demagnetizing field or stray magnetic field:

$$\vec{B}_s = \mu_0\vec{H}_s = -\mu_0 N\vec{M}. \tag{7.15}$$

We see that the demagnetizing field is proportional to the geometric demagnetization factor N. To plot the magnetization M as a function of B_i, B_0 must be corrected by subtracting $B_s = \mu_0 NM$. This is most simply done geometrically by shearing back the magnetization curve. The principle is shown in Fig. 7.11(a). The demagnetization field is plotted as an inclined purple line with slope $(\mu_0 N)^{-1}$. For a given magnetic field B_x, we draw a horizontal line intersecting the magnetization curve at $B = B_x$. The position of the line on the vertical axis at $B = 0$ is $M(B_x)$. In addition, the horizontal line intersects the inclined line of the demagnetizing field at $B = B_s$. The distance B_s represents the demagnetizing field at B_x, that is, $B_s = B_s(M(B_x))$. The horizontal distance (red line) between the intersection with the line of the demagnetizing field and with $M(B_x)$ is the internal magnetic field $B_i = B_x - B_s$. Therefore, by shifting the point $M(B_x)$ of the magnetization curve along the horizontal line to the magnetic field B_i, we get the corrected position of the magnetization. With respect to the horizontal axis, we have shifted the magnetization curve from $M(B_x)$ to $M(B_i = B_x - B_s)$. This is the same as pulling each point of the magnetization curve along the horizontal direction until the inclined line of the demagnetization field coincides with the vertical axis as depicted in Fig. 7.11(b). This procedure is called shearing back the magnetization curve. It reveals the true remanence magnetization M_R (always larger than

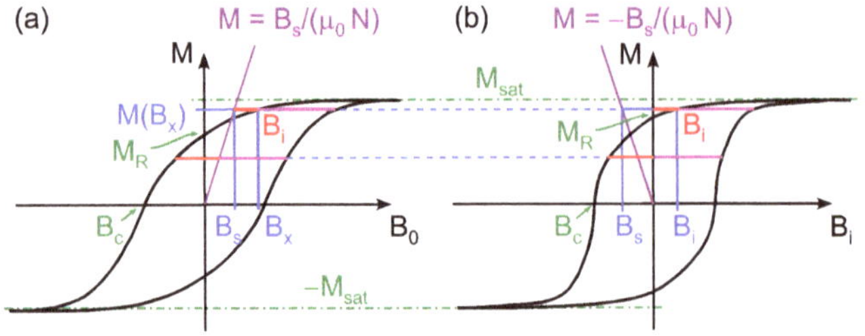

Fig. 7.11: Sketch of the principle of shearing back a magnetization curve. The inclined purple line is the demagnetizing field. (a) Horizontal line at $M(B_x)$ defines the demagnetizing field B_s (intersection with inclined purple line) and the internal magnetic field B_i which is the distance between B_s and B_x (red horizontal line). (b) Magnetization curve sheared back. Each point of the magnetization curve has been shifted along a horizontal line by the distance $B_s(M)$.

the measured one). The remanence magnetization M_R is defined by the intersection between the straight line of the demagnetizing field and the magnetization curve. Note that the saturation magnetization M_{sat} and the coercive field B_c are not affected by the back shearing.

(b) As the polar Kerr rotation is proportional to the magnetization, the polar Kerr hysteresis curve must also be sheared back. Therefore, one must determine the proportionality factor N_θ for the demagnetizing field:

$$B_s = N_\theta \theta_K = \mu_0 \, N \, M. \tag{7.16}$$

This can be done by measuring θ_{sat} and M_{sat}, or θ_{rem} and M_{rem}, and using

$$\theta_{sat} = C_{MO} \, M_{sat} \tag{7.17}$$

or

$$\theta_{rem} = C_{MO} \, M_{rem}, \tag{7.18}$$

where θ_{sat} and θ_{rem} are the polar Kerr rotation at saturation and at remanence, respectively. The constant C_{MO} is the proportionality factor between magnetization and polar Kerr rotation. In general, we have

$$\theta_K(B) = C_{MO} \, M(B), \tag{7.19}$$

Inserting this into eq. (7.16) yields

$$B_s = N_\theta \theta_K = \mu_0 N M = \mu_0 N C_{MO}^{-1} \theta_K. \tag{7.20}$$

So, we finally get the proportionality factor, N_θ, for the demagnetizing field:

$$N_\theta = \mu_0 C_{MO}^{-1} N, \tag{7.21}$$

where C_{MO} is determined by eq. (7.17) or (7.18).

Appendix
Alternative representation of magneto-optic effects

In this book, we have defined the magneto-optic effects as a magnetic-field-induced anisotropy in the dielectric tensor according to eq. (3.33) and subsequently in the optical conductivity tensor according to eq. (3.34). The experimentally observed quantities, Faraday and polar Kerr rotation and ellipticity, were then related to this anisotropy and, finally, also the transition rates in a quantum-mechanical approach. We chose the optical conductivity as the key material property to describe magneto-optic effects. This was motivated by the fact that the optical conductivity directly relates to the absorbed power which is easily accessible in experiment and which directly relates to the transition rates. We could have also taken the dielectric tensor as the key material property to describe magneto-optic effects because there exists a direct relation between optical conductivity tensor and dielectric tensor expressed in eq. (3.35).

This is, of course, not the only way to describe magneto-optic effects and we would like – for completeness and comparability – to mention an alternative representation for magneto-optic effects. While the approach used in this book is motivated by the classical Lorentz theory and corroborated by a quantum-mechanical approach, other representations rely on a purely phenomenological approach. The most widely used one is the representation using the Voigt material constant Q_V for effects linear in the applied magnetic field or magnetization (such as Faraday effect and polar Kerr effect) and the constants B_1 and B_2 for effects that are quadratic in the magnetic field (such as the Voigt effect) (131). The relation between the two representations is then simply defined via the dielectric tensor:

$$\overleftrightarrow{\varepsilon} = \begin{pmatrix} \tilde{\varepsilon}_{xx} & \tilde{\varepsilon}_{xy} & \tilde{\varepsilon}_{xz} \\ \tilde{\varepsilon}_{yx} & \tilde{\varepsilon}_{yy} & \tilde{\varepsilon}_{yz} \\ \tilde{\varepsilon}_{zx} & \tilde{\varepsilon}_{zy} & \tilde{\varepsilon}_{zz} \end{pmatrix} = \tilde{\varepsilon} \begin{pmatrix} 1 & -iQ_V m_3 & iQ_V m_2 \\ iQ_V m_3 & 1 & -iQ_V m_1 \\ -iQ_V m_2 & iQ_V m_1 & 1 \end{pmatrix} + \begin{pmatrix} B_1 m_1^2 & B_2 m_1 m_2 & B_2 m_1 m_3 \\ B_2 m_1 m_2 & B_1 m_2^2 & B_2 m_2 m_3 \\ B_2 m_1 m_3 & B_2 m_2 m_3 & B_1 m_3^2 \end{pmatrix}, \tag{A.1}$$

where $\tilde{\varepsilon}$ is the dielectric function in the absence of magnetic field or magnetization, and m_1, m_2, m_3 are the directions of the magnetization vector $\vec{M}$, defined as follows:

$$\vec{m} = \begin{pmatrix} m_1 \\ m_2 \\ m_3 \end{pmatrix} = \frac{\vec{M}}{M_{\text{sat}}}. \tag{A.2}$$

https://doi.org/10.1515/9783110635300-008

M_{sat} is the saturation magnetization. The Voigt parameters B_1 and B_2 are usually two orders of magnitude smaller than the Voigt material constant Q_V (131) and can be neglected in many cases except if the contribution of Q_V vanishes.

Now, let us neglect B_1 and B_2 and assume that the magnetization $\vec{M}$ is aligned along the z direction. So, the z component is

$$m_3 = |\vec{m}| = \left|\vec{M}\right| \Big/ M_{sat} = M/M_{sat}, \tag{A.3}$$

and $m_1 = m_2 = 0$. Then, we get according to eq. (A.1) a very simple relation between the off-diagonal element of the dielectric function and the Voigt material constant Q_V:

$$\tilde{\varepsilon}_{xy} = -i\tilde{\varepsilon}Q_V m_3 = -i\tilde{\varepsilon}Q_V \frac{M}{M_{sat}}. \tag{A.4}$$

Using eq. (3.35), we can express the off-diagonal element in terms of the Voigt material constant:

$$\tilde{\sigma}_{xy} = \tilde{\varepsilon}\varepsilon_0\omega\, Q_V m_3 = \tilde{\varepsilon}\varepsilon_0\omega\, Q_V \frac{M}{M_{sat}}. \tag{A.5}$$

The dielectric displacement $\vec{D}$ can be written in a simple form as well (131):

$$\vec{D} = \tilde{\varepsilon}\varepsilon_0 \left(\vec{E} + iQ_V\vec{m} \times \vec{E}\right). \tag{A.6}$$

From eq. (A.6), we see that the effect of the Voigt material constant Q_V is simply to add an additional vector to the nonmagnetic dielectric displacement $\tilde{\varepsilon}\varepsilon_0\vec{E}$. The additional vector is perpendicular to the magnetization $\vec{M}$ and to the electric field vector $\vec{E}$.

References

(1) É. Verdet, Leçons d'Optique Physique **1**, Victor Masson et fils, Paris (1869).

(2) F. Arago, and A. Fresnel, Mémoire sur l'action que les rayons de lumière polarisés exercent les uns sur les autres, *Annales de Chimie et de Physique* **10**, 288–305 (1819).

(3) A. Fresnel, Sur la diffraction de la lumière, *Annales de Chimie et de Physique* **11**, 246–296 (1819).

(4) A. Fresnel, Suite du mémoire sur la diffraction de la lumière, *Annales de Chimie et de Physique* **11**, 337–378 (1819).

(5) A. Fresnel, Note sur le calcul des teintes que la polarisation développe dans les lames cristallisées, *Annales de Chimie et de Physique* **17**, 102–112 (1821).

(6) A. Fresnel, 2e Note sur la coloration des lames cristallisées, *Annales de Chimie et de Physique* **17**, 167–196 (1821).

(7) A. Fresnel, Addition à la 2e note insérée dans le cahier précédent, *Annales de Chimie et de Physique* **17**, 312–315 (1821).

(8) M. Faraday, On the magnetization of light and the illumination of magnetic lines of force, *Philosophical Transactions* **136**, 1–20 (1846).

(9) Lord Rayleigh (J. W. Strutt), On the constant of magnetic rotation of light in bisulphide of carbon, *Philosophical Transactions* **176**, 343–366 (1885).

(10) É. Verdet, Recherches sur les propriétés optiques développées dans les corps transparents par l'action du magnétisme, première partie, *Annales de Chimie et de Physique* **41** (série 3), 370–412 (1854).

(11) É. Verdet, Recherches sur les propriétés optiques développées dans les corps transparents par l'action du magnétisme, troisième partie, *Annales de Chimie et de Physique* **52** (série 3), 129–163 (1858).

(12) É. Verdet, Recherches sur les propriétés optiques développées dans les corps transparents par l'action du magnétism, deuxième partie, *Annales de Chimie et de Physique* **43** (série 3), 37–44 (1855).

(13) A. Fresnel, Extrait d'un mémoire sur la double réfraction particulière que présente le cristal de roche dans la direction de son axe, *Annales de Chimie et de Physique* **28**, 147–161 (1825).

(14) É. Verdet, Recherches sur les propriétés optiques développées dans les corps transparents par l'action du magnétisme, quatrième partie, *Annales de Chimie et de Physique* **69** (série 3), 415–491 (1863).

(15) G. B. Airy, On the equations applying to light under the action of magnetism, *Philosophical Magazine* **28** (3rd series), 469–477 (1846).

(16) E. Becquerel, Expériences concernant l'action du magnétisme sur tous les corps, *Annales de Chimie et de Physique* **17** (série 3), 437–451 (1846).

(17) J. C. Maxwell, A dynamical theory of the electromagnetic field, *Philosophical Transactions* **155**, 459–512 (1865).

(18) H. A. Lorentz, Ueber die Theorie der Reflexion und Refraction des Lichtes, *Zeitschrift für Mathematik und Physik* **22**, 1–219 (1877).

(19) H. Helmholtz, Ueber die Bewegungsgleichungen der Elektricität für ruhende leitende Körper, *Journal für die reine und angewandte Mathematik* **72**, 57–129 (1870).

(20) H. Becquerel, Recherches expérimentales sur la polarisation rotatoire magnétique, *Annales de Chimie et de Physique* **12** (série 5), 5–87 (1877).

(21) H. Becquerel, Pouvoir rotatoire magnétique des gaz à la température et à la pression ordinaires, *Comptes rendus hebdomadaires des séances de l'Académie des sciences* **88**, 709–712 (1879).

(22) H. Becquerel, Recherches expérimentales sur la polarisation rotatoire magnétique dans les gaz, *Comptes rendus hebdomadaires des séances de l'Académie des sciences* **90**, 1407–1410 (1880).

https://doi.org/10.1515/9783110635300-009

(23) A. Kundt, and W. C. Röntgen, Nachweis der elektromagnetischen Drehung der Polarisationsebene des Lichtes im Schwefelkohlenstoffdampf, *Sitzungsberichte der mathematisch-physikalischen Classe der königlich bayerischen Akademie der Wissenschaften zu München* **8**, 546–549 (1878).
(24) A. Kundt, and W. C. Röntgen, Nachtrag zu Drehung der Polarisationsebene in Schwefelkohlenstoff-Dampf, *Sitzungsberichte der mathematisch-physikalischen Classe der königlich bayerischen Akademie der Wissenschaften zu München* **9**, 30 (1879).
(25) A. Kundt and W. C. Röntgen, Ueber die electromagnetische Drehung der Polarisationsebene des Lichtes in den Gasen, *Sitzungsberichte der mathematisch-physikalischen Classe der königlich bayerischen Akademie der Wissenschaften zu München* **9**, 148–170 (1879).
(26) H. Becquerel, Sur la propagation inégale de la lumière polarisée circulairement, dans les corps soumis à l'action du magnétisme, suivant le sens de l'aimantation et le sens des vibrations lumineuses, *Comptes rendus hebdomadaires des séances de l'Académie des sciences* **88**, 334–336 (1879).
(27) A. Righi, Sulla velocità della luce nei corpi trasparenti magnetizzati, *Il Nuovo Cimento* **3** (3rd series), 212–234 (1878).
(28) J. Kerr, On rotation of the plane of polarization by reflection from a magnetic pole, *Report of the British Association for the Advancement of Science* **46**, 40–41 (1876).
(29) J. Kerr, On rotation of the plane of polarization by reflection from the pole of a magnet, *Philosophical Magazine* **3** (5th series), 321–343 (1877).
(30) J. E. H. Gordon, A Physical Treatise on Electricity and Magnetism **2**, 296, Sampson Low, Marston, Searle & Rivington, London (1883).
(31) E. H. Hall, On the "rotational coefficient" in nickel and cobalt, *Philosophical Magazine*. **12** (5th series), 157–172 (1881).
(32) G. F. Fitzgerald, On the rotation of the plane of polarization of light by reflection from the pole of a magnet, *Proceedings of the Royal Society of London* **25**, 447–450 (1876).
(33) J. Kerr, On reflection of polarized light from the equatorial surface of a magnet, *Philosophical Magazine* **5** (5th series), 161–177 (1878).
(34) A. Kundt, On the electromagnetic rotation of the plane of polarization of light by means of iron, cobalt, and nickel, *Philosophical Magazine* **18** (5th series), 308–327 (1884).
(35) A. Kundt, Über die elektromagnetische Drehung der Polarisationsebene des Lichtes im Eisen, *Sitzungsberichte der königlich preussischen Akademie der Wissenschaften zu Berlin*, 1055–1064 (1885).
(36) P. Zeeman, Ueber einen Einfluss der Magnetisirung auf die Natur des von einer Substanz emittirten Lichtes, *Verhandlungen der Physikalischen Gesellschaft zu Berlin* **7**, 128–130 (1896).
(37) W. Voigt, Ueber den Zusammenhang zwischen dem Zeeman'schen und dem Faraday'schen Phänomen, *Nachrichten von der Königlichen Gesellschaft der Wissenschaften zu Göttingen – Mathematisch-physikalische Klasse*, 329–344 (1898).
(38) W. Voigt, Magneto- und Elektrooptik, B. G. Teubner, Leipzig (1908).
(39) P. Drude, Zur Elektronentheorie der Metalle, *Annalen der Physik* **306**, 566–613 (1900). Formerly: vol. 1 (4th series).
(40) P. M. Oppeneer, Theory of the Magneto-optical Kerr Effect in Ferromagnetic Compounds, Habilitation Treatise, Technische Universität Dresden (1999).
(41) O. Halpern, Notiz über die Rotationspolarisation ferromagnetischer Körper, *Annalen der Physik* **404**, 181–184 (1932). Formerly: vol. 12 (5th series).
(42) H. R. Hulme,The Faraday effect in ferromagnetics, *Proceedings of the Royal Society of London A* **135**, 237–257 (1932).
(43) P. N. Argyres, Theory of the Faraday and Kerr effects in ferromagnetics, *Physical Review* **97**, 334–345 (1955).

(44) Q. Majorana, Il fenomeno Kerr nei metalli non ferromagnetici, Il *Nuovo Cimento* **2** (8th series), 1–13 (1944).
(45) B. R. Cooper and H. Ehrenreich, The ferromagnetic Kerr effect in nickel, *Solid State Communications* **2**, 171–174 (1964).
(46) B. R. Cooper, Theory of the interband ferromagnetic Kerr effect in nickel, *Physical Review* **139**, A1504–A1514 (1965).
(47) P. C. Hohenberg and W. Kohn, Inhomogeneous electron gas, *Physical Review* **136**, B864–B871 (1964).
(48) W. Kohn and L. J. Sham, Self-consistent equations including exchange and correlation effects, *Physical Review* **140**, A1133–A1138 (1965).
(49) C. S. Wang and J. Callaway, Band structure of nickel: spin-orbit coupling, the Fermi surface, and the optical conductivity, *Physical Review B* **9**, 4897–4907 (1974).
(50) M. Singh, C. S. Wang, and J. Callaway, Spin-orbit coupling, Fermi surface, and optical conductivity of ferromagnetic iron, *Physical Review B* **11**, 287–294 (1975).
(51) R. Kubo, Statistical-mechanical theory of irreversible processes. I. General theory and simple applications to magnetic and conduction problems, *Journal of the Physical Society of Japan* **12**, 570–586 (1957).
(52) G. Dresselhaus, A. F. Kip, and C. Kittel, Cyclotron resonance of electrons and holes in silicon and germanium crystals, *Physical Review* **98**, 368–384 (1955).
(53) S. Zwerdling and B. Lax, Oscillatory magneto-absorption of the direct transition in germanium, *Physical Review* **106**, 51–52 (1957).
(54) B. Lax, L. M. Roth, and S. Zwerdling, Quantum magneto-absorption phenomena in semiconductors, *Journal of Physics and Chemistry of Solids* **8**, 311–318 (1959).
(55) H. Ebert, Theoretical investigations of the magneto-optical Kerr effect by means of the spin-polarized relativistic linear muffin-tin orbital method, *Physica B* **161**, 175–178 (1989).
(56) Y. A. Uspenskii and S. V. Khalilov, Electron structure and magnetooptics of ferromagnetic 3d metals, *Soviet Physics Journal of Experimental and Theoretical Physics* **68**, 588–596 (1989).
(57) G. H. O. Daalderop, F. M. Mueller, R. C. Albers, and A. M. Boring, Theory of the magneto-optic Kerr-effect in NiUSn, *Journal of Magnetism and Magnetic Materials* **74**, 211–218 (1988).
(58) P. M. Oppeneer, T. Maurer, J. Sticht, and J. Kübler, Ab initio calculated magneto-optical Kerr effect of ferromagnetic metals – Fe and Ni, *Physical Review B* **45**, 10924–10933 (1992).
(59) G. Q. Di and S. Uchiyama, Temperature dependence of Kerr rotation for Ni film, *Journal of Applied Physics* **75**, 4270–4272 (1994).
(60) G. S. Krinchik and V. A. Artem'ev, Magneto-optical properties of Ni, Co, and Fe in the ultraviolet, visible, and infrared parts of the spectrum, *Soviet Physics Journal of Experimental and Theoretical Physics* **26**, 1080–1085 (1968).
(61) K. H. J. Buschow, Magneto-optical properties of alloys and intermetallic compounds, in: *Handbook of Ferromagnetic Materials* **4**, 493–595, K. H. J. Buschow and E. P. Wohlfarth, Elsevier (1988).
(62) J. Schoenes, Magneto-optical Properties of Metals, Alloys and Compounds, in: *Materials Science and Technology* **3A**, 147–255, R. W. Cahn, P. Haasen, and E. J. Kramer, VCH Weinheim (1992).
(63) W. Reim and J. Schoenes, Magneto-optical spectroscopy of f-electron systems, in: *Handbook on Ferromagnetic Materials* **5**, 133–236, K. H. J. Buschow and E. P. Wohlfarth, Elsevier (1990).
(64) H. J. Williams, F. G. Foster, and E. A. Wood, Observation of magnetic domains by the Kerr effect, *Physical Review* **82**, 119–120 (1951).
(65) C. A. Fowler and E. M. Fryer, Magnetic domains on silicon iron by the longitudinal Kerr effect, *Physical Review* **86**, 426–426 (1952).
(66) H. J. Williams, R. C. Sherwood, F. G. Foster, and E. M. Kelley, Magnetic writing on thin films of MnBi, *Journal of Applied Physics* **28**, 1181–1184 (1957).

(67) R. L. Conger and J. L. Tomlinson, Magneto-optic readout for computer memories, *Journal of Applied Physics* **33**, 1059–1060 (1962).
(68) E. J. Supernowicz, Magneto-optical readout from a magnetized nonspecular oxide surface, *Journal of Applied Physics* **34**, 1110–1111 (1963).
(69) P. Chaudhari, J. J. Cuomo, and R. J. Gambino, Amorphous metallic films for magneto-optic applications, *Applied Physics Letters* **22**, 337–339 (1973).
(70) J. L. Erskine and E. A. Stern, Calculation of the M_{23} magneto-optical absorption spectrum of ferromagnetic nickel, *Physical Review B* **12**, 5016–5024 (1975).
(71) B. T. Thole, G. van der Laan, and G. A. Sawatzky, Strong magnetic dichroism predicted in the $M_{4,5}$ x-ray absorption spectra of magnetic rare-earth materials, *Physical Review Letters* **55**, 2086–2088 (1985).
(72) G. van der Laan, B. T. Thole, G. A. Sawatzky, J. B. Goedkoop, J. C. Fuggle, J.-M. Esteva, R. Karnatak, J. P. Remeika, and H. A. Dabkowska, Experimental proof of magnetic x-ray dichroism, *Physical Review B* **34**, 6529–6531 (1986).
(73) G. Schütz, W. Wagner, W. Wilhelm, P. Kienle, R. Zeller, R. Frahm, and G. Materlik, Absorption of circularly polarized x-rays in iron, *Physical Review Letters* **58**, 737–740 (1987).
(74) H.-Ch. Mertins, C. Jansing, M. Krivenkov, A. Varykhalov, O. Rader, H. Wahab, H. Timmers, A. Gaupp, A. Sokolov, M. Tesch, and P. M. Oppeneer, Giant magneto-optical Faraday effect of graphene on Co in the soft x-ray range, *Physical Review B* **98**, 064408-1 – 064408-9 (2018).
(75) L. E. Hargrove, R. L. Fork, and M. A. Pollack, Locking of He–Ne laser modes induced by synchronous intracavity modulation, *Applied Physics Letters* **5**, 4–5 (1964).
(76) T. Brabec and F. Krausz, Intense few-cycle laser fields – frontiers of nonlinear optics, *Reviews of Modern Physics* **72**, 545–591 (2000).
(77) A. Kirilyuk, A. V. Kimel, and T. Rasing, Ultrafast optical manipulation of magnetic order, *Reviews of Modern Physics* **82**, 2731–2784 (2010).
(78) W. Voigt, Doppelbrechung von im Magnetfelde befindlichem Natriumdampf in der Richtung normal zu den Kraftlinien, *Nachrichten von der Königlichen Gesellschaft der Wissenschaften zu Göttingen. Mathematisch-physikalische Klasse*, 355–359 (1898).
(79) A. A. Cotton and H. Mouton, Sur les propriétés magnéto-optiques des colloïdes et des liqueurs hétérogènes, *Annales de Chimie et de Physique* **11** (série 8), 145–203 (1907).
(80) Q. Majorana, Sur la biréfringence magnétique, *Comptes rendus hebdomadaires des séances de l'Académie des sciences* **135**, 159–161 (1902).
(81) L. D. Landau and E. M. Lifshitz, Electrodynamics of Continuous Media, *Course of Theoretical Physics* **8**, 251–253, Pergamon Press, Oxford (1960).
(82) P. S. Pershan, Magneto-optical effects, *Journal of Applied Physics* **38**, 1482–1490 (1967).
(83) A. Fresnel, Mémoire sur la loi des modifications que la réflexion imprime à la lumière polarisée, *Annales de Chimie et de Physique* **46**, 225–264 (1831).
(84) F. Goos and H. Hähnchen, Ein neuer und fundamentaler Versuch zur Totalreflexion, *Annalen der Physik* **436**, 333–346 (1947). Formerly: vol. 1 (6th series).
(85) D. J. Griffiths, Introduction to Electrodynamics, 4th Edition, Pearson Education (2013).
(86) R. F. Potter, Germanium (Ge), in: *Handbook of Optical Constants of Solids*, 465–478, E. D. Palik, Academic Press (1985).
(87) D. F. Edwards, Silicon (Si), in: *Handbook of Optical Constants of Solids*, 547–569, E. D. Palik, Academic Press (1985).
(88) H. S. Bennett and E. H. Stern, Faraday effect in solids, *Physical Review* **137**, A448–A461 (1965).
(89) S. Richard, F. Aniel, and G. Fishman, Energy-band structure of Ge, Si, and GaAs: a thirty-band k·p method, *Physical Review B* **70**, 235204-1 – 235204-6 (2004).
(90) E. D. Palik, J. K. Furdyna, IR and microwave magnetoplasma effects in semiconductors, *Reports on Progress in Physics* **33**, 1193–1322 (1970).

(91) E. B. Hagen and H. Rubens, Über die Abhängigkeit des Emissionsvermögens der Metalle von der Temperatur, *Sitzungsberichte der königlich preussischen Akademie der Wissenschaften zu Berlin*, 478–492 (1909).

(92) R. C. Jones, A new calculus for the treatment of optical systems, I. Description and discussion of the calculus, *Journal of the Optical Society of America* **31**, 488–493 (1941).

(93) H. Feil, Magneto-optical Kerr effect. The relation between optical, transport and electronic properties in magnetic compounds, PhD Thesis, University of Groningen (1987).

(94) P. Fumagalli, T. S. Plaskett, T. R. McGuire, R. J. Gambino, N. A. Bojarczuk, Magneto-optical and magnetic study of amorphous U-As, U-As-Cu, and U-As-Ti films, *Physical Review B* **46**, 6187–6194 (1992).

(95) W. Reim, Magneto-optical Kerr effect of uranium compounds: a study of the electronic structure, *Journal of Magnetism and Magnetic Materials* **58**, 1–47 (1986).

(96) W. Reim and D. Weller, Dielectric enhancement and passivation layers for magneto-optical thin films, *Journal de Physique* **49**, C8-1959 – C8-1960 (1988).

(97) T. Katayama, Y. Suzuki, H. Awano, Y. Nishihara, and N. Koshizuka, Enhancement of the magneto-optical Kerr rotation in Fe/Cu bilayered films, *Physical Review Letters* **60**, 1426–1429 (1988).

(98) J. C. Suits and K. Lee, Giant magneto-optical Kerr effect in EuO, *Journal of Applied Physics* **42**, 3258–3260 (1971).

(99) H. Brändle, New materials with large magneto-optical polar Kerr effects at room temperature, PhD Thesis no. 9252, ETH Zurich (1990).

(100) W. Schütz, Magnetooptik, in: *Handbuch der Experimentalphysik* **16** (1), Akademische Verlagsgesellschaft, Leipzig (1936).

(101) L. H. Thomas, The motion of the spinning electron, *Nature* **117**, 514–514 (1926).

(102) L. Uba, S. Uba, and V. N. Antonov, Magneto-optical Kerr spectroscopy of noble metals, *Physical Review B* **96**, 235132-1 – 235132-11 (2017).

(103) J. L. Erskine and E. A. Stern, Magneto-optic Kerr effects in gadolinium, *Physical Review B* **8**, 1239–1255 (1973).

(104) J. Schoenes, Magnetooptik und elektronische Struktur der magnetisch ordnenden Europiumchalkogenide, *Zeitschrift für Physik B* **20**, 345–368 (1975).

(105) P. Schwob, Die magnetischen Eigenschaften der Europium-Chalkogenide unter hydrostatischem Druck, *Physik der Kondensierten Materie* **10**, 186–218 (1969).

(106) A. Kornblit, G. Ahlers, and E. Buehler, Heat capacity of $RbMnF_3$ and EuO near the magnetic phase transitions, *Physics Letters A* **43**, 531–532 (1973).

(107) W. Zinn, Microscopic studies of magnetic properties and interactions recent results on europium-monochalcogenides, *Journal of Magnetism and Magnetic Materials* **3**, 23–36 (1976).

(108) P. Wachter, The optical electrical and magnetic properties of the europium chalcogenides and the rare earth pnictides, *Critical Reviews in Solid State and Material Sciences* **3**, 189–241 (1972).

(109) D. D. Berkner, Magneto-optic study of the critical region of EuS, *Physics Letters A* **54**, 396–398 (1975).

(110) H. G. Bohn, W. Zinn, B. Dorner, and A. Kollmar, Neutron scattering study of spin waves and exchange interactions in ferromagnetic and paramagnetic EuS, *Journal of Applied Physics* 52, 2228–2230 (1981).

(111) R. Griessen, M. Landolt, and H. R. Ott, A new antiferromagnetic phase in EuSe below 1.8 K, *Solid State Communications* **9**, 2219–2223 (1971).

(112) G. Busch, P. Junod, and P. Wachter, Optical absorption of ferro- and antiferromagnetic europium chalcogenides, *Physics Letters* **12**, 11–12 (1964).

(113) R. Reisfeld, and A. Glasner, Absorption and fluorescence spectra of Eu^{2+} in alkali halide crystals, *Journal of the Optical Society of America* **54**, 331–333 (1964).

(114) J. Schoenes and P. Wachter, Exchange optics in Gd-doped EuO, *Physical Review B* **9**, 3097–3105 (1974).
(115) P. Wachter, Europium chalcogenides: EuO, EuS, EuSe and EuTe, in: *Handbook on the Physics and Chemistry of Rare Earth* **2**, 507–574, K. A. Gschneidner and L. Eyring, North Holland, Amsterdam (1979).
(116) T. Kasuya, s–f exchange interactions and magnetic semiconductors, *CRC Critical Reviews in Solid State and Material Sciences* **3**, 131–164 (1972).
(117) W. Müller and W. Nolting, Temperature-dependent quasiparticle band structure of the ferromagnetic semiconductor EuS, *Physical Review B* **66**, 085205–1 – 085205–9 (2002).
(118) T. Jeong, Investigation on the electronic structure and magnetic properties of EuS, *International Journal of Modern Physics B* **26**, 1250150–1 – 1250150–7 (2012).
(119) J. Ferré, B. Briat, C. Paparoditis, S. Pokrzywnicki, and R. Suryanarayanan, Magneto-optical properties of EuS thin films: magnetic circular and linear dichroism, *Solid State Communications* **11**, 1173–1177 (1972).
(120) J. Schoenes, P. Wachter, and F. Rys, Magneto-optic spectroscopy of EuS, *Solid State Communications* **15**, 1891–1895 (1974).
(121) Wikipedia: Table of Clebsch–Gordan coefficients, https://en.wikipedia.org/wiki/Table_of_Clebsch-Gordan_coefficients, cited: October 8, 2021
(122) J. Schoenes and P. Wachter, High field magneto-optical study of EuSe and EuTe, *Physica B* **89**, 155–158 (1977).
(123) S. C. P. van Kooten, P. A. Usachev, X. Gratens, A. R. Naupa, V. A. Chitta, G. Springholz, and A. B. Henriques, Converting Faraday rotation into magnetization in europium chalcogenides, *Journal of Applied Physics* **126**, 095701-1-095701-8 (2019).
(124) C. Müller, H. Lippitz, J. J. Paggel, and P. Fumagalli, Evidence of exchange-induced spin polarization in the magnetic semiconductor EuS, *Journal of Applied Physics* **95**, 7172–7174 (2004).
(125) W. B. Zeper, F. J. A. M. Greidanus, P. F. Carcia, and C. R. Fincher, Perpendicular magnetic anisotropy and magneto-optical Kerr effect of vapor-deposited Co/Pt-layered structures, *Journal of Applied Physics* **65**, 4971–4975 (1989).
(126) Z. Q. Qiu, J. Pearson, A. Berger, and S. D. Bader, Short-period oscillations in the interlayer magnetic coupling of wedged Fe(100)/Mo(100)/Fe(100) grown on Mo(100) by molecular-beam epitaxy, *Physical Review Letters* **68**, 1398–1401 (1992).
(127) S. T. Purcell, W. Folkerts, M. T. Johnson, N. W. E. McGee, K. Jager, J. aan de Stegge, W. B. Zeper, W. Hoving, and P. Grünberg Oscillations with a period of two Cr monolayers in the antiferromagnetic exchange coupling in a (001) Fe/Cr/Fe sandwich structure, *Physical Review Letters* **67**, 903–906 (1991).
(128) K. Yosida, Magnetic properties of Cu-Mn alloys, *Physical Review* **106**, 893–898 (1957).
(129) P. Fumagalli, R. J. Gambino, T. S. Plaskett, and T. R. McGuire, Magneto-optical evidence of phase separation in amorphous U-As films, *Applied Physics Letters* **60**, 258–260 (1992).
(130) A. Hubert and R. Schäfer, Magnetic Domains – The Analysis of Magnetic Microstructures, Springer, Berlin (1998).

Index

https://doi.org/10.1515/9783110635300-010

www.ingramcontent.com/pod-product-compliance
Lightning Source LLC
Chambersburg PA
CBHW061403210326
41598CB00035B/6083
* 9 7 8 3 1 1 0 6 3 5 2 2 5 *